This book is dedicated to

His Highness, The Amir, Shaikh Isa bin Salman Al-Khalifa

citizens of the State of Bahrain

and

to all those who have contributed to the development of Bahrain's oil industry

The Bahrain refinery (*Joe C. Torres*).

BAHRAIN
OIL AND DEVELOPMENT
1929 - 1989

Angela Clarke

**International Research Center for Energy
and Economic Development (ICEED)**

Text Copyright ©1990 THE BAHRAIN PETROLEUM COMPANY B.S.C. (C)
Design Copyright ©1990 IMMEL PUBLISHING. London

International Standard Book Number: 0-918714-25-7

Library of Congress Catalog Card Number: 90-084725

United States publisher:
The International Research Center for Energy and Economic Development (ICEED)
206 Business Building, Campus Box 418
University of Colorado
Boulder, Colorado 80309-0418 U.S.A.

Author: Angela Clarke
Copy Editor: Judith Wardman
Design and Illustration: Jane Stark
Indexing: Pamela Le Gassick
Jacket Illustration: Derek Musgrave
Commissioned Photography: Anthony Nelson, Gulf Colour Laboratories
UK Picture Research: Michael Nicholson
UK Library Research: Geoffrey Lumsden, Ph.D.
Translation Consultant: Husain Dhaif, Ph.D.
UK Publishing Director: Peter Vine, Ph.D.
USA Publishing Director: Dorothea El Mallakh, Ph.D.

Phototypeset in New Baskerville by Shirley Kilpatrick, Icon Publications Ltd
Printed and bound in the Netherlands by Roto Smeets

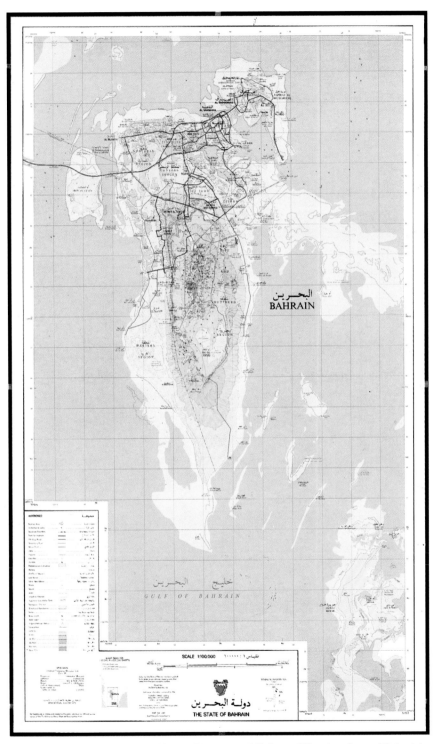

The State of Bahrain – 1988 (*Survey Directorate, Ministry of Housing, State of Bahrain*).

CONTENTS

PART II

CHRONOLOGY

DIRECTORY:

Companies and Organizations associated with the evolution of the oil industry in Bahrain

PART III

ABBREVIATIONS and ACRONYMS

GLOSSARIES

Economic and Financial

Language

Legal

Place-names

Technical

NOTES

BIBLIOGRAPHY

INDEX TO PART I

FOREWORD

The early 1930s were unkind years for the people of Bahrain. With the depression cutting deeply into the economy of the world, and the cultured pearl industry threatening, then destroying, the natural pearl industry of Bahrain, the economy of the islands, employment and Government revenue suffered heavily. No one realized at that time, that the advent of the cultured pearl industry would destroy completely an integrated technology which had thrived for over four thousand years, encompassing shipbuilding, pearl diving and trading.

The discovery of oil in Bahrain in 1932 brought a dramatic change in the economic activities of the nation. The new industry demanded a new technology, trades and skills that were previously unknown on the islands. Although the Sitra wharf and tank farm, the refinery and Awali township are quite separate geographically from the islands' other communities, for three decades Bapco was Bahrain's only provider of industrial training, jobs and new technology. Gradually, increasing oil revenues generated the means to finance Bahrain's modern economic growth. Nevertheless, even today, two wrong impressions about Bahrain and Bapco are often mentioned.

Many people believe that prior to the discovery of oil, no modern infrastructure existed in Bahrain, and that Bapco evolved and operated quite independently of the islands' communities. Both are misconceptions.

In the early decades of this century, before Bapco established its own infrastructure, a municipality system had been created in Bahrain; boys and girls were being educated in purpose-built schools; legislation to establish a Board of Trade and to improve the human and social conditions of the pearl industry had been passed; health centres, hospitals and a hospital ship provided health care both on land and at sea during the diving season; a nucleus of modern government had been formed, based on the three principles of administration, finance and security; and Bahrain had become a transshipment centre for entrepôt trading, as well as a communications centre for steamer traffic and aviation services from Europe to Africa and Australia.

Throughout my career I have been involved in Bahrain's oil industry. From many years of observation and participation, two themes immediately come to my mind. The oil industry and modern Bahrain have grown together. With the advent of independence, under the wise leadership of His Highness The Amir and the excellent relationship maintained with Bapco and its owners, the transition towards partnership in the nation's oil industry was relatively painless, natural and productive, hence my insistence that Bapco's name should not be changed.

Today, the responsibility for the development, training and financing of Bahrain's modern economic growth does not fall on one company. It was my vision of the future that Bahrain should not have a huge monolithic oil company. Instead, we have several separately organized companies, responsible for many aspects of Bahrain's hydrocarbons industry – oil and natural gas exploration and production, and such downstream operations as refining, aviation fuelling and petrochemicals. Other industries have developed such as aluminium smelting, as well as offshore banking. Tourism is a product of the 1980s and by the end of the century I am sure that more economic enterprises will have been created.

The evolution of the original Bapco over the last six decades has been, and continues to be, a remarkable example of a successful partnership between the Bahrain Government and foreign investors. Many people would argue that the State of Bahrain has coped better than most countries which have faced dramatic economic and social changes caused by the rapid development of their oil resources. The sixtieth anniversary of Bahrain's oil industry was celebrated with dignity and maturity.

This book is a work of reference blended with the story of human beings, who, through foresight, devotion and loyalty, allowed Bahrain to evolve in material as well as spiritual values. It is a tribute to that achievement.

Khalifa bin Salman Al-Khalifa
Prime Minister
State of Bahrain
June 1990

ACKNOWLEDGEMENTS

Some four years ago, His Excellency Yousuf A. Shirawi, Minister of Development and Industry and Acting Minister of State for Cabinet Affairs, recognized that valuable historical documents relating to Bahrain's oil industry were disappearing rather quickly, mostly the result of company reorganizations and relocations in both Bahrain and the USA. The ageing and passing of many of the pioneers and observers who participated in and witnessed the changes that took place in Bahrain during the last sixty years of the islands' oil industry were seen to be of real concern. Rich memories were fading and disappearing too. So a book to record the history of the oil industry in Bahrain was conceived in His Excellency's mind. Without his inspiration, and the patronage of the Bahrain Petroleum Company, this book would not have become a reality.

I am also indebted to His Excellency Tariq A. Almoayed, Minister of Information, whose encouragement and support for almost a decade has been a major contribution to the research and development of my historical studies of Bahrain. In this respect, I also thank Ahmed Al-Sherooqi, Director of Public Relations and Media, for his continuing guidance.

In the course of creating this volume, many people have provided administrative and technical support, academic and legal advice, contributed to the research, and participated in the book's production, all of whom are recognized in the appreciation which follows. However, such a list does not lend itself to embellishment and so by way of an introduction, I should like to mention certain people without whom, quite unequivocally, this book would not have been produced.

In particular, I thank Abdul Hussain Faraj (General Manager, Administration Division, Bapco), who sought me out in London and encouraged me to undertake this assignment; Abdul Hussain Mirza (General Manager, Finance and Legal Division, Bapco), who spared much time unscrambling the complexities of financial and legal agreements, and his colleague, Rahat Siddiq (Legal Adviser, Bapco), whose frequent sorties to the vaults enabled the jigsaw puzzle of financial and legal history to be pieced together.

The chemistry and physics to which I should have paid more attention as a student were retaught to me by Majeed Shafea (General Manager, Logistics and Marketing, Caltex Bahrain), who also spared many hours explaining the technical and marketing aspects of the oil industry in Bahrain. Additional assistance was provided by the Engineering

Services, Maintenance, Refining and Services Divisions of Bapco. A. M. Center and Derek Clark provided important technical assistance. Jalil A. Samahiji (Manager, Exploration and Development, Banoco) advised on geological matters. Official maps of the island were graciously supplied by Ebrahim Al Jowder (Director of the Survey Directorate, Ministry of Housing) and his colleagues, particularly Michael Richards.

The project "engine room" could not have functioned without the technical support of many people. Special thanks are due to Ken Foxall and Peter Holland. John Sammons, Sawsan Aryan and their colleagues (Bapco's Computer Services Department), together with Leonyl Ignalaga and P. T. Joseph – better known to us as Leo and Joseph – maintained remarkable equilibrium throughout the hardware and software challenges which frequently arose. Mahmood Sh. Abdul Latif Al-Mahmood (Public Relations Manager), Samuel Knight and his colleagues in the Public Relations department; Rashid Mubarak Abdulla and Hassan Essa Gattami (Photographic Laboratory, Engineering Services); Ali Ismail Al-Awadi and his team in Awali Services; Abdul Rahman Mahmeed, Ali Ahmed Bubshiet, Khalifa Ali Rashid (Awali Post Office) and Abdulla Ebrahim Ali (Bahrain Refinery Postroom) who took care of all the mail and international courier despatches; Mahmood A. Al-Alawi (Manager, Financial Accounting, Bapco), A. R. Ferdousi and their colleagues; Mohammed A. Al-Kawarie, Ahmed Saleh Al-Mannai and Ali Ramadhan (Office Services); Moosa Abdul Aziz, Hassan Haji and Milton Zuzarte, who maintained the overworked photocopying machine; Akbar Akbar and the Transport Department staff; Hassan Thawady and Najma Sadiq Abdul Rahman in the Bapco Immigration and Travel Department, all deserve special mention. Francis Afonso, Pio Silveira (Al Dar Guesthouse) and their colleagues took care of me during my own stays at the guesthouse as well as visiting contributors, catered for gatherings, prepared for countless meetings and above all kept the electronic whiteboard supplied with paper and pens. Jameela Al-Khooheji and Jamila Jamal will recognize the help they have given me, as will Saleem Mohamed Ali Abdulla. The Chief Executive's secretary, known to us all as Ram (alias A. P. Ramanathan) and Ruth Anderson, my assistant for almost a year, know the true worth of their considerable contributions.

Special thanks are due also to the staff of Caltex who, in Bahrain, Dallas and London, have given significant assistance over many months. In Bahrain, I thank Lilias Picken and Maggie Lazell; Ali Fadhlulla, who delivered courier packages and FAX messages to me; Albert Mansingh and Jothi Samuel for their help with the despatch of seemingly endless facsimiles, as well as George Clements (Majeed Shafea's secretary). In Dallas, Parker Monroe Jr and the staff of the Public Relations Department provided invaluable facilities during the North American research and follow-up phases, as did John Graham (Travel Department), who orchestrated the complex task of hotel, travel and visa arrangements throughout the USA and Canada. In London, Ellen Price and Tom Smythe took care of my travel and courier despatches through Caltex UK. I thank Hazel Pattinson, also in London, for transcribing more than 100 hours of recorded conversations, the foundation of the oral history selected for this book.

The research for this project evolved into an immense undertaking which, from a gentle flow of books and paper, generated into a tidal wave of boxes and files. To all of those people who volunteered to assist this process, as well as those who were caught unwittingly in its ground swell, I extend my most sincere appreciation. Apart from those people who were kind enough to talk to me about their memories, their thoughts and observations, there were almost as many who helped with the technical process of research. In Bahrain, I thank Helen Crombie and the Awali Library staff, Lally Varghese (Bapco Archive), Adel Sharif and Ibrahim Abdul Muhsin (Translators). Dr Husain Dhaif, Director of the English Language Centre and Assistant Professor of English, Bahrain University, interpreted for me on several occasions, translated the conversations from Arabic and advised on transliteration.

In England, I thank Ted Anstey (Technical Librarian, Caltex UK) for arranging the acquisition and loan of essential books and periodicals. Dr Geoffrey Lumsden spent three months ensconced in the British Library, India Office Library, Foreign and Commonwealth Office Library and Public Records Office Library in London research- ing documents which became crucial ingredients to the narrative. J. Peter Tripp, CMG (Political Agent Bahrain 1963-5), and Lt-Col. Arnold C. Galloway, CIE (Political Agent Bahrain 1945-7 and UK-based Bapco representative 1950-68), were kind enough to lend me documents from their private collections of papers.

In America, three libraries provided major research support. Meg Linden (Corporate Librarian) and Marie Tilson (Senior Reference Librarian) of Chevron Corporation, San Francisco; Ruth Baacke (Librarian) and Monica Stein (Assistant Librarian) of the Middle East Institute, Washington DC; and Muriel Hummell (Corporate Librarian, Caltex Petroleum Corporation, Dallas), all assisted me quite beyond the call of duty, and without their enthusiastic efforts this book would be sadly lacking. Louis C. Lenzen in Belvedere, California, and Thomas E. Ward Jr in Summit, New Jersey, were kind enough to lend me documents from their fathers' private collections, and Cdr Hugh G. Story, now retired in San Diego, California, loaned documents essential to an understanding of specific periods in Bapco's history.

The accuracy and coherence of the final manuscript would not have been possible without the guidance of several specialists. In Bahrain, I thank His Excellency Habib A. Kassim (Minister of Commerce and Agriculture, State of Bahrain), His Excellency John A. Shepherd, CMG (British Ambassador to the State of Bahrain), and Dr Henry T. Azzam (Vice-President and Head of the Economics Unit, Gulf International Bank in Bahrain). In America, I thank Michael Ameen Jr, Dr Irvine Anderson and George Ballou, whose knowledge of the Middle East oil arena through their associations with Mobil Oil Corporation, Aramco and Chevron Corporation respectively steered me towards a mature, if not vintage, perspective to the manuscript. I owe an equal debt to Birney Van Benschoten, formerly Attorney to the Caltex Petroleum Corporation in New York, who was kind enough to loan me documents from his private collection and who devoted many retirement hours in Massachusetts to reviewing the manuscript and diverting me from legal pitfalls. Ralph Buck Jr also performed a similar role, with specific reference to the history of the last twenty years. Especially, I wish to thank Dr

Edith Penrose (Emeritus Professor of Economics, University of London) for reading the manuscript, spending several days in discussion with me in Bahrain and for her suggestions which have improved my presentation immeasurably. Dr Dorothea H. El Mallakh (Director, International Research Center for Energy and Economics Development [ICEED], University of Colorado) also reviewed the manuscript and identified errors which might otherwise have been published. I also thank Ian Seymour (Editor of the Middle East Economic Survey, Cyprus) for his critical appraisal of the final text. Any errors which may remain are my own.

For more than two years, Dr Peter Vine (Immel Publishing) to whom I introduced the idea of publishing the book internationally, waited patiently for the manuscript, yet proved that the age of miracles is not yet over by producing the book under extreme pressure in just a handful of months. His task force included Ann Johnson in Immel's London office, and Judith Wardman who painstakingly copy-edited the manuscript to ensure continuity and consistency. Credit for the book's design goes to Jane Stark who took my words on a creative journey through more than a century of change, assisted by the imaginative talent of Bapco's artist, Joe Torres.

Finally, but not least, I thank Don Hepburn (Chief Executive, Bapco) for his unfailing support as my mentor, and his wife, Roma, for her companionship and encouragement during the long and isolated days of writing.

ROLL OF APPRECIATION

To everyone who shared the vision,
contributed to the complex research and review process,
gave academic and legal advice, provided administrative and technical support and
participated in the creation of this book,
the author extends her sincere gratitude

Ali bin Ebrahim Abdul Aal
Rashid Mubarak Abdulla
Saleem Mohamed Ali Abdulla
Haji Abdul Razak Abdullah
Dr M. Morsy Abdullah
Francis Afonso
Akbar Akbar
Adel Hassan Al-A'ali
Abdul Ghaffar Al-Alawi
Mahmood A. Al-Alawi
Hussain Al-Ansari
Ali Ismail Al-Awadi
Yousuf Ali Al-Awadhi
Mohammed Ali Mohammed Hassan Al-Dhaif
Abdulla A. A. Al-Emadi
Ali Ahmed Al-Emadi
Abdul Nabi Kadhim Al-Fardan
Rasool Al-Fardan
Mohammed A. Al-Kawarie
His Excellency Shaikh Abdulla bin Khalid Al-Khalifa
Shaikh Ebrahim bin Rashid Al-Khalifa
Jameela Al-Khooheji
Mahmood Shaikh Abdul Latif Al-Mahmood
Ahmed Seleh Al-Mannai
Saleh Yousif Al-Najjar
Abdul Karim J. Al-Sayed
Fatima Al-Sayed
Khalfan Al-Sayeedi

Ahmed A. Al-Sherooqi
Mohammad Al-Sherouqi
Doris Alford
Starley Alford
Abdulla Ebrahim Ali
Amina Ali
Jaffar Akbar Ali
Mohamed Saleh Shaikh Ali
Thoraya Ali
His Excellency Tariq A. Almoayed
Michael M. Ameen Jr
Jaffar G. Ameeri
Saleh bin Mohammed An-Naimi
Dr Irvine H. Anderson
Ruth Anderson
Celia Andresen
Ray Andresen
C. Michael Anglin
Ted Anstey
Sawsan Aryan
Moosa Abdul Aziz
Dr Henry T. Azzam
Ruth K. Baacke
Jean Bair
Guy H. Baldwin
George T. Ballou
Irene Barkhurst
Robert Bartlett

Ambassador Lucius D. Battle
Rose Nykerk Battleson
Jennie Bell
Dennis M. Berdine
Thomas E. Berry
Irene Biggar
John R. Billingsley
Arthur B. Brown
Derek J. S. Brown
Ali Ahmed Bubshiet
Ralph D. Buck Jr
George M. Calder
A. M. Center
Peter D. Chaplin
Archibald H. T. Chisholm
Derek F. Clark
George Clements
John Creecy
Helen Crombie
Jim Crombie
Russell Davis
Janis Dawkins
Dr Husain Dhaif
Dr Dorothea H. El Mallakh
Ali Fadhlulla
Isabella Fairbairn
Bahia Juma Fakhro
Dr Hassan A. Fakhro
Abdul Hussain G. Faraj
A. R. Ferdousi
Richard G. Fernie
James Foley
R. Gwin Follis
Molly A. G. Ford
Kenneth E. W. Foxall
The Hon. Richard Funkhouser
Helen P. Gallagher
Lt-Col. Arnold Galloway, CIE
Mary Galloway
Hassan Essa Gattami
Raymond Gerhart
Abdul Aziz Abdul Ghaffar
Mohamad Hassan Ghaith
Peter Gilbert
Edward F. Given, CMG, CVO
Kathleen Given
S. Kenneth Gold
John Gornall
John Gourlay

John P. Graham
Hassan Haji
Simon C. Harrison
Ali Hassan
Harold J. Haynes
Dr Frauke Heard-Bey
Donald F. Hepburn
Roma Hepburn
Peter Holland
Howard H. Holm
Marjorie Holm
John Holman
Edwin Howard
Florence Howard
Muriel Hummell
Leonyl Ignalaga
Ebtisan Issa
Abdul Reda bin Jaffar
Majeed Habib Jaffar
Jamila Jamal
Ann Johnson
P. T. Joseph
L. D. Josephson
Ebrahim Al Jowder
His Excellency Habib A. Kassim
Isabel Kerley
Abdullah Ahmed Khalil
Yousif Khalil
Samuel Knight
Edward N. Krapels
Maggie Lazell
Louis C. Lenzen
William H. Leopold
Professor George Leuczowski
Margaret J. Linden
Jack W. Loveless
Dr Geoffrey Lumsden
Laurence (Bud) Machin
Abdul Rahman Mahmeed
Mohammed A. Mannai
Albert Mansingh
Abdul Nabi A. Mansoor
Ahmed Mohamed Mansoori
Philip C. McConnell
Richard H. Meeker
Marie Miller
Abdul Hussain Ali Mirza
Zahra Mirza
Juma Abd Mohamed

David Moloney
Joan Moloney
Parker Monroe Jr
Alan E. Moore
Canon Alun Morris
Ibrahim Abdul Muhsin
Grant Murray
Peter N. Nash
Anthony Nelson
Dr Dennis O'Brien
Gerard O'Donnell
Muriel O'Donnell
Mary Osborne
Hazel Pattinson
Dr Edith T. Penrose
John E. Pepper
Barry Phillips
Lilias Picken
Ellen Price
Innes Rae
Najma Sadiq Abdul Rahman
Ali Ramadhan
A. P. Ramanathan
Khalifa Ali Rashid
Claire Raven
Vern A. Raven
Stuart A. Reynolds
Hussain R. H. Reza
Ruqaya Reza
Michael Richards
Charles A. Rodstrom
Henrietta Rodstrom
A. Karim Salimi
Jalil A. Samahiji
M. John Sammons
Jothi Samuel
Imtiaz I. Sarwani
Thomas P. Scott
Ian Seymour
Majeed A. Shafea
Shamsi Shafea
Adel Sharif
Akbar A. Shefi'ee
His Excellency John A. Shepherd, CMG

Dr Mae Shirawi
His Excellency Yousuf A. Shirawi
S. M. Rahat Siddiq
Pio Silveira
Thomas Simon
Leslie A. Smith
Muriel Smith
Tom Smythe
W. Douglas Sprague
Jane Stark
Monica Stein
William L. Stine
Walter O. Stolz
Cdr Hugh G. Story
Hussain Tadayon
Mohammed Tadayon
Burton W. Teague
René Temoin
John D. Thackwray
Hassan Thawady
Anthony Thomas
Marie Tilson
Joe C. Torres
J. Peter Tripp, CMG
William E. Tucker
Harry F. Tyner
Birney M. Van Benschoten
Lally Varghese
Dr Peter Vine
Barbara Vosberg
Don Kent Wallace
Fred P. Walstow
Muriel Walstow
Patrick J. Ward
Thomas E. Ward Jr
Judith Wardman
Noel Watson
Philip H. G. Watson
John Weir
Rodney E. Willoughby
Aisha Yateem
Hussain Yateem
Dr Daniel Yergin
Milton Zuzarte

USE OF NAMES

Company Names

Company names are contemporary with the period in which they are mentioned in the narrative. Corporate lineages are identified in the Company Guide (Part II).

Country Names

Until the *1st of January 1953*, officially Bahrain was spelt "Bahrein". To remain faithful to historical documents written prior to this date, the former spelling has been retained in quotations. Otherwise, for the sake of consistency, the modern spelling using "a" has been adopted throughout.

It was not until the *21st of May 1935*, that Persia became known as Iran. These country names are used throughout this book in their correct historical context. Thus, from that date until the 17th of December 1954, the Anglo-Persian Oil Company (APOC) became the Anglo-Iranian Oil Company (AIOC).

In the same manner, the body of water which is known today in Arabia and Bahrain as the Arabian Gulf is often referred to in historical documents as the Persian Gulf. This latter name only appears in this book in the context of contemporary quotations. Otherwise the modern preference of Arabian Gulf has been adopted.

Place Names

Many place-names which appear in this book have been transliterated from Arabic. As several Arabic sounds are not heard in the English language, the notes in the Language Glossary (Part III) will help readers to pronounce the transliterated words as closely as possible to the phonetic sounds heard generally in Bahrain.

Names in Quotations

Following correct procedure, the spelling used in contemporary documents has been retained in the quotations which appear in this book.

For example,
>*Gassari* (Qassari),
>*Manameh* (Manama),
>*Bellad-i-Kadim* or *Billad-i-Kadim* (Bilad Al-Qadim) and
>*Avari* (A'ain Adhari).

Technical Names

English-English vocabulary has been used throughout this book. However, to avoid confusion the following anomalies are identified in particular:

>*gasoline* (American) = *petrol* (English)
>*kerosene* (American) = *paraffin* (English)

The Technical Glossary in Part III assists with other expressions.

PREFACE

In much the same way as an archaeologist is captivated by the excavation of artefacts, my search for Bahrain's oil industry history produced many rewards. A fascinating story, beginning with international rivalry, which many years later gave way to the blossoming of Bahrain as a modern independent state, emerged from the wealth of material which accumulated over many months. Fortunately, the temptation to rely heavily on other people's published work, often caused by limited research resources and tight deadlines, was never a choice: few secondary sources specifically relevant to Bahrain's oil industry exist. Thus the quest for first-hand knowledge became an essential requirement if not a compulsion. The contents of corporate library archive boxes, old trunks in neglected attics and precious memories, all of which had remained locked away for years, became the focus of the project's research.

Essentially, this book is the product of primary sources, unpublished material in the form of letters, telegrams, diaries, reports, memoranda and conversations with over 150 people in Bahrain, Canada, Cyprus, the United Kingdom and the United States of America. Tales of intrigue and innovation enthralled me as much as the transition from oil pioneering to modern industrial development. Distilling the volume, discarding the irrelevant and putting the rest to work as a dramatic narrative, remained the challenge.

The result is not a thesis or a technical book, but a celebration. History is not just a study of documents and dates, it is about human beings too. In the case of Bahrain's oil industry, many people of different nationalities, age-groups and walks of life participated in and observed the events which moved and influenced the shaping of modern Bahrain. To enrich and balance the essential aspects of historical detail, a selection of collected thoughts and memories has been woven into the narrative.

In my search for Bahrain's oil industry history, fascination often led me along seemingly unexplored trails or caused me to linger on diversions from the main story line. Similarly, it was inevitable that in a work of this kind one should ask when contemporary history ends and current affairs begin. Predictably, there is no precise answer. One American historian who has faced the same problem suggests that "to rush in upon an event before its significance has had time to separate from the surrounding circumstances may be enterprising, but is it useful?" [Tuchman, 1981, p. 24] Eventually I concluded that enthusiasm is no substitute for sticking to the main track and allowing the material to speak for itself without the author's intrusion.

This is the story of a major episode in Bahrain's history, albeit incomplete, since no historical record can be exhaustive. There are several aspects of the oil industry which cannot be dealt with in detail in a book of this size, but nevertheless I hope that the broad picture of the historical record is clear.

PART I, the *narrative*, introduces the host, the guests and the players. The concession negotiations, preliminary surveys of Bahrain and the terms under which Bapco Limited was eventually allowed to operate are explained. The scene then set, the later chapters track the evolution of Bahrain's oil industry from the day on which oil was discovered on the 1st of June 1932 up until the present day.

PART II, the *reference* section, comprises a chronology of events and a guide to the companies and organizations directly and indirectly associated with the history of the oil industry in Bahrain.

PART III, the *appendices*, complement Parts I and II.

If, as Barbara Tuchman also suggests, history is a collaboration between author and thoughtful readers, the latter making of a book what they choose, not what has been imposed upon them, then I hope that it will be with such spirit that readers will enjoy and make their own interpretations of this celebratory history.

Angela Clarke
Awali
June 1990

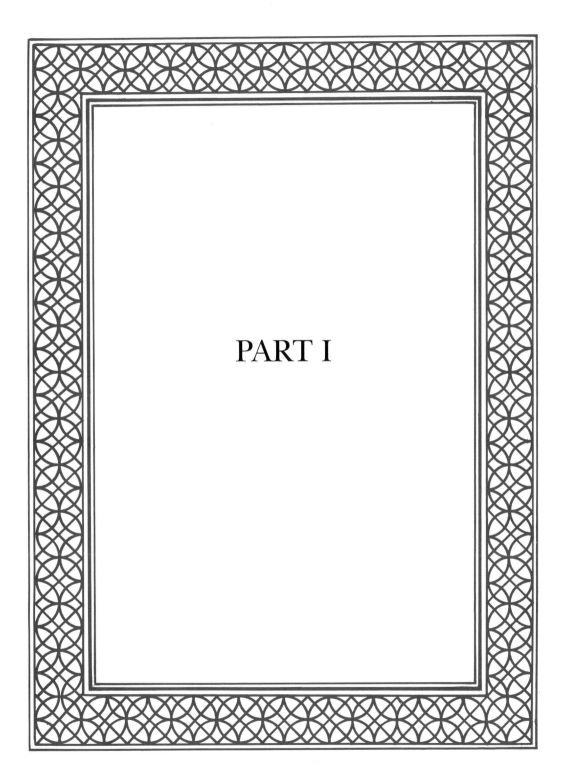

PART I

PROLOGUE

Not long before oil was struck in Bahrain, Dr George Lees, then Chief Geologist of the Anglo-Persian Oil Company, made a striking statement. He declared that he

> "would drink any commercial oil found in Bahrain"
> [Chisholm, 1975, Note 59, p. 162].

Little did Dr George Lees know that he had just ordered the longest drink of his life.

Masjid Al-Khamis.

Al-Khamis mosque (*Barry Phillips*).

CHAPTER ONE

The Host

"On looking out to sea on the morning of a clear sky and a fresh norwester, it would seem as if nature, at all times lavish of effect, had here, however, exhausted every tint of living green in her paint box; and then, wearying of the effort, had splashed an angry streak of purple into the foreground. The water itself is so clear that you can see far down into the coral depths, while springs of fresh water bubble up through the brine, both near the entrance of the harbour and at several other places along the coast" [Durand, *JRAS*, 1880].

Poetic rhetoric was often the hallmark of nineteenth century chroniclers and Captain Durand, a Government of India official,[1] was no exception. If he were to see Bahrain today, he would be confused and incredulous at the landscape changes which have taken place and the increased pace and pressures of life. The medieval twin minarets of *Masjid Al-Khamis* (Al-Khamis Mosque), which had guided seafarers into *Khor Tuubli* (Toobli Bay) for centuries, no longer dominate the horizon. Now, the minarets of Bahrain compete for attention with high-rise buildings and the products of twentieth century technology. Nowhere is this more apparent than in the new diplomatic area of Manama, the modern capital of Bahrain, and some twenty kilometres south where the grey columns of Bahrain's refinery stand proud. Even *Jebel Ad-Dukhan*, Bahrain's highest peak, which Captain Durand referred to as the "Hill of Smoke", now hosts a tropospheric scatter station. In less than a century, the change has been immense.

Travellers, as well as administrators, were moved to write with equally poetic phrases. In 1905, Aubrey Herbert observed that "Bahrain is not antagonistic to life. ... The golden dusted roads which cross it are broad and shaded on either side by long forests of date palms, deepening into an impenetrable greenness, cool with the sound of wind among the great leaves and the tinkle of flowing water."

Dilmun – The Land of the Rising Sun

Bahrain comprises a welcoming group of thirty-three islands, immortalized in the myth that they were the original Garden of Eden, a fertile paradise adjacent to the Arabian desert where Adam and Eve are reputed to have been exiled. No wonder that more than 4000 years ago, traders believed they had arrived on the shores of the islands they called the Land of the Rising Sun, the Land of the Living. Travellers then were convinced that this was a kind of Elysium to which sages and heroes were transported to live eternally in a land which was then, reputedly, the home of the gods. It seems almost certain that the island "abundant with fresh water and lying some two days' sail with a following wind from Mesopotamia" was Bahrain. From the third millennium BC until the time of Cyrus of Persia, there is repeated mention in cuneiform texts of a land somewhere to the south of Babylonia called Dilmun. In the Sumerian language the name is often written as *Ni-Tukki*, sometimes deciphered as "the place of the bringing of oil", for the Dilmuns are thought to have used petroleum seepages to caulk their ships and export cordage to their trading partners in the Arabian Gulf and the Indian Ocean.

Gilgamesh (Joe C. Torres).

Interestingly, modern excavation in Bahrain revealed a unique bitumen cup attributed to the Dilmun period.[2] The name also occurs as *Kur-Ni-Tuk*, deciphered as "the mountain of Dilmun", and as the place-name *Niduk-ki* which appears in the Babylonian *Myth of the Creation*.

Believed to have been the centre of the Dilmun civilization from the end of the third or the beginning of the second millennium BC to as recently as 400 BC, Bahrain was steeped in the traditions of Eridu, the most southerly and reputedly the oldest Sumerian city. Within the Barbar Temple complex which the Dilmuns built on the northern shore of their paradise, they worshipped Enki, the God of Wisdom and the Sweet Waters under the Earth, at a subterranean shrine interpreted as a symbolic temple *abzu* or sacred well. Reputedly, Enki lived in the *abzu*, believed to be the abyss of fresh water upon which the whole world rests. Enki became the father of the patron divinity of Dilmun, and his city was Eridu.

The oldest epic poem in the world, *The Epic of Gilgamesh* [Sandars, 1972], is thought to be Sumerian. Written on cuneiform clay tablets which were excavated in Mesopotamia more than 100 years ago, subsequently pieced together and more recently deciphered, it describes Dilmun as a land of immortals.

Utnapishtim, a wise king and priest of Shurrupak, was taken by the gods to live there forever. Gilgamesh, fifth king of Uruk and hero of the epic, sought and found the flower of eternal youth in Dilmun, so that when he returned to Eridu he could make the old men of his city young again. Interpretations vary as to how his vision was destroyed: either as he swam in an enticingly cool pool on his way home or as he lay asleep by the water's edge, a serpent devoured the flower of eternal youth. Whichever version one accepts, there seems little doubt that Gilgamesh returned to Eridu forever a mortal.

For millennia, Bahrain has been blessed with a rich source of underground springs, both beneath the sea and land. Many Arabs considered in ancient times that the source of these springs was an underground river which was an extension of the River Euphrates flowing south beneath the coastal region of Arabia.

"The principal springs are the *Gassari*, on the road from Manameh to *Bellad-i-Kadim*. The *Umm Shaoom*, a mile to the eastward of Manameh, the *Abu Zeidan* in *Billad Al-Qadim* and the *Avari* which last supplies many miles of date-groves through a canal of ancient workmanship with a perfect river of fast-running water, some 10 feet [3 metres] broad by 2 feet [0.6 metre] in depth. The spring itself is some 30 to 35 feet deep [9 to 10.6 metres], and rises so strongly that a diver is forced upwards on nearing the bottom. I do not mean that you cannot reach it, but merely that the force of the water is felt against you. The water where it rises in this deep spring… is some 22 yards across by 40 long [20.1 by 36.5 metres], is as clear as crystal with a slightly green tint and very beautiful. It holds a shoal or two of large fish and many water-tortoises" [Durand, *RGIFD*, 1880, p. 15].

Port of Pearls

Since the days of the Dilmun civilization, Bahrain's importance as a strategic and flourishing entrepôt is well documented. Traders stopped to replenish their supplies of fresh water, dates and salted fish before continuing on their way. New evidence of such commercial activity is found regularly by archaeologists, such as Chinese coins, thought to be 800 years old, discovered at Ras Al-Qala' (Headland of the Fort).[3] But perhaps the Arabian Gulf's most recorded maritime activity is pearl-fishing; for millennia Bahrain has been famous for its oyster beds and harvests of pearls.

An ancient cuneiform tablet found at Ur, dating to circa 2000 BC, mentions "a

Opening oyster shells and harvesting pearls at the end of a day's diving (*Joe C. Torres*).

parcel of fish-eyes" from Dilmun. Some scholars have interpreted this to mean pearls from Bahrain's waters. Arabian Gulf pearls, reputed to be the finest in the world, are referred to in another ancient cuneiform inscription from Nineveh: "In the sea of changeable winds [the Arabian Gulf] his merchants fished for pearls".

Pliny, writing in the first century AD, states that Tylos (the classical name for the island of Bahrain) "is famous for the vast number of its pearls". The Arabian Gulf pearl fisheries were celebrated in *The Periplus of the Erythraean Sea*, written by a Greek sailor about AD 60. The Holy Koran refers to pearls as the property of paradise.

By the fifteenth century, the Arabian Gulf had attracted the attention of Portuguese explorers such as Duarte Barbosa. In 1485 he wrote of Barem (Bahrain): "Around it grows much seed pearl, also large pearls of good quality. The merchants of the island itself fish for these pearls and have therefrom great profits". Just less than a century later, an English merchant, Ralph Fitch, visited the island of Hormuz at the mouth of the Arabian Gulf and remarked on the "great stores of pearls which came from the Isle of Bahrain … the best pearls of all others".

This chronicle of international attention and economic success continued for more than four centuries. Although, "it is a matter of great difficulty to arrive at anything approaching to a correct estimate of the amount and value of the pearls that are now yearly harvested, as they are carried to many different markets to suit the varying tastes of the nations" [Durand, 1878, p. 27], it is thought that about 90-per-cent of the Arabian Gulf's pearls passed through Bahrain. In the earliest recorded figure, the value of Bahrain's pearl production was estimated to be about two thirds of the total value of the Arabian Gulf fisheries for that year [Wilson, 1883, p. 284]. The last individual return on the principal exports from Bahrain was for the year 1889. Thereafter, the annual return was consolidated in a statement which included all ports along the Arab Coast of the Gulf.

The oscillating fortunes of the pearl merchants were especially apparent during the 1899 season. Anticipating that the opening of the Paris Exhibition would create an abnormal demand, the merchants "keenly competed for these gems and raised the price from 40 to 60-per-cent. Their expectations, however, were not fulfilled, and many of them sustained heavy losses and some became bankrupt" [Calcott Gaskin, 1901, p. 112]. This state of affairs only intensified the pressures on the pearl-diving fleets, whose seasonal harvests already varied according to the weather conditions at sea, the divers' health and the condition of the pearl banks. To make matters worse, the Political Assistant to Bahrain reported in 1900:

"The prosperity of the Bahrein Islands primarily depends upon the pearl fishery, in which about one-half of the male population is occupied. The fishery in the year under report opened on the 12th May … and closed on the 17th September. One of the principal pearl banks situated to the north of the islands, where the oysters were found to be diseased and producing no pearls, was abandoned in the early part of the season … the chief cause in the falling off in the quantity of pearls obtained" [Calcott Gaskin, 1901, p. 112].

The harvesting of pearls required great skill, endurance and courage on the part of the divers. From Hidd village on Muharraq island to Jasra on the west coast of Bahrain island, the anticipated 300 to 400 rupees a month which *might* be earned if the diving was favourable barely compensated for the emotional farewells at the beginning of each season. As Abdullah Ahmed Khalil, a veteran puller on a diving dhow, commented: "The worst aspect of it was parting from your families." Mohammed Mannai, a Bahraini jeweller, recalled one of his grandfather's comments: "… the people were very close to each other … even today, some Bahraini ladies don't like to wear pearls or anything related to the sea, because it took their husbands, fathers, brothers and sons away from them. … *Inshallah*, they returned safely, but everyone knew of the hazards." In later years, when a hospital boat was available, it was usual for it to make four trips a season to the pearl banks and treat more than 400 patients at sea.

It was not an easy life, yet at the beginning of the twentieth century there was little alternative, either as a substitute to or a supplement for the unpredictable pearl-divers' and pullers' incomes. Although it was conventional for the dhow captains to pay the crews an advance at the beginning of each season, their ability to repay the loans depended very much on the fortunes of the forthcoming season. Agriculture was not developed sufficiently at that time to create additional employment and income. Trade and merchant activity are not labour intensive and provide little employment for the local people. The primary supplementary activities were textile and mat weaving, basket making, pottery and lime manufacturing, as well as boat building, which in 1903 employed 200 carpenters, whose wages had doubled in recent years. Nearly 130 boats, ranging from 300 rupees to 8000 rupees, were sold along the Arab Coast during that year.[4] This was good news, but with 50-per-cent of the male population employed in the pearl industry, it could hardly absorb more than a fraction of that percentage who might wish to seek safer, healthier, more secure and year-round work.

Foreign Trade

For thousands of years travellers and traders have visited the islands of Bahrain, established communities on them and used the entrepôt facilities they had to offer. As long ago as the Formative Dilmun Period, now considered to have existed between 3200 and 2200 BC, Bahrain was a trading station, part of an international network which included the region east of the River Tigris, Dilmun, Makan (Oman) and Meluhha (the Indus Valley). Bahrain's link in these ancient trade routes was found during excavations at Ras Al-Qala' and Diraz, believed to have been trading sites on Bahrain island circa 2400 BC. Between 200 BC and AD 200 the entrepôt trade through Bahrain was immense, particularly because the demand by the rich Roman and Parthian empires for Arabia's incense seemed to be inexhaustible.

During recent centuries, commodities were transhipped between Bombay and Basrah, with vessels calling in at Muscat, Bandar Abbas, Bahrain and Bushire en route both up and down the Arabian Gulf. Eventually, wind power gave way to coal-burning engines, as square-rigged ships and lateen-sailed dhows were superseded by steamers. Today, sea and air cargo shipments are handled by ocean-going container ships and aircraft, fuelled by refined oil products.

Whilst there was no refining capability in the Middle East at that time, refined oil products were no strangers to the Arabian Gulf region. In the 1860s more than half of the kerosene refined in the United States was exported.

By the 1880s petroleum and its products, mostly sent to Europe, ranked fourth in the value of the United States' exports. Standard Oil had secured a dominant position in the export trade, achieving a near-monopoly in Asian markets. Although it is not certain whether Standard Oil's products reached Bahrain at that time, it is known that kerosene was transhipped to Bahrain in tins, bearing the then familiar lion trademark.

As Rockefeller's Standard Oil interests were expanding in the United States during the early 1880s, the Europeans were widening their commercial interests and exporting manufactured goods to the Middle East. The Suez Canal, opened in 1869, had quickened the journey and made otherwise inaccessible ports along the Red Sea available for trade. The Canal had made the passage of square-rigged ships to the Indian Ocean and the Arabian Gulf safer, avoiding the hazards of sailing round the African continent. Overland routes from the eastern end of the Mediterranean across Iraq and down to the port of Basrah were well-established, and from Basrah four steamer companies [5] competed for business in the transhipment of goods to the Gulf's coastal ports and onwards to India.

In 1879 Emil Tietze travelled east from Europe. His destination was Arabia, Mesopotamia, and Persia. If, when preparing for his survey trip, Tietze had seen the draft Abstract Table showing the total estimated value of articles imported into Bahrain during the previous year, it is possible that his eyes would have scanned the pages until they rested on the oil entry [*ARPGPR*, 1879, Table 15]. Fish oil had been imported from Aden, the Red Sea, ports along the southern coast of the Arabian Gulf, as well as Kuwait, Basrah and Baghdad. Lesser amounts of gingelly (jinjali), coconut and sesame oil had been imported too.

Interestingly, just one person is mentioned in the entire document (apart from the compiler), the oilman. He rates a mention because his store provisions, such as flour, ghee, salt and saltpetre, are meticulously itemized.

No lamp oil or naphtha were specifically imported that year, although it is likely that a small amount was included in the shipment of "other oils". Imported candles would have illuminated the islands' homes for the year. Firewood and charcoal serviced heating and cooking fuel requirements. Food, fabric and frankincense were among the many other listed commodities. Donkeys, horses, cows and sheep arrived from India and Persia by steamer and dhow, along with cargoes of cotton bales, gold lace, thread and twist. Medical consignments of quinine, camphor and salammoniac; food shipments of rice, nuts, vegetables, sugar and spices, together with metal for blacksmiths, timber for dhow and house builders and canvas for sail makers were part of the inventory too.

The former long distance dhow harbour, Manama, now reclaimed land (*Caltex*).

What was not needed for Bahrain's requirements was re-exported to other countries in the region. The port of Bahrain was a busy place for traders, with the emphasis firmly fixed on merchant and entrepreneurial activities. The export of indigenously manufactured products was conspicuously absent; even locally produced pearls and shells were the products of nature and not man.

At the beginning of the twentieth century, pearl exports were the largest on record and of better than average quality. [6] The demand for small pearls grew, as did the price of larger ones. Business was brisk, stimulated by the approaching coronation of the British monarch, King Edward VII, which created a bullish European market for the gems. Bahrain's role as an entrepôt was buoyant, although it had to be remembered that the good overall trade figures were the result of considerable commodity transhipment, such as coffee, tea, rice and shells, an activity which created little indigenous employment. The fluctuations in the rates of exchange were less violent than usual. The Austrian dollar and the Turkish lira, of which a considerable quantity was used, passed hands in exchange for Indian rupees. As yet, sterling and the American dollar were not the usual trading currencies in the region. Even the Gulf rupee, the predecessor of the Bahraini Dinar, had not yet come into being. [7]

Although import and export statistics were not kept by the customs authorities, published figures were compiled from the manifests of steamers, courteously lent by the agents, records of local merchants and other sources. They were regarded as being fairly representative of the islands' trade and provided some indication of market trends and the growing demand for all descriptions of piece-goods which, Calcott Gaskin reported [1901, p. 112], "ought to be of interest to manufacturers and merchants in India as well as in the United Kingdom, whose attention I would particularly draw to the study of the requirements of this important market, which, in time, it is expected will develop considerably."

Already the German firm of Traun, Sturken and Co. of Hamburg had opened a branch office in Bahrain to do business in mother-of-pearl shells. [8] Commenting particularly on the lack of transport for the estimated 600 tons of remaining shell stock at the close of the following year, the Assistant Political Agent remarked that:

"The advantages of this trade, as well as of the general business to be done in Bahrein, appears to be better appreciated by foreign trading houses than British merchants. The French firm of M. M. Dumas and Guien of Marseilles has recently opened a branch office. It is hoped that some energetic British firm will realise the importance of this growing trade and will endeavour to obtain a share in it before it is too late" [Calcott Gaskin, 1903, p. 35].

He added that the stability of the Indian rupee was being appreciated by the Arabs and, as a consequence, the circulation of the Maria Theresa dollar was decreasing. Exchange transactions with Europe were arranged principally through Bombay banking houses and in some instances through Bushire, where the rates were higher than at the ports where banks existed.

Market and Mosque

Like most towns in the Middle East, those in Bahrain developed with a market and mosque as their focal points and homes were within walking distance. Commercial life was integrated with daily prayer. A member of a prominent business family [9] reflected on his grandfather's reminiscences:

"The *suuqs* [markets] developed themselves around mosques, so that when the men were called for prayer they stopped their shopping or trading and went to pray. Due to the longer time taken in travelling on donkeys, the working hours were shorter and would start much earlier than now – 5 or 5.30 in the morning until praying time at noon."

Al-gahwa (the coffee shop) was, and remains, an integral part of suuq business and social activity. Before the days of offices and telephones people would meet there to discuss business. Some coffee shops were frequented mostly by the *Tawawiish* (pearl merchants) and *nowaakhdha* (pearling dhow captains), and they were known accordingly. But most of them were simply places of relaxation where dominoes and other games were played or people just sat and talked. Suuq trade usually took place in the mornings, since many people had to travel long distances, often on foot, to buy or sell. Unlike now, when many of the shops in the suuqs open in the evenings, traders would ensure that they were home by sunset. A resident of Karzakhan,[10] which until a few years ago remained a small village on the west coast of Bahrain island, describes his routine journey to Manama, twenty kilometres away as the crow flies, the actual distance varying according to the route taken:

"I used to walk from Karzakhan village to Manama. I used to leave here at dawn and arrive in Manama in time to take dates, tomatoes and fish to the market. Sometimes we would leave here earlier, about three o'clock in the morning and arrive in Manama by dawn. Of course, in the past, there was no gas for fuel or even stoves. There was only wood from the palm trees, so when the vegetable season was over I used to take wood to the market instead. We used to walk with the donkeys which carried the load. Before, we used to take the shortest way, but when the road was made, we followed that. We left Karzakhan, passed Buri, on to A'ali, then Salmabad, then Radm Al-Kawara [Kawara Bridge]. From there we would walk until we reached what is now called Baqer Furniture. From there we continued through the Water Garden to the centre of Manama. Sometimes we would be home before noon. Then we would go to work in the palm fields or in the garden."

A Bahraini shop vendor taking a break to smoke a *gidu* and drink *chai* (*Barbara Vosberg*).

Each market specialized in a particular commodity or opened on a specific day of the week. *Suuq Al-Khamis* (Thursday market), a fine example, was, until the 1960s Bahrain's donkey market. The Khamis neighbourhood is one of the oldest in Bahrain, for it is here that the first phase of *Masjid Al-Khamis* (Al-Khamis Mosque) is believed to have been constructed during the second half of the eleventh century. Nearby, the former

Water being drawn from the well at A'ain Hunainiiyah, circa 1930s (*Bapco archive*).

medieval capital of Bahrain, *Bilad Al-Qadim* (the Old Town), was built on the habitation mounds of an earlier Islamic city. Travellers from Manama with an interest in either the ancient burial mounds near A'ali village or the more distant Jebel Ad-Dukhan would have journeyed through this medieval town and past the equally historic mosque.

Spring water was brought to market too, as well as to people's homes, from as far away as *A'ain Hunainiiyah*, below the rimrock of East Rafa'. There the sweet water, reputed to be twenty fathoms deep at one time, was raised by a goatskin-bucket let down over a pulley and walked up to the cistern level by cattle or donkeys pulling down an incline from channels, generally levered by a palm-trunk lightly swung by ropes to a frame. This was balanced at one end by a basket of earth so that little exertion was required to lift the water. Every day water sellers would walk with their donkeys or camels, each animal carrying between six to eight goatskins for many kilometres at a time. Once at his destination, the water carrier might call out "Yo, Yo, Yo", which means "Bring the water". Buyers decanted the water into clay pots made in A'ali village, so that it cooled. But, as one Bahraini recalls, the contents of one goatskin would cost twelve annas which meant that this sweet well water was too expensive for most people to buy.

The Rhythm of Life

Many people relied on the *oyuun* (natural springs) near their homes in the villages as a focal point to gather, talk, bathe, launder clothes and collect water. Some springs contained *helow* (sweet water) suitable for drinking. The water in other springs, known as *khariij*, is a mixture of salt and fresh water which, although too brackish for drinking, was used for bathing and laundering clothes. The islands of Muharraq, Sitra and Nabiih Saleh, in particular, were blessed with many fresh water springs, as was the northern coastal region of Bahrain island. A Bapco employee [11] who has lived all his life on Sitra island remembers his childhood:

> "On Sitra we had *A'ain Al-Raha*, *A'ain Abdan* and *A'ain Mahazza*, but now they are no longer there. During the summer nobody could swim in the night because the water was too cold. In the winter time it was very hot. This is because of the heat transfer in the soil which takes place faster than the liquid. During the summer when it is hot, the water is colder. In the winter, if the atmosphere is cold, the water feels warm. The water came from underneath the ground. People filled their clay containers with water and washed their clothes in the *oyuun*. Children could not swim in these springs as they were deep and fast-flowing. The water at Nabiih Saleh [island] was the most dangerous, particularly *A'ain Saffahiiyah*, where children could swim inside the big holes and disappear completely. Now everything is safe as there are concrete sides and steps."

The spring water supplies were used not only by the residents of Bahrain. For centuries, trading vessels have sheltered in the bays around the northern area of Bahrain and replenished their fresh water, dried fish and date supplies. In recent times, the two-weekly mail steamers were no exception. Yet to any traveller or trader who arrived in Bahrain the rhythm of life would have been immediately obvious. Prayer and family life are central to Bahraini tradition and culture, based on the words of the Koran. Islam has permeated business and family communities in Bahrain for generations. One lady [12] from a prominent Bahraini family and now well into the eighth decade of her life remembers:

> "After sunrise prayers [*Salat Al-Fajr*] people would rest a bit or read the Koran. The men would have their breakfast and then go to work. The women would do what had to be done, then guests would arrive around 9.30 in the morning. Family friends brought their sewing with them and big trays of refreshments would be brought in. About 11.30 the guests went home, just before noon prayers [*Salat Al-Thuhr*]. Then lunch, the main meal of the day, was served. We did not have much fruit in those days, except water melon which was brought by steamer from Iraq and Iran. There were always dates, fresh or dried. After lunch was a time when no visitors came. We had our privacy. According to the Koran you are supposed to

have three times of privacy a day: morning, after lunch and at night. Therefore there would be no visitors until after the afternoon prayers [*Salat Al-A'asr*] about four o'clock. Between sunset prayers [*Al-Maghrib*] and the final prayers of the day [*Salat Al-A'isha*], one-and-a-half hours apart, we'd have dinner, sit and tell stories. In those days, a few books were available which we could read by kerosene lamp, but not everyone could read, so tales were remembered and repeated. In the days before television and car travel, children would often play blind man's buff or statues before going to bed, especially on moonlit nights."

Although it would have been less obvious to the newcomer, after a short time visitors to the islands would have become aware of the tribal system of government which operated in the country. Traditionally, throughout the entire Arabian peninsula region, the head of each tribe was not a despotic ruler of his subjects but the "father" of his own people. The Rulers of Bahrain [13] were no exception. Even today they look after the welfare of their people and are accessible to listen to their concerns personally. As one cabinet minister [14] now expresses the concept, "every system in the world allows a child to see his father. So it is with the Bahraini system." Whereas in the West a firm leader will make decisions which do not suit every constituent, the ability to do this is not necessarily one which qualifies a man to be a great leader in the Arabian Gulf region. The paternal approach is as much a concern of the Amir of Bahrain today as it was for his forefathers.

Life in Bahrain may be conducted more slowly and less directly than often seen elsewhere in the world, but frequently it is done with more restraint and politeness. When the first oil explorers and developers arrived in Bahrain, they would have found few similarities with the way and pace of life they knew in America or Europe. The fabric of Bahraini society may be likened to an intricate carpet: it comprises many patterns woven with many threads. Like any complex carpet design, it is impossible to appreciate it fully at the first viewing. It takes time to absorb the work in its entirety, if it is possible ever to do so.

The relationship of Bahrain's oil resources with its people and those international participants who helped to shape the industry's early development is not straightforward. Perhaps an appropriate analogy is the suggestion that oil revenues are the warp threads on Bahrain's loom. The weft patterns of modern economic development and social change have been woven during the last sixty years or more to represent an achievement that many countries and individuals may admire. This is its story.

Water carrier in Manama (*Joe C. Torres*).

Edwin L. Drake and Peter Wilson at Drake Well in 1861 (*Drake Well Museum, Titusville, Pennsylvania*).

CHAPTER TWO

Early Oil Rivalries

Crude oil is not a wonder confined to the twentieth century. The hanging gardens of Babylon were one of the seven wonders of the ancient world and, reputedly, the walls of that city were cemented with bitumen. As recently as 1200 years ago, crude oil was produced from hand-dug wells in Burma and in Japan. The use of the drill to dig wells came much later and the commercial development of crude oil resources did not occur until the nineteenth century, following the industrial revolution. Steam engines and boilers, capable of generating power, formed part of the appropriate technology necessary for deep oil well drilling.

Edwin Drake was partly responsible for this. Although he did not invent drilling, the effective beginning of modern crude-oil production is often dated from 1859 when he first applied the technique in America. Using a six-horsepower steam engine, he managed to drive a pipe down into bedrock and thereby demonstrate convincingly that large deposits of oil could be extracted from deep below the surface by drilling. Drake never benefited from his achievement financially, but his legacy to the world's oil industry was a technique which became known as the "Pennsylvania" standard rig, 65-85 feet (20-26 metres) tall. Interestingly, the designed drill depth was 2000 feet (609.6 metres), just eight feet (2.4 metres) less than the depth at which oil was discovered in Bahrain 73 years later.

With the use of Drake's technique, oil production boomed in both the United States and Europe, prompting the establishment of refining, transportation and distribution networks. In the United States, the industry came under the domination of a great entrepreneur, John D. Rockefeller, who in the 1860s, together with business associates, started to build or acquire control over several refineries, transport facilities, railroads, pipelines and ancillary services across the continent.

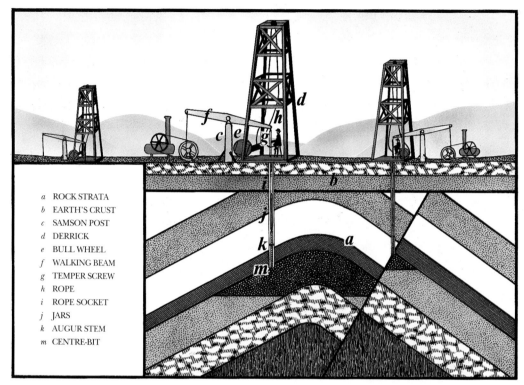

a	ROCK STRATA
b	EARTH'S CRUST
c	SAMSON POST
d	DERRICK
e	BULL WHEEL
f	WALKING BEAM
g	TEMPER SCREW
h	ROPE
i	ROPE SOCKET
j	JARS
k	AUGUR STEM
m	CENTRE-BIT

"Drake's Folly", an innovative oil derrick devised by Edwin Drake to improve the system of oil exploration (*illustration: Jane Stark; source: Drake Well Museum, Pennsylvania*).

International Rivalry Begins

In 1882 Rockefeller formed the Standard Oil Trust, bringing more than thirty oil-related companies together under this organizational structure. Before long, Standard Oil not only dominated the United States oil industry, but also had established itself as the world's most important supplier of oil products, particularly kerosene, to markets in Europe, India, China and elsewhere in the Far East. Nevertheless, Standard Oil did have competition. By the 1880s, the Russian oilfields at Baku were in full production, as had been those in Romania. Apart from the Russian independents, the Nobel brothers of Europe and the Rothschild interests in Paris were of great importance. Rivalry became intense, particularly because transport from Europe to the Mediterranean, the Near and Middle East, as well as Asia was much cheaper than from North America.

In Indonesia, a small company was formed in 1890 to exploit the oil of the Dutch East Indies. A decade later, Henri Deterding became its manager. Matching Rockefeller's commercial aggressiveness and brilliance, Deterding soon brought Standard's largest foreign rivals into one marketing organization. Thus, Royal Dutch Petroleum, Shell Transport and Trading, and Rothschilds became Asiatic Petroleum, supplying Asian markets.

John D. Rockefeller (*Rockefeller Archive Center*).

At the beginning of the twentieth century, three other rival organizations were formed: the Anglo-Persian Oil Company (APOC), Burmah Oil Company and Gulf Oil Corporation, all of which became part of the Bahrain oil industry story in one way or another.

The origins of APOC began in 1901 when an English entrepreneur, William Knox d'Arcy, decided to spend some of the money he had made in the Mount Moyan Goldfield in Australia. Instead of gold, he chose oil exploration. Having obtained a concession in Persia and financial backing in London he developed his plan. In 1908 oil was found, and the Anglo-Persian Oil Company (APOC), partly financed by Burmah Oil Company, was formed in 1909 to develop the newly discovered oilfield.

At more or less the same time, British interests in the Burmah Oil Company successfully blocked Standard Oil's efforts to explore for oil concessions in Burma and India, as Dutch interests had done in the Dutch East Indies. By 1905 Burma supplied over one-third of the kerosene trade in India. Although the trade records do not specifically say so, it is quite probable that some of the kerosene exported from Bombay to ports along the Arabian Gulf coasts, including Bahrain, originated from the Burmah Oil Company.

Gulf Oil Corporation, an American company, was formed in 1907, the year before oil was discovered in Persia. The purpose was to acquire the stock of two other American companies which had their roots in the early history of the oil discoveries in Texas and Oklahoma. Financed by the Mellon banking group, the company grew to become a major player in the international oil industry, particularly in South America. Eastern Gulf Oil Company, one of its subsequent subsidiaries, later acquired an option on the Bahrain oil concession which ultimately it assigned to its competitor, Standard Oil Company of California.

While these oil companies were staking out their territorial interests, J. Calcott Gaskin (by now Assistant Political Agent in Bahrain) was beginning to wonder about the strange tales he had heard from local fisherman. Fascinated and curious, he decided to find out more about them:

"At the end of March, when some of the headmen of the Dawasir Arabs came to pay me a friendly visit, amongst other things in their conversation they mentioned that on the conclusion of the pearling season in September 1902, a diving boat owned by a friend of theirs was beating up the Gulf from one of the southern pearl-beds. At about ten to fifteen miles [16 to 24 kilometres] north of Ha'lul island, the attention of the people aboard was attracted by an agitation of the surface confined within a circle of some yards in diameter. As the locality was known to have deep water, they sailed up to the spot to satisfy their curiosity and were surprised to find on reaching it, that liquid bitumen was being thrown upwards to the surface which smeared the sides of the boat as they passed through it. If these statements are true it would appear that a natural spring of liquid bitumen or crude petroleum which occasionally is found in eruption, exists somewhere in the locality indicated and may be worth exploiting" [Calcott Gaskin, 23 April 1904].

Calcott Gaskin's holograph letter reached the desk of the Political Resident in Bushire. To carry out a proper investigation, PRPG would need the services of a geologist and the British Navy. The Surveyor General of the Government of India, Lt-Col. F. B. Longe, was given instructions to commission a survey of Arabia and the islands of Bahrain for the purpose of determining the oil exploration potential in that region. HMS Lapwing was seconded for geological duties and the Deputy Superintendent of the Geological Survey of India, Guy E. Pilgrim, was conveyed to the vicinity.

On the 9th of June 1905, Pilgrim reported to the Political Resident in Bushire that he had failed to find any traces of bituminous material or to detect any bituminous smell. He had carefully examined Halul island and Diyina island but:

"I cannot recommend that any serious thought be given to possible mining operations in the sea North of Halul, at all events not until mining operations of a less expensive nature in other localities in the Gulf shall have given us a reasonable hope that an exploitable reserve of petroleum exists anywhere in the Gulf, and shall have proved that such local indications are something more than mere surface manifestations connected with a deep-seated and fitful volcanic activity" [Pilgrim, 1905].

The Political Resident Persian Gulf promptly communicated Pilgrim's findings to the Government of India in Simla. The PRPG at this time was a 40-year old diplomat named Percy Zachariah Cox who was to become both a linchpin in Britain's search for Middle East oil concessions and a threat to the Bahrain concession negotiations.

It was therefore an event of considerable interest to the British when, in 1908, oil was discovered in Persia and G. E. Pilgrim published the geological account of his 1904-5 reconnaissance of the Arabian Gulf region. In his analysis of Bahrain, he discussed the occurrence of bitumen in relation to the possibility of finding a petroleum supply.

"... The strata which overlie the Jebel Dukhan beds are admirably adapted lithologically for storing petroleum. ... In the present instance, ... not only the whole of the 'cover' but also the whole of these porous beds have been removed

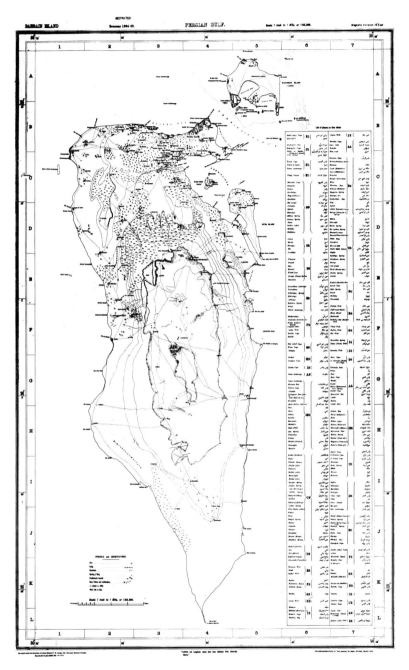

Bahrain Island, as surveyed during the 1904-05 seasons under the direction of Lt-Col. F. B. Longe, Surveyor-General of India (India Office). It should be noted that Longe did not include Umm An-Nassan island nor the Hawar archipelago on his geological map although they are part of the State of Bahrain (*India Office, London*).

from the anticline by the operation of atmospheric forces, so that any petroleum that ever existed in these beds must long ago have drained away. There is, of course, a possibility that beneath the limestone that now forms the topmost beds of the anticlinal crest there exists another porous stratum" [Pilgrim, 1908, pp. 113-14].

It was a prudent account, but one which gave hope that oil *might* be found in Bahrain. Six years later, in 1913, S. Lister James (Chief Geologist of the Anglo-Persian Oil Company) accompanied a tour of the Arabian Gulf by an Admiralty Commission to inspect the d'Arcy Concession area, which did not include Bahrain. In January 1914, Lister James noted in his report that: "In view of the definite occurrence of asphalt and the ideal nature of the structure [on Bahrain], it appears inadvisable to ignore the area before testing with a fairly deep well" [Owen, 1975, p. 1322].

Whilst the Anglo-Persian Oil Company never seriously pursued this interest, naturally it did not look with approval on attempts by others to secure competitive oil rights in Bahrain.

Bahrain and Britain – A Treaty Relationship

Rivalry and international competition are no strangers to Bahrain. In the sixteenth century, Portuguese colonial aspirations in the Arabian Gulf region were neither discreet nor disguised, as the remains of their monumental fortifications at Qala'at Al-Bahrain (Bahrain Fort) bear witness. British interest in the area can be traced back to the East India Company's trading contacts in 1616. [1] Later the Ottoman Turks extended their ambitions south and, particularly during the latter part of the nineteenth century, harassed the Arabian Gulf's pearl-diving fleets. On more than one occasion, the Ruler of Bahrain, Shaikh Isa bin Ali Al-Khalifa, found himself reluctantly having to ask the British Political Resident in the Persian Gulf to despatch a gunboat from Bushire to protect the islands' dhows and maintain order at sea.

As late as the second decade of the twentieth century, the Ottoman Turks were still entrenched in part of the southern flank of their empire, Hijaz, now incorporated into the western province of the Kingdom of Saudi Arabia. The Ottoman Empire extended eastwards across the route to India, which included Bahrain. For much of the nine-teenth century, the British considered this route essential for the safety of imperial communications from London to Bombay and supported the Ottomans against the threatening power and increasing influence of Russia. Eventually, Britain lost interest in maintaining that friendship and allowed Constantinople to look to Berlin rather than London for help. [2]

In 1861, Bahrain and Britain signed a Treaty of Perpetual Peace and Friendship concerning matters of maritime aggression, slavery and trade. In accordance with the treaty terms, Britain established an administrative network from Bombay to Basrah, which included Bandar Abbas, Bahrain and Bushire, all coastal ports of the Arabian

Portuguese bastion, Qala'at Al-Bahrain (*Department of Tourism and Antiquities*).

Gulf. After 1873, when responsibility for Arabian Gulf affairs was transferred from the Government of Bombay to the Government of India, printed administration reports were produced annually by the Resident, compiled from individual reports submitted from the agents and consuls resident at the various administrative centres. The Political Resident Persian Gulf (PRPG), as he was called in contemporary documents, was based in Bushire. (After Lord Louis Mountbatten relinquished his post as Viceroy of India on the 15th of August 1947, signalling the end of 163 years of British rule, the British Residency reported to the Foreign Office in London.)

Whilst the Political Resident Persian Gulf was responsible overall for the area, the day-to-day administration was attended to by Political Agents based at the administrative centres and coastal ports. In 1900, the first incumbent of that post in Bahrain was appointed. Thereafter, all communications with the Ruler of Bahrain concerning administrative matters were dealt with in the first instance by the Political Agent, who either referred matters directly to the Ruler or, as was the case with the oil concession negotiations in particular, discussed them with the PRPG. During the 1920s, this administrative arrangement between Bahrain and Britain proved to be the critical counterpoise in the Anglo-American negotiations which focused on the Bahrain oil concession and oil company operations in Bahrain.

Forces of Change

In May 1911, the Sherman Antitrust Act, which had become United States law on the 2nd of July 1890, was enforced upon Rockefeller's monopolistic Standard Oil Trust. A United States Supreme Court judgment upheld a Circuit Court decision which found Standard Oil Company of New Jersey in violation of the Act. This ruling forced the dissolution of Standard Oil Company within six months. As part of the break-up, thirty-four subsidiary companies of Standard Oil Company of New Jersey (Jersey Standard) were distributed to individual shareholders of the New Jersey corporation. Standard Oil Company (California), of which the Bahrein Petroleum Company Limited later became a wholly-owned subsidiary, was organized as a separate company. [3]

Calouste Gulbenkian, also known as Mr Five Per Cent (*Associated Press*).

Meanwhile, in Eastern Europe there was another major development. In 1912 Calouste Sarkis Gulbenkian, an Armenian financier, formed the Turkish Petroleum Company (TPC). The shareholders were the National Bank of Turkey (35-per-cent), the Deutsche Bank (25-per-cent), the Anglo Saxon Petroleum Company, today a subsidiary of Royal Dutch Shell (25-percent), with Gulbenkian himself holding 15-per-cent. In the course of validating the petroleum rights in Mesopotamia, TPC found itself in conflict with the Anglo-Persian Oil Company, and the result was deadlock.

After a revised shareholding had been negotiated, the impasse was resolved whereby the Deutsche Bank and Anglo Saxon Petroleum Company held 25-per-cent each and the Anglo-Persian Oil Company 50-per-cent, with the understanding that Gulbenkian would have a 5-per-cent lifetime beneficiary interest, contributed equally by Anglo Saxon and d'Arcy Exploration Company (in effect, the Anglo-Persian Oil Company). Gulbenkian objected to this arrangement, took legal action and ended up with a 5-per-cent share in the Turkish Petroleum Company. However, once the First World War was declared, it was clear that this arrangement would no longer work since it included British and German shareholders. The political map of Europe and the Near East changed dramatically after the end of World War I, resulting in the disintegration of the Austro-Hungarian, Ottoman and Russian empires, as well as the dissolution of the Habsburg and Romanov dynasties.

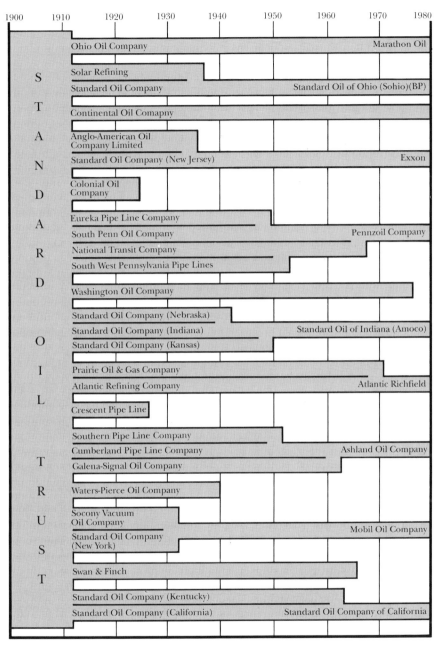

The descendants of the Standard Oil Trust. Three became "majors", including the Standard Oil Company of California (*from Fundamentals of the Petroleum Industry, by Robert O. Anderson. Copyright© 1984 Robert O. Anderson, courtesy University of Oklahoma Press*).

More than a decade later, on the 3rd of February 1928, the Near East Development Corporation (NEDC) was incorporated as a holding company in respect of 23.75-per-cent of the shares of the Turkish Petroleum Company (TPC). When NEDC was formed, five companies held 20-per-cent of its shares each: Standard Oil Company (New Jersey), today known as Jersey Standard, Esso or Exxon; Socony-Vacuum Oil Company Inc (now Mobil Corporation); Atlantic Refining Company; [4] Pan-American Petroleum and Transport Company; and Gulf Oil Corporation.

Meanwhile, negotiations were under way to set up the Iraq Petroleum Company (IPC) as the successor to the Turkish Petroleum Company (although this did not occur formally until the 8th of June 1929). In the new arrangement, Turkish Petroleum Company shareholders comprised the Anglo-Persian Oil Company (APOC), Compagnie Française des Pétroles (CFP), Royal Dutch Shell and the newly-formed Near East Development Corporation (NEDC), each holding a 23.75-per-cent share, with the Germans no longer shareholders. The remaining 5-per-cent was allocated to Participations and Investments Limited (Partex), which took care of Gulbenkian's interests.

The Red Line Agreement

It was against this background that the famous Red Line Agreement of 1928 was mutually approved by the partners of the Turkish Petroleum Company. Each of the participating companies undertook not to conduct independent operations in a large area comprising most of the territory of the former Ottoman Empire. The object was to prevent the shareholding companies competing against each other, both in seeking concessions and in the purchase or refining of crude oil produced in the area. The agreement was also intended to ensure that all members of the Turkish Petroleum Company would gain equally from the activities of any of them in the area. The inclusion of this restriction in the agreement which admitted the American companies into TPC following the formation of NEDC was insisted upon by Gulbenkian and the French. They wished to make sure that the other participants of TPC could not hinder its expansion while at the same time being in a position to develop oil outside the agreement's territory to the disadvantage of French interests.

There are various accounts as to how, rather than why, the boundary for the Red Line Agreement was drawn up. One version [5] suggests that when, at a meeting in July 1928 of the TPC participants, the conference looked like foundering, Gulbenkian called for a large map of the Middle East, took a thick red pencil and drew a line round the central area. He is reputed to have said: "That was the Ottoman Empire which I knew in 1914, and I ought to know. I was born in it, lived in it and served it." From that moment, everyone could see that all of the countries of the Arabian peninsula region, including Bahrain, but *excluding* Kuwait, were within the Red Line boundary.

Immediately this posed a problem for Gulf Oil Corporation, which had a 23.75-per-cent shareholding in NEDC and, therefore, was automatically a minority shareholder of the newly-organized Turkish Petroleum Company. According to the terms of the Red

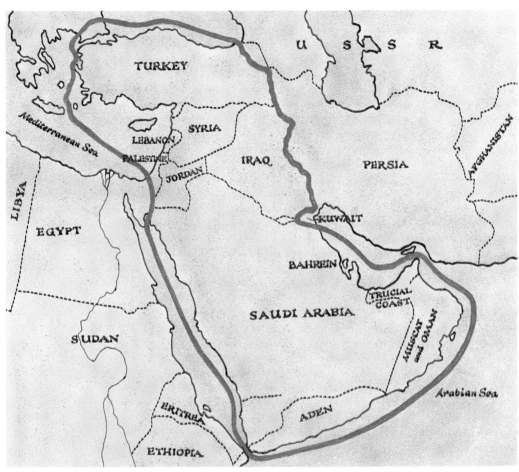

The boundaries of the Ottoman Empire as drawn by Calouste Gulbenkian in 1928, to define the area of the Red Line Agreement (*Jane Stark*).

Line Agreement, Gulf Oil found itself with a conflict of interests between its Kuwait concession negotiations and, more precisely and relevant to the Bahrain oil story, the involvement of Eastern Gulf Oil Company (a wholly-owned subsidiary of Gulf Oil Corporation) in the Bahrain concession.

Competitive Interest in Arabia Awakens

Prior to the First World War, the oil potential of Arabia had been viewed with caution. At the end of the war, one person declared his conviction that a major oilfield lay undisturbed far below the desert sands of Arabia. His name was Major Frank Holmes. This rugged New Zealander, who was later dubbed by the Arabs *Abu Al-Neft* (Father of Oil), wrote home to his wife in 1918:

"I personally believe that there will be developed an immense oilfield running from Kuwait right down the mainland coast" [Bapco archive].

To many, this was a preposterous statement. To others, it was a glimmer of hope on an apparently gloomy horizon. At the beginning of the 1920s it was felt, particularly in America, that there might be a shortage of oil. Although within a few years this proved to be an erroneous assumption, the immediate reaction was another scramble for further oil concessions in the Middle East.

Following the success of their oil exploration in the Near East, the British *were not* anxious to explore the Arabian Gulf region, mostly because the bitumen seepages there did not provide the same encouraging signs which had been so apparent in Mesopotamia and Persia. Nevertheless, they did not wish to see foreign competition in such close proximity. The Americans, on the other hand, *were* keen to obtain an oil concession in the Arabian peninsula region. But, at the same time, they recognized that the kudos of such an achievement had to be balanced with practical considerations. As Moore commented in his thesis many years later [1948, p. 22]:

"… It is necessary to emphasize that the mere occurrence of an oil seepage or conviction that an oil pool exists underground, is not enough to justify the major expense of drilling, roadbuilding, personnel care, storage and marine terminal facility construction, and other investments which attend production in remote foreign lands. Not only must there be a good market, but also there must be reasonable expectation that a sizable oilfield exists."

These economic factors became the crux of the matter regarding Bahrain. Whilst, in time, several people came to the conclusion that there was a reasonable expectation of finding oil somewhere on the islands, no-one expected that this oilfield *alone*, if it existed, would be big enough to support the development costs. Thus, the successful outcome of the Bahrain oil concession negotiations hinged on what other exploration opportunities might be secured on the mainland, particularly in Al-Hasa province, in Kuwait, and in what became known as the Neutral Zone.

As a result, attempts by "outsiders" to secure competitive oil rights in Al-Hasa province did not meet with the Anglo-Persian Oil Company's approval. Although APOC never overtly blocked concession negotiations, it is fair to say that, together with the British Government, it successfully frustrated other British attempts to secure a concession in the region. When, in 1923, a concession was awarded to the Eastern and General Syndicate, the way became clear for this British syndicate of businessmen to pursue a concession in Bahrain, and then to exercise any option which it might be granted to explore and to prospect for oil on the islands. However, the same frustrations featured throughout these protracted negotiations, reaching a crescendo in the late 1920s when Gulf Oil Corporation and Standard Oil Company of California became the American protagonists in the early exploration and development of Bahrain's oilfield.

Al-Hasa Concession

On the 6th of August 1920, a consortium of British businessmen and mining engineers registered the Eastern and General Syndicate Limited in London with the object of obtaining concessions and investigating business opportunities in Arabia. The group included Sir Edmund Davis and Mr Percy Tarbutt, pioneers in the international trade of the British Empire, particularly regarding mining ventures. Now demobilized from the British army in which he had served as a Major during the First World War, Frank Holmes became the Syndicate's Middle East representative. Capitalizing on his hunch and sharpening his latent skills as an opportunist businessman, Holmes ventured forth. Although he never reaped great financial rewards from his endeavours, he was instrumental in securing several oil concessions for the Syndicate, first in Al-Hasa province, then in Bahrain and later in Kuwait and the Neutral Zone.

Major Frank Holmes
(*Chevron archive*).

Born in New Zealand in 1874 and trained as a mining and metallurgical engineer, he had gained wide experience working mostly in gold and tin mining enterprises in Africa, Asia, South America and even Russia. During the 1914-18 War, while arranging supplies from Abyssinia (now Ethiopia) through Aden for the army stationed in Mesopotamia, Holmes visited the Arabian Gulf for the first time. As an engineer he became interested in the oil seepages and water problems along the Arabian Gulf coast and in what he saw of the Anglo-Persian Oil Company's refinery in Abadan. Once Holmes became the Eastern and General Syndicate's representative, this middle-aged figure made a marked impression.

"Burly, amiable, loquacious and unpredictable he cut a fantastic figure as he traveled on foot, ass, camel and steamer along the coast of the comparatively unknown Arabian subcontinent and its environs. He was accompanied only by an interpreter and a Somali servant. … With chance acquaintances along the road he chattered expansively, but always he posed as one traveling for reasons of health" [Moore, 1948, p. 10].

Holmes' eccentric personality aroused curiosity at every capital and port of each state he visited from the Red Sea to the Arabian Gulf, where the omnipresent British Political Agent was on hand to clear the course of his excursions. Not only were Holmes' activities monitored closely by the British Government officials, but also the Anglo-Persian Oil

Company's interested parties were anxious to track his movements, if not his discussions.

In the late summer of 1922, Holmes visited His Highness, Abdul Aziz bin Abdul Rahman bin Faisal bin Saud, Sultan of Nejd and the Dependencies in Arabia. Already, he had acquainted himself with officials in Bahrain, the geological structure of the islands and the oil seepages which occurred upon them. During his first visit to Al-Hasa province, Holmes talked with Ibn Saud and drafted a tentative agreement for his consideration. According to Ameen Rihani, Holmes' friend at court:

"On the day I arrived at Ojair, Saiyed Hashem, by the usual Oriental circumlocution, mentioned the name of my fellow-traveller in connection with a certain document which His Highness the Sultan wanted me to examine. Ha! So the Major is travelling for his health! The Saiyed brought the document with him, which I was to criticize if necessary, and give my opinion on the subject matter.

"Now, there being but one decent room in the Qasr, [6] that which I occupied – the Saiyed [7] lived downstairs with the Ameer; [8], and there being but one lamp, which was in my room; and I having come out of the *jalbout* exhausted and feeling the need of rest and sleep, we agreed to read the document together in the evening. But in the afternoon, just in time for tea, who should pop in, but Major Holmes? He had just arrived in a *jalbout* from Bahrein. … He wore over his conventional European clothes, a thin *aba* which concealed nothing; and over his cork helmet, a red kerchief and *ighal* which made his head appear colossal. But in this attempt to combine good Arab form with comfort and hygiene, he certainly looked funny. He was no longer a mystery to me. The document and this second visit to the Sultan indicated concessions and economic schemes" [Rihani, 1928, p. 81].

Since there were no spare rooms in Ojair, Holmes was urged to proceed immediately to Al-Hasa. If he rode all night, he would arrive in time to have an interview with the Sultan before he started out for Ojair. Rihani recalled:

"Later in the evening, I regretted his going; for I certainly should have enjoyed the Major more than his document, on the twenty pages of which were his twenty signatures in full. Of what interest to me, in sooth, is a concession to drill for oil and minerals and salt in the Province of Al-Hasa?

"For the sake of the Sultan, however, I read the document and execrable Arabic translation, clause by clause; and, summoning from the past my long neglected business sense, I was able, I think, without hurting the prospects of Major Holmes and his Company, to make a few suggestions" [Rihani, 1928, p. 82].

During October 1922, the city of Ojair, a port of Al-Hasa, was preparing to host a regional treaty conference under the direction of the British High Commissioner and the British Government's senior representative in the Persian Gulf, Sir Percy Cox. The objective was to determine the borders between Kuwait and Saudi Arabia, the result of which was the creation of the Neutral Zone.

Rihani recorded in his diary on the 30th of November that Holmes "loomed up on the horizon unexpectedly as usual, and incorporated himself into the Ojair Conference". The Major "pitched his tent between the camps of the British officialdom and the retinue of Ibn Saud, but a little nearer Saud than the British" [Moore, 1948, p. 12]. He "eats with his own people, ... although he does not share their confidence" [Rihani, 1928, p. 83].

The Anglo-Persian Oil Company also had agents on hand at the conference for, on the 29th of November, Rihani was shown a letter by Abd'ul-Latif Pasha which Pasha had just received from his friend, Sir Arnold Wilson (President of APOC in Abadan). The communication indicated that Sir Arnold would be arriving shortly to visit the Sultan with the idea that maybe they could strike an oil deal. With swift footwork on the 1st of December, Holmes solicited favour with Ibn Saud through his "friend at court", Ameen Rihani. The anxious Major was assured by Rihani that, already, he had advised Ibn Saud to favour the Eastern and General Syndicate's approaches for a concession. Rihani's rationale was, that the less a company applying for a concession had to do with politics, the better it would be for the Sultan.

By the evening of the 2nd of December, Holmes' contract was in the hands of the British High Commissioner. However, before the close of the day, Rihani received a messenger bearing some disturbing news about the proposed Al-Hasa concession. Sir Percy Cox had asked the Sultan to write a letter to Major Holmes explaining that he, Ibn Saud, could not give his decision until he had made certain enquiries of the British Government. According to Rihani's diary for that evening, the messenger had added that the Sultan had refused three times to comply with the British High Commissioner's request, and that three times Sir Percy had insisted. The letter *was* written, 1922 came to a close and Major Holmes departed for Baghdad.

The winter passed, as did the following spring. Holmes still waited for a decision from somebody – the Sultan, the Residency or the Colonial Office – about the concession's fate. Nothing had arrived. His misgivings intensified. Believing that the British Government was against him, he decided to pack his bags and leave for his home in England. But Rihani persuaded him otherwise, knowing that Ibn Saud was well disposed towards Holmes and the Eastern and General Syndicate, which he represented.

"I told the Major to change his mind and go back to Al-Hasa. The Sultan will soon be there. 'I will give you a letter to him, and I'm certain you'll get the Concession. Never mind what Sir Percy Cox says. ... By all means, accept the invitation of Lady Cox to tea, and tell her you are going back home. ... Say Good-bye, too, to Sir Percy. For if he suspects that you are going back to Al-Hasa, he might get ahead of you to the Sultan with one or two of those lead-pencil notes of his. ... I leave in a few days for Damascus and thence to Freiké [Mount Lebanon]. Good-bye and good luck'" [Rihani, 1928, p. 86].

King Ibn Saud of Saudi Arabia (*Popperfoto*).

Holmes did return to Al-Hasa and, as Rihani had shrewdly predicted, Ibn Saud granted an exclusive option to the Eastern and General Syndicate for the exploration of oil and mining rights in the province. In agreeing to the terms of the document, for and on behalf of the Syndicate, Holmes confirmed in writing to Ibn Saud that the syndicate would not sell to the Anglo-Persian Oil Company Limited either as to the whole or part thereof, any oil or mineral concession or concessions that may be granted by His Highness to the Eastern and General Syndicate Limited.

Focus on Bahrain

The intrigues of the Al-Hasa episode proved to be the hallmark of subsequent negotiations concerning the *Bahrein Island Concession* which was granted to the Syndicate on the 2nd of December 1925, and eventually assigned to the Bahrein Petroleum Company Limited on the 1st of August 1930. Throughout this period, the British Government, the Anglo-Persian Oil Company, the Eastern and General Syndicate and Major Holmes were caught in a complex web of cables, conferences, legal counsel, financial dispute, diplomatic debate, even suspicious surveillance. Once the interest of Gulf Oil Corporation and Standard Oil Company of California was aroused, and the British became aware of the possibility that American capital might finance oil exploration in Bahrain, Anglo-American rivalries became ardently contested issues. The next few years were frustrating times for Major Holmes. Nevertheless, he kept his vision in focus, one which was soon shared with His Highness Shaikh Hamad bin Isa Al-Khalifa, Deputy Ruler of Bahrain.

During his visits to Bahrain, Holmes had noticed that, although there was an excellent network of underground springs, there was no evidence of water wells on Bahrain's islands. So, in February 1924, the Eastern and General Syndicate commissioned Dr Arnold Heim, a Swiss geologist and Docent at the University of Zurich, to report on the oil possibilities in eastern Arabia, including Bahrain. In his report (dated 5th September 1924, but not published for two years), Heim wrote: "Bahrein forms a large and very gentle *anticlinal dome.* ... To drill on the dome of Jebel Dukhan not only would be extremely expensive, but also a pure gamble" [*Chisholm*, 1975, Note 25, p. 109].

Not deterred, Holmes returned to Bahrain later in 1924. During that year prolonged storms had made the collection of spring water from the submarine springs very difficult, resulting in a shortage of fresh drinking water. Anticipating an emergency, Holmes, with the co-operation of Muhammad Yateem (with whom he had made his Arabian headquarters in Bahrain), approached Shaikh Hamad bin Isa Al-Khalifa, and offered to drill two water wells. The cost to the Ruler, Shaikh Hamad's father, would be nothing if he failed; but should he succeed, he suggested a fee of $15,000 per well and consideration of his application for an oil concession. [9]

While these negotiations were going on, the Anglo-Persian Oil Company was considering Bahrain too. At a management committee meeting held on the 9th of September 1924, Charles Greenway (later Lord Greenway) stated:

"Although the geological information we possess at present does not indicate that there is much hope of finding oil in Bahrain or Kuwait, we are, I take it, all agreed that even if the chance be 100 to 1 we should pursue it, rather than let others come into the Persian Gulf and cause difficulties of one kind and another for us" [*Ward*, 1965, pp. 19-20].

As the Anglo-Persian Oil Company and the British Colonial Office corresponded on the matter, Holmes reached an agreement with the Ruler of Bahrain to drill twelve to sixteen water wells at different points on the islands of Bahrain. Once the contract was signed, Holmes returned to London and the offices of the Eastern and General Syndicate to make the necessary drilling arrangements. With the help of Captain Albert H. Farley, a director of Phoenix Oil and Transport Company Limited, two drillers were engaged. The necessary equipment was procured, and T. George Madgwick, Professor of Oil Mining at Birmingham University, was persuaded to oversee operations in Bahrain. Once on the island, Madgwick chose a site for the first water well, assembled the drilling equipment and set to work. Meanwhile, anticipating no hitches, Holmes confidently set off for the Red Sea area again.

In May 1925, the drilling engine broke down. Madgwick, having cabled Holmes, left Bahrain to rendezvous with the Major at the Marina Palace Hotel in Port Said and to collect the engine spare parts which were being shipped from Basrah. During their meeting, Madgwick was pressed to agree that in addition to his water well drilling activities in Bahrain, he would examine the geological surface structure of the island. The plan was that if his findings were favourable, permission would be sought from the Ruler of Bahrain to drill another well, this time to test for oil.

Six months later, in November, Holmes cabled the Syndicate that a strong flow of water had been struck on the main island of Bahrain at the rate of 110 tons per day from a depth of 142 feet (43 metres) from one well and 140 tons per day from a depth of 121 feet (36.8 metres) from another. The drilling had been a complete success. Shaikh Hamad bin Isa Al-Khalifa, the Deputy Ruler of Bahrain, was "immensely pleased with the assurance of an excellent supply of potable water from conveniently located wells. Sixteen wells were drilled and all proved to be good producers" [*Ward*, 1965, p. 25].

The Eastern and General Syndicate's interest in Bahrain was rewarded on the 2nd of December 1925. Shaikh Hamad, acting on behalf of his elderly father, granted the Syndicate an exclusive exploration licence, the First Schedule of the *Bahrein Island Concession*, for a period not exceeding two years, with the right of renewal for two more years. An annual fee of 10,000 rupees was payable throughout the currency of the agreement, and the Syndicate was not liable to pay more unless it was awarded a Mining Lease from the Ruler of Bahrain. A stipulation was that the rights of the agreement could not be transferred to a third party without the Ruler's consent, acting on the advice of the British Resident in the Persian Gulf.

It was this stipulation which was to cause Holmes, and many others, considerable concern and delays during the coming years, for it was not until August 1930 that the concession was assigned, at last, to the Bahrein Petroleum Company Limited. Only then could arrangements be made to drill a test oil well in Bahrain. However, the political issues relating to the vexed question who should control Bahrain's oil industry development were secondary to the primary question. Who would provide the venture capital and the technical skills to exercise the rights secured in the 1925 Bahrain agreement?

Since the Syndicate did not have adequate funds, it had no choice but to sell its *option* on the Bahrain concession. But unfortunately for the Syndicate, the international commercial climate regarding oil exploration had begun to change. The American panic of the early 1920s had mellowed towards a more cautious interest in overseas oil reserves. By the mid-1920s, immense domestic fields, such as Huntingdon Beach, "flooded the United States market with an apparently limitless supply of oil and, indeed, at times drove the prevailing price to disastrously low levels. Because of the apparently excessive supply available domestically, there was little occasion for an aggressive foreign policy oriented toward the acquisition of foreign concessions and supplies" [Rayner, 1944, p. 72].

Who will back Bahrain?

Many American companies which had sought oil abroad during the perceived crisis curtailed or suspended their operations. Nevertheless, the Department of State insisted on an open-door principle of equal opportunity for United States oil interests in areas which became available for concessions. While the British rigidly maintained the validity of the pre-First World War Turkish Petroleum Company contract, which did not include American participation, American diplomatic pressure was being exerted to "open the door" on terms acceptable to the United States of America. These terms took time to negotiate, as, for example, in 1928, when the Americans secured a shareholding in the British-controlled Near East Development Corporation. Meanwhile:

"The British and British-Dutch companies, while continuing to explore most of the areas favorable for the occurrence of petroleum, appeared to relax their exploratory efforts in a time of plenty, as is natural with the industry, and concentrated on developing the concessions already in hand and showing great promise, such as Iraq, Persia, Venezuela, and the East Indies" [Moore, 1948, p. 18].

It was amidst such a combination of international rivalry and financial reserve that Archibald Chisholm, who was involved closely with the Kuwait concession negotiations during the 1930s and was acquainted with Major Holmes, recalled that for over five years the Major "sought in vain to find a British or American oil concern to back his fancy and exploit the [Bahrain] concession" [Chisholm, 1975, Note 15, p. 95].

Eventually he succeeded. To the chagrin of the British Government, it was not British but American capital and technical skills which were to back Holmes' "fancy".

Water well drilling engine in Bahrain, circa 1925 (*Chevron archive*).

Although Standard Oil Company of California became the key player, the British Government maintained its rigid stance that the Bahrein Petroleum Company should be British. The deadlock was broken by a convoluted compromise. The Bahrein Petroleum Company Limited was registered in the Dominion of Canada (a member of the British Commonwealth of Nations), as a wholly-owned subsidiary of the American company, Standard Oil Company of California.

On the 1st of June 1932 no one could have been more delighted than the Ruler of Bahrain, Shaikh Isa bin Ali Al-Khalifa (then in the twilight months of his long life), to learn that oil had been discovered in Bahrain at Oil Well Number One. However, the intervening years between the signing of the *Bahrein Island Concession* on the 2nd of December 1925 and this victorious day some six-and-a-half years later, generated some of the most extraordinary and exciting chapters in the history of modern Bahrain.

Drilling a water well in Bahrain, circa 1925 (*Chevron archive*).

CHAPTER THREE

The Concession Catch

Commercial Caution

In 1926, Professor Madgwick returned to London and informed the Eastern and General Syndicate directors of the "perfect dome" which he had seen in Bahrain. However, despite his encouraging remarks, he found the directors "lukewarm to undertaking the risk of oil developments on their own" [Moore, 1948, p. 22] and more inclined to consider the unfavourable opinion of Dr Heim, who had concluded that such a venture would be both very expensive and a gamble. Madgwick's case was not helped by the fact that he had no supporting documents. Since the Political Agent had not allowed him to undertake any oil survey work during his assignment in Bahrain, the diplomatic argument being that the purpose of his stay was solely to drill for water, Madgwick could offer the Syndicate directors no more than an Admiralty Chart and his professional confidence.

The Syndicate faced two problems. It was not an oil exploration company but a dealer in negotiating for oil concessions from host countries and selling options on such concessions to interested companies. Since it had neither the funds nor the technical experience to exercise its option on the Bahrain concession, the directors decided to sell the option instead.

At that time, one of Madgwick's clients was Cory Brothers and Company, who had for many years been engaged in the coal business and had developed an interest in prospecting for oil. He hoped they would show interest in drilling on the Bahrain anticline, since the company had ample funds for oil development. But, in the end, Cory Brothers and the Syndicate could not agree on terms.

The Syndicate then approached the Anglo-Persian Oil Company, Royal Dutch Shell, Burmah Oil Company and other London oil and financial groups, but each declined to pursue discussions. It appeared to Madgwick that the main reason for this negative attitude centred on the commercial viability of oil exploration in the Arabian peninsula region as a whole. No one seemed willing to consider Bahrain on its own, the general feeling being that oil exploration on the islands would only be attractive to an oil company, if conducted in association with similar exploration on the mainland. Even without this proviso, no company was prepared to go to the expense of drilling a test oil well in Bahrain, despite the favourable surface indications on the main island, until a detailed geological survey was available.

Matters were further complicated by the fact that Iran and Iraq were pressing for increased crude and refined oil output in their own countries, upon which it was believed that their incomes, present or prospective, depended. The Arabian Gulf region was considered to be a special province and foreign companies were reluctant to create a commercial situation where Bahrain might find herself competing with these two larger countries. The oil companies who had plenty of oil, such as the Anglo-Persian Oil Company, were not in a hurry to explore new fields, but at the same time they did not want other people to compete and spoil their markets.

This commercial climate created a major problem for the Eastern and General Syndicate. Its three Arabian concession agreements were already close to, if not in, default. Urgently, the Syndicate required a substantial sum of money to honour royalty payments and to prevent its options from lapsing, therefore rendering them unsaleable. The anxious directors recognized, however, that British oil companies might choose to bide their time, explore the territories concerned at a leisurely pace and negotiate more satisfactory concessions, unhampered by conditions imposed by a third party. With characteristic grit, Major Holmes was not deterred. He is reputed to have toured the London clubs and offices relentlessly throughout 1925 and 1926 and became "an interminable bore" in his search for an organization to relieve the Syndicate of its financial burdens in the Arabian Gulf region. [1]

Another potential catch was the existence of restrictive agreements made between 1913 and 1917 relating to oil exploration and development which meant in effect that any American company considering purchase of the options had to accept that diplomatic intervention might be necessary. [2] It was against this setting of polite regrets in London, and potential diplomatic impediments, that Madgwick set sail for New York in August 1926, en route for his new post as petroleum consultant to the Canadian Government, based in Calgary, Alberta.

Assessing the Risk

When Madgwick arrived in America, he went to see his friend and former colleague, Thomas Ward, who had recommended him for his new appointment. They discussed the Syndicate's predicament and resolved to find a solution. Ward, already familiar with Madgwick's geological reports concerning Trinidad, had concluded that it was characteristic of the Professor to express reservation when making geological prognoses. He

The Windsor Hotel on Dominion Square, Montreal, Canada, 1926 (*National Archives of Canada*).

recognized, therefore, that although Madgwick had had no maps to guide him apart from an Admiralty Chart, and while it was impossible to fix the exact area of the Bahrain dome, he had reported with more boldness than usual that he had never seen so striking an oil structure as that which he had observed in Bahrain.

Ward's interest was aroused. Before Madgwick left New York for Calgary, he and Ward lunched with John Alvin Young, who at that time was associated in Mexico with Ward and had worked with William T. Wallace, then in charge of the Gulf Oil's New York office. He suggested that Ward should discuss the Bahrain concession with Wallace who might show an interest. Although Wallace did not discount the idea of Gulf Oil Corporation becoming involved in the oil exploration of Bahrain, the lack of convincing geological data on the islands' structure did not stimulate his interest immediately. However, he did indicate that he might become enthusiastic if Ward could produce some encouraging facts. The only way that this could be achieved would be for Madgwick to write a report once he had settled in Calgary and his trunk containing all his research notes had arrived. Meanwhile, Ward wrote to Edmund W. Janson, a director of the Syndicate, requesting a copy of the Bahrain concession. In his reply, Janson intimated that the Syndicate had decided to drill in Bahrain, in association with another group, and that Major Frank Holmes as the Syndicate's representative was empowered to conclude a deal, if such an opportunity should arise in America.

While Madgwick was busy working on his report in Calgary, Holmes had arrived in Montreal and installed himself at the Windsor Hotel. His purpose was to discuss

Arabian Gulf oil possibilities with Edwin B. Hopkins, a geologist with whom Ward had worked in Mexico. Hopkins knew Dr Kenneth C. Heald, Chief Geologist of the Gulf Oil Corporation. Like Heald, Hopkins was at that time disinclined to consider any new oil prospects, other than those in Venezuela. Although Madgwick's earnest remarks on the Bahrain structure had made a good impression, the general feeling, yet again, was that little could be done without more convincing geological data.

An urgent cable was despatched to the Professor urging him to report without delay. In response, on the 23rd of September 1926, Madgwick offered what he described as a hurried opinion, but nevertheless a considered one. With Pilgrim's 1908 survey at hand Madgwick reported:

> "In the centre [of Bahrain island] is a depression 12 miles long [19.3 km] and four [6.4 km] wide. ... This is enclosed by an almost perfect low escarpment of Middle Eocene [3] rocks so that we have an ideal 'dome'. In the centre of this depression rise the only prominent hills in the Islands, the principal one known as Jebel Dukhan whose height is marked as 440 feet [134 m]. ... There is some faulting ... but so far as can be judged not of great magnitude. ... Where the southern steepening of the dip begins to make itself felt ... about where the crest of the fold would be, is an occurrence of asphaltic material connected with vertical fissuring. ... The structure is so striking that in conjunction with the asphalt – which is regarded as a desert seepage – there can be no hesitation in saying that test drilling is called for. ... Bahrain and the adjacent coast must stand on its own merits and may not be regarded as a possible extension of the fields in Persia. ... In my opinion, any oil geologist gifted with reasonable optimism ... would advise as I do, that a deep test is warranted. A deep test is essential for, in the absence of data as to where the seepages came from, it will be necessary to be prepared to go to possibly considerable depth. On the other hand, with such a fine structure it ought to be possible to make a location which should definitely determine whether or not oil does occur in payable quantities" [Ward, 1965, pp. 29-32].

Holmes and Ward, very much encouraged by Madgwick's report, immediately arranged an itinerary of visits to potential backers and exploration companies. Among them was Major Thomas R. Armstrong, Vice-President and a Director of the Creole Petroleum Corporation (the Venezuelan subsidiary of the Standard Oil Company (New Jersey)), who also happened to be a specialist in South American concession negotiations and was second-in-command of Standard's producing department in New York City. Ward's son recalls that, in preparation for the meeting, his father had purchased the only available map in New York which showed the Red Sea and Arabian Gulf areas. Ward kept his appointment with Armstrong who, with succinct and blunt economy, declared:

> "Since Bahrain, when viewed on a world map, is small enough to fit under the tip of my pencil, it is not big enough for us". [4]

Despite this abrupt dismissal, Ward and Holmes continued their search. Before long, Norval Baker, geologist of the Standard Oil Company (New Jersey), showed interest and expressed a wish to secure more information. Captain C. Stuart Morgan, who was connected with the Standard Oil Company (New Jersey) on Middle East matters was even more encouraging although his principals "were not inclined to immediate interest" [Ward, 1965, p. 34].

Survey Solution

The prospects of securing commercial interest in Bahrain's oil potential now looked more hopeful, although it was unlikely that anyone would be prepared to make a swift commitment. This was especially unfortunate for the Syndicate since payment on the Al-Hasa option was in arrears, £14,000 to £15,000 was needed to validate the Neutral Zone agreement, and a payment was needed to honour the terms of the Bahrain concession. [5]

Clearly, no progress could be made towards easing the Syndicate's financial worries until a favourable geological report was available. To this end, in January 1927, Ward decided to approach Captain Morgan about the possibility of undertaking a thorough survey of Bahrain. Morgan, as secretary of the group responsible for negotiating for American participation in the Turkish Petroleum Company, had considerable personal knowledge of the Arabian Gulf and Mesopotamia and, to Ward's relief, agreed to push the idea of Standard Oil Company (New Jersey) conducting a survey. He suggested that this might be discussed with C. F. Bowen, Standard's Chief Geologist, and his assistant, Norval Baker.

Upon re-reading Madgwick's report, Baker showed renewed and greater interest and asked if it might be feasible for the Syndicate to arrange for Standard's survey party, then in Kenya, to visit Bahrain. Ward cabled London seeking an immediate and affirmative response. But the Syndicate directors were not to be hurried. Eventually, Ward pressed for an answer. "Morgan's people insist upon having a reply one way or another, for they are not interested unless they can make arrangements to use the geological survey party which is at present available" [Ward, letter of 21 January 1927].

The Syndicate's reply was not helpful. Not only would it be impossible to secure the necessary permit in the time available, but also the Syndicate repeated, Ward's negotiations must still be based on Madgwick's report. [6] It is curious, too, that by the end of January 1927, the Syndicate would have received a letter dated the 16th of January from Holmes in Bahrain in which he wrote:

"Since my arrival I have put up tents near the oil seepage and also have done some rough road making. It is now possible to take a motor car to the oil seepages. I have also had some wells sunk, and find that the area over which bitumen can be found is greater than was ever hinted at before. I have had a good look over the area surrounding the seepages, and the dome formation is exceedingly attractive. …

The oil indications in places Dr Heim never saw are more promising and extensive than I had any conception of."

The question arises as to why the Syndicate chose not to pursue the survey opportunity now presented to it, even though it knew that oil companies were reluctant to commit themselves without further favourable geological evidence. These extracts from Holmes' 16th of January letter to the Syndicate were not forwarded to Ward in New York until the 9th of June, almost six months later. Ward does not recount in his monograph whether Holmes sent him a copy of the letter separately. Meanwhile, one of the Syndicate directors, Edmund W. Janson, had been in touch with the Colonial Office. He reported that: "… we should get no support if we introduced American capital for boring in the Persian Gulf. … I am afraid nothing can be done at the present time with regard to your investigating the geological formation at Bahrein, but I am now at work trying to get the position cleared up" [Janson, letter to Ward, 23 February 1927].

As the storm clouds gathered over the issue of American capital, the possibility of securing financing via Canada was considered as a way of overcoming objections raised by the British Colonial Office. At the same time Ward fulfilled his promise to keep Wallace of Gulf Oil Corporation informed of developments, but added that Norval Baker of the rival Standard Oil Company had now expressed an interest in surveying Bahrain. Wallace agreed to reconsider the Bahrain prospect and instructed Dr J. Volney Lewis, a member of Gulf Oil's New York staff, to review the relevant information already available to the company which pertained to Bahrain, Al-Hasa, the Neutral Zone and Kuwait.

On the 28th of March 1927, Holmes wrote to the Syndicate: "I am extremely glad that the Board of Directors have decided to drill for oil. I hope for success. I have the huts erected at the oil seepage site, a fresh water well sunk and everything (including a motor road to the seepage) ready for the drillers to take up their residence so soon as the plant arrives". Again, this extract was not sent to Ward until the 9th of June and is not referred to in his monograph. It raises the question whether or not Ward knew that Anglo-Persian Oil Company geologists had been active in Bahrain while Holmes was busy making preparations:

"Since I have made the road to the seepage and built the huts there, the APOC geologists have been three times to Bahrein port. On two of these visits, they (the geologists, four) have visited the oil seepage and taken away samples of the bitumen and oil sands. … I hear that the geologists have already reported that Bahrein is well worth boring. … The work at the seepage has exposed some interesting things, among them a larger showing of oil bitumen" [Holmes, letter to Eastern and General Syndicate, 9 April 1927].

Whether or not Holmes was under APOC surveillance, is open to debate. Nevertheless, it seemed possible that Gulf Oil and Standard Oil (New Jersey) were not the only companies moving towards the view that Bahrain might be worth drilling after all and that a comprehensive survey would settle the matter.

Bizarre Bargaining

Whatever correspondence may or may not survive, and however lacking the Syndicate may have been in communicating effectively with Thomas Ward, its New York representative, all parties who had read the Bahrain concession document would have known one inescapable fact. The Syndicate's two year exclusive Exploration Licence was due to expire at the end of 1927. By late April, inertia was replaced with action. On the 20th of April Ward received an unprecedented cable from 18 St Swithin's Lane, the Syndicate's London office:

"HOLMES WIRES THAT FROM A LETTER FROM YOU MELLONS [7] APPEAR STILL MUCH INTERESTED OUR OIL BUSINESS STOP AUTHORISE YOU OFFER THEM OUR ENTIRE INTEREST IN BAHREIN CONCESSION FOR ONEHUNDREDTHOUSAND DOLLARS CASH PAYABLE HALF ON SIGNING AGREEMENT REMAINDER IN TWO YEARS STOP THIS OFFER IS SUBJECT TO MELLONS DECIDING ON BASIS OF MADGWICKS REPORT AS THERE IS NOT TIME FOR THEM TO MAKE SPECIAL EXAMINATION OF TERRITORY IT BEING IMPERATIVE THAT DRILLING MUST COMMENCE FORTHWITH STOP IF THIS DEAL IS COMPLETED WE WILL GIVE MELLONS THE FULLEST CONSIDERATION REGARDING ANY OF OUR OTHER OIL INTERESTS THE POSITION OF WHICH HOLMES DISCLOSED TO YOU BUT BAHREIN MUST BE DEALT WITH FIRST AND OWING NECESSITY IMMEDIATE DRILLING THERE MELLONS DECISION MUST BE TELEGRAPHED QUICKLY EASGENSYND" [radiogram from Eastern and General Syndicate, 20 April 1927].

No doubt Ward was delighted. On the 22nd of May he met Volney Lewis, Staff Geologist of the Gulf Oil Corporation, who reported back to the Syndicate. Two days later, Ward took

"great pleasure in confirming that I have been authorized by my principals, the Eastern and General Syndicate, to offer you the entire interest of the Eastern and General Syndicate, in their Bahrein concession for the sum of $100,000, payable in cash, one half on signing and the balance in two years from that date. ... This offer is made on the understanding that you will decide on the basis of the report of Professor T. G. Madgwick, in as much as there is no time for a special examination of the territory. It is imperative that drilling commence forthwith" [Ward, letter to Volney Lewis, 24 May 1927].

Volney Lewis' reply could not have been more disappointing. Having reconsidered Professor Madgwick's report and other material in Gulf Oil's files relating to Bahrain island and mainland Arabia, and noting that under the urgent circumstances no time had been allowed for what he described as "even a cursory geological examination, I find that I cannot recommend the offer of the Eastern and General Syndicate to the serious consideration of our officials" [letter [8] to Ward, 27 May 1927].

This might have been the end of the story, except that Ward was ever mindful that various American oil men had told him repeatedly that Bahrain was too small a venture to justify the necessary risk and expense *on its own*. Could it be that if he interested an American oil company in *more* than just the Bahrain concession, their attitude might change? It was a notion worth exploring. Thomas Ward's son recalls that "it was my father's idea to make two agreements – one for Bahrain and one which would include the mainland". With this in mind, Thomas Ward Snr decided that the time had come to visit London and talk things over with Holmes and the Syndicate directors. However, before leaving New York he had other matters to attend to. Volney Lewis' letter required a reply. By return, on the 28th of May, Ward, who was determined to keep "the door open", asked for ideas regarding a geological survey of Bahrain so that he could advise the Syndicate accordingly.

He added that as plant was being shipped to fulfil drilling obligations, he would be glad to do everything possible to help Volney Lewis in this matter. To this end, Ward submitted a list of questions to the Syndicate.

Having acquired drilling equipment, were the drilling obligations in Bahrain being fulfilled? Since the Gulf Company was prepared to negotiate for the purchase of the Syndicate's interest in Bahrain, would the Syndicate accept Gulf's provisions for doing so, namely that time had to be allowed for a geological survey and that Al-Hasa and Kuwait had to be included? Was the Syndicate willing to negotiate on this basis? If yes, then how long would it take to arrange for proper escorts to accompany a geological party to the three concession areas? Would Major Holmes be available to co-operate with the Gulf Oil geological party?

Ward emphasized that Mr William T. Wallace "has very plainly stated that unless our negotiations cover Al-Hasa and Kuwait, in addition to Bahrain, the proposition is of no interest to his people" [Ward, letter to Eastern and General Syndicate, 1 June 1927]. However, the 9th of June reply (which included extracts from Holmes' letters to the Syndicate earlier in the year) was curiously unrealistic, particularly since the Bahrain concession was due to lapse in six months: "… our offer was not intended to be, nor can it be, kept open until the drilling has demonstrated the value of the concession. … We must therefore reserve the right to withdraw our offer at any time after the end of this month if the negotiations have not been completed by then". A week later the Syndicate added that it could not usefully add to what it had written already and that it was sorry that Ward's friends could not agree to deal with the Bahrain concession first.

As planned, Ward departed for London and contacted Syndicate director Edmund

A drilling rig on Sitra island (*Bapco archive*).

W. Janson. He also cabled Wallace in America seeking confirmation that Gulf Oil was still prepared to negotiate according to Ward's proposal. Wallace answered yes, but with the caveat that Gulf would require 100-per-cent control. The Syndicate directors were briefed to authorize the sale of the Bahrain, Al-Hasa and Neutral Zone concessions and undertake to deliver a concession for Kuwait, as and when negotiated, with an overriding royalty of 25 cents (then equal to one shilling) per ton in addition to cash payments. The overriding royalty was to apply to an excess of minimum daily production of 250 tons of oil per day. This minimum was to be allowed as a contribution to refining operations.

The Syndicate directors required some convincing, as they did regarding the proposed agreement covering the mainland areas of Al-Hasa, the Neutral Zone and Kuwait. After several meetings, agreement was reached which resulted in Ward drafting a memorandum on behalf of the Syndicate. Specifically relating to Bahrain, the suggestion was:

"Bahrein Concession

"£750 per annum to be paid until oil in commercially exploitable quantities found, then 3R. 8A. (= 5/3) [9] per ton of net crude oil Royalty. Royalties not to be less than 30,000 rupees (say £2250) per annum.

"Two years exploration licence from 2 December 1925 with a possible extension for a further two years, then two years prospecting licence over a definite area with a possible extension for a further two years, then a Mining Lease for 55 years up to 100,000 acres in not more than 3 blocks" [Ward, 1965, p. 39].

The recommended terms were US$250,000 for all properties (Bahrain, Al-Hasa, Kuwait and the Neutral Zone), and it was suggested that payments should be made in $50,000 tranches, the first being due when the geological party reported favourably on Bahrain. Outstanding obligations, including 10,000 rupees (approximately US$3500), were to be paid by the purchaser, meaning Gulf Oil Corporation, by February 1928.

Additionally, arrangements for a geological survey should be complete by the end of November 1927, with nine months allowed for the geological work and the submission of a report. This period included the best season for exploration work and giving enough time for a preliminary survey upon which a final decision could be reasonably made. Meanwhile, it was expected that the Eastern and General Syndicate Limited would assist by negotiating with the Colonial Office in London. In the event of the Gulf Oil Corporation not making a decision by the 1st of November to despatch a geological party to Bahrain, then the Syndicate would make its own arrangements to drill for oil on the islands, with the equipment already being used in Kuwait for artesian well drilling.

Averting an Option Crisis

Ward left for New York on the 15th of September, carrying an authoritative memorandum from the Syndicate which enabled him to reopen negotiations with William Wallace immediately upon his arrival. Despite the fact that within two months the Bahrain option would lapse, which in turn would cause the Syndicate even greater financial hardship, the Syndicate directors rigidly maintained their position. The Bahrain concession should be dealt with first, and not in association with Al-Hasa, the Neutral Zone or Kuwait.

The Syndicate's inflexibility and procrastination continued to frustrate efforts to bring the option discussions to a successful conclusion. Thomas Ward Jr recalled, "the Syndicate was very lax in responding to my father – the records stand for themselves". Dr Volney Lewis introduced another difficulty by suggesting that Ward's estimate of nine months gave insufficient time to arrange, complete and report on a geological survey of Bahrain. In his opinion eighteen months would be more appropriate. Gulf Oil made its position clear too. The Syndicate had to understand that Gulf's interest in Bahrain was but a portion of a bigger picture.

It was now the 26th of October 1927 and time was running out. The recommended deadline was a matter of days away, as was the due date for the Syndicate's next option payment on Bahrain. Commercial risk was not the only consideration. It had been known all along that the British were reluctant to allow American capital to be invested in Bahrain. Expressed precisely, the requirement that any oil operation in Bahrain should have British nationality was the crucial issue, as yet unresolved by the Syndicate and the Colonial Office in London.

Some time later, when riding home on the train reading the *New York Times* magazine section, Ward saw an article on the British Commonwealth of Nations. He became convinced that the outcome of the Imperial Conference in 1926 offered the answer to the problem. The conference had described the Dominions as "autonomous communities … equal in status, in no way subordinate one to another … though united by a common allegiance to the Crown" [10]. Ward put his suggestion to Wallace, who instructed his legal adviser James M. Greer to investigate. The conclusion was that registration of "The Bahrein Petroleum Company Limited" as a subsidiary company of Gulf Oil Corporation in Ottawa, under the laws of Canada, should fulfil the British nationality clause. There was additional merit to this proposal: the company laws of the Dominion of Canada permitted 100-per-cent stock ownership by American citizens, a fact which, Ward recalls, strongly appealed to Wallace and Greer [Ward, 1965, p. 42].

In New York, Wallace and Ward (representing Gulf Oil and the Syndicate respectively) had come to an arrangement. Gulf Oil Corporation was prepared to take over the Eastern and General Syndicate's option on the Bahrain concession. Also, Gulf Oil agreed to an overriding royalty of one shilling per ton on production in excess of 750 tons of oil per day, the Syndicate applying sterling rather than the American equivalent of 25 cents. Before leaving for Venezuela on the 12th of November, Wallace instructed his assistant Charles W. Hamilton (Vice-President of the Gulf Oil Corporation) and James M. Greer (Gulf Oil attorney) to draw up two legal contracts. One was to cover a completed concession for Bahrain, and the other for the mainland, the concession for which still had to be negotiated. The signing date for both documents was set for the 30th of November 1927.

In London, just two days before the due signing date, Syndicate director E. W. Janson informed Ward:

"Surely the Gulf Oil people must understand that it is quite impossible for us to guarantee that any Government department will sign a statement to be operative in the future. That they will agree when the time comes, I have no doubt whatever. We have now done everything we possibly can in the matter, and I hope the Gulf Oil people will be satisfied. … I hope the deal comes off, because I think it is in all our interests that it should, and you have done a lot of hard work on it, but you must remember that we cannot handle our Government offices like you can yours in America" [Janson, letter to Ward, 28 November 1927].

The deal did come off. On the 30th of November 1927, just three days before the option payment for Bahrain was due, Frank A. Leovy, on behalf of the Eastern Gulf Oil Company (a subsidiary of Gulf Oil Corporation), and Thomas E. Ward, on behalf of the Eastern and General Syndicate, signed the two contracts [Moore, 1948, pp. 29-30].

Specific to the Bahrain contract, the Syndicate granted to Gulf, or its nominee, an exclusive option until the 1st of January 1929 to acquire from the Syndicate the Bahrain concession (Option, 30 November 1927, Clause II). The immediate consideration for the option was one dollar ($1.00) cash in hand to be paid by Gulf Oil to the Syndicate. Further considerations were that Gulf undertook to pay the Syndicate 10,000 rupees not later than the 2nd of December 1927 (Clause III). If the option still existed a year later, on the 2nd of December 1928, meaning that the Syndicate remained obliged under the terms of the concession to pay the Ruler of Bahrain 10,000 rupees, Gulf Oil undertook to pay the Syndicate a further 10,000 rupees by that date. During the term of the option, Gulf was required also to undertake a geological survey of the island of Bahrain (Clause V).

The Syndicate's obligations were to maintain the concession in "full force and effect" (Clause IV); to secure (effective as of the 2nd of December 1927) a valid extension to the 2nd of December 1929 of the Exploration Licence granted in the First Schedule of the Concession (Clause IV); and to make the services of Major Holmes available to Gulf's geological party, when in Bahrain (Clause V). Clause VII addressed the controversial nationality issue by requiring the Syndicate to obtain effective written consent from the Ruler of Bahrain, and effective assurances from the British Colonial Office, that no objections would be raised to the assignment and transfer of the option to Gulf's nominee company. In the event that Gulf Oil decided to exercise this option, then it was required to establish a British corporation under the laws of England or of Canada.

Thomas Ward in New York City and Frank Leovy in the City of Pittsburgh, Allegheny County, Pennsylvania State, could breathe sighs of relief. The payments due to the Ruler of Bahrain on the 2nd of December 1927 could be made. Thirteen days later, Syndicate director E. W. Janson wrote to Ward from London: "I am very glad that you pulled off the business for us, and I hope your crowd will make money out of the proposition" [Ward, 1965, p.43].

The Lie of the Land

Immediately, Gulf Oil sent two of their geologists, Ralph O. Rhoades and Dr J. Volney Lewis, to see Professor Madgwick in Calgary and to acquaint themselves with Bahrain's local conditions. As time was critical, they had no choice but to visit Madgwick in hospital where he was recovering from an appendectomy. Although his report was not cut-and-dried with maps and dimensions, which is what they really wanted, it was persuasive enough and they believed in his opinion as a geologist. Upon their return to New York, Dr Kenneth C. Heald, Chief Geologist of Gulf Oil, appointed Rhoades to head up the geological party to Bahrain. W. F. Eastman and C. F. Shalibo were chosen as assistants, and preparations began.

Information on travel itineraries was obtained from London. The advice was that, as it was sometimes difficult to secure steamer passages, it would be better to make provisional arrangements with Messageries Maritimes (through Thomas Cook in London) to sail from Marseilles direct to Beirut, then travel through Syria, across the desert to Baghdad by Tawil motor cars, thence by rail to Basrah and finally, by boat to Bahrain. The next best way, should there be any delay in catching the Messageries steamer at Marseilles, was to go by P & O steamer to Port Said, railway to Haifa, thence by motor car to Beirut, across the desert to Baghdad and so on. Fourteen days should be allowed to travel from London to Basrah by either line of steamers.

December was a busy month in New York. Meetings were held and great care taken to perfect the arrangements for the smooth operation of the geological expedition. Messrs Thomas Cook and Sons were paid £217 16s. 9d. for the steamer fares and Rhoades was briefed to look out for Cooks' representatives at Cherbourg and Paris. He was also equipped with two letters. One he should present to Messageries Maritimes in Marseilles, and the other, a letter of introduction from the National Bank of Commerce, should be handed to the Eastern Bank Limited in Bahrain. According to the Syndicate's estimates and providing that there were no delays, the geological party should arrive in Bahrain on the 17th of January 1928. The only doubtful transit point was Basrah. However, Cooks had communicated with the steamship owners there, and Major Holmes had been asked to ensure that the boat waited for the party. On the 24th of December 1927, Rhoades and his assistants sailed from New York for Cherbourg. Their task was clear: determine whether Bahrain was likely to be a commercially attractive proposition for Gulf Oil to explore for oil.

New Year news from Bahrain was that Messrs Rhoades, Shalibo and Eastman had missed their Basrah connection. Nonetheless, on the 19th of February 1928, Holmes reported to Wallace from his office at Messrs A. M. Yateem Bros, Yateem Building, Bahrain, that the geological party had arrived in early February, had visited the British Political Agent and then on the following day had dined with the Deputy Ruler, Shaikh Hamad bin Isa Al-Khalifa.

"I can congratulate you on the three men you have sent. They have made a good impression and have been well received by both the local European people and the Arab Ruling family. I feel that they have been well launched and have fallen into the right groove in the place" [Holmes, 19 February 1928 and Ward, 1965, p. 45].

Nevertheless, Holmes told Wallace that it could not be stressed enough how important it was to ensure the standing of Gulf Oil's geologists. As the geologists were American, the British Political Agent was anxious to know all about them and the Eastern and General Syndicate's relationship with the American company. Although Shaikh Hamad felt comfortable with this American presence in Bahrain, Holmes emphasized that

"it should not be supposed that we are not being watched and our competitors have begun a campaign to annoy us. … They have guessed wrongly and are quite convinced that your men came from the Standard Oil Company. The Chief Political Agent of the Gulf [11] is coming to Bahrein next week with the avowed intention of finding out from me what we are doing and what our connections are. I think I can hold my own with him, but we shall see. All this questioning and agitation will die down in a little while, but we want to keep it under control until our campaign [12] has been brought to a successful issue" [Holmes, 19 February 1928 and Ward, 1965, p. 46].

Rhoades and his team were ahead of schedule. Based on the fieldwork they conducted between the 6th of February and the 19th of March, Rhoades reported to Wallace on the 10th of April that with the complete co-operation and able assistance of his two colleagues, Eastman and Shalibo, the work had been completed much sooner than the anticipated nine months. The nine-page report, with accompanying sketch maps, was encouraging. In his covering letter Rhoades informed Wallace that he believed Bahrain offered two decidedly favourable advantages. One was the remarkably well-developed anticline with surface evidence that petroleum was contained in the subsurface; the other was that Bahrain is on a well established trade route. The key issue was the extent, if any, of petroleum possibilities. In this respect, the only surface indication of petroleum in Bahrain was a bitumen deposit several kilometres slightly east-of-south from Jebel Ad-Dukhan.

"At that point there is a small area of locally accumulated sand, thru which oil has, in the past, seeped to the surface. Under the arid desert conditions the seepage has become practically inactive and only the dried asphaltic residue remains. Some years ago, several pits were dug over an area of about half an acre, and each encountered the bitumen. One of these pits was reopened a few years ago by the Eastern and General Syndicate's geologists. It is about 18 feet deep and exposes the asphalt quite well. ... In the sides and bottom of the pit, the tarry sands, protected from air, are fresher, a bit spongy, and give off a strong odor of petroleum. ..."

"The character of the Cretaceous formations which possibly underly the Bahrein region, their thicknesses and probable relation to the generation of oil in the latter region, can be determined by study of the mainland of Arabia. The question of reservoir beds, their character and probable depth below the surface in the Bahrein anticline must also await study on the mainland. ... The presence of asphalt deposits and seepage along the western flank of the Arabian Gulf geosyncline strongly suggest that oil bearing formation underly [sic], more or less continuously, the western part of the Gulf and the adjacent mainland as far south as Bahrein. ... It is concluded that a test well favorably located on the Bahrein anticline may reasonably be expected to encounter oil" [Rhoades, 1928, pp. 7-8].

Five factors favoured the development of Bahrain's oil potential: easy access to deep-sea shipping from the fold, political security, an abundance of available local labour, the probability of purchasing fuel oil from the APOC refinery at Abadan and a convenient

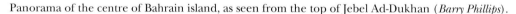

Panorama of the centre of Bahrain island, as seen from the top of Jebel Ad-Dukhan (*Barry Phillips*).

water supply from shallow wells which could be drilled on the flanks of the anticline.

Rhoades made four recommendations. The third, in principle, could be carried out with relative ease, namely a geological study of the mainland. The fourth was that the cable tool method should be used for drilling. Recommendations one and two were not so straightforward in practice. The first was that the Eastern and General Syndicate Limited's option on the Bahrain concession should be exercised and the rights thereto acquired by the company. The second was that, as a precaution against "unpleasant competition", an option on the area not covered by the present concession, should be negotiated with the Bahrain Government. [13]

Beyond 100,000 Acres

In early March, Colonel Haworth, Political Resident of the Persian Gulf, arrived in Bahrain. Contrary to Holmes' expectations, Haworth did not appear aggressive, but seemed anxious to learn what was being done regarding financial arrangements for the current geological work, and to ascertain Holmes' view of the prospects for finding oil in Bahrain. Nevertheless, there was a sharp edge to the meeting. Haworth revealed that the Anglo-Persian Oil Company was "dreadfully annoyed" that the Syndicate had found strong financial backing outside the United Kingdom. Holmes, suspecting that the real purpose of Haworth's visit was intelligence gathering, politely told the PRPG that he would communicate his views to Gulf Oil and think it over himself, which he did. Not long afterwards, Holmes learned of APOC's interest in Bahrain, perhaps the prompt for

Rhoades' reference to "unpleasant competition". APOC was interested in an area of the Bahrain islands (later known as the Additional Area) other than that which Gulf Oil had obtained the right to select under the terms of its agreement.

From Rhoades' perspective, this was most unsatisfactory. By the end of March, he had had a chance to assess the entire geological formation of the Bahrain archipelago. He and Holmes concluded that Gulf Oil should attempt to secure the remaining Additional Area of the Bahrain territory, after the 100,000 acres provided for in the Bahrain concession had been located. As Bahrain island was a huge geological dome, and even if promising results were secured within the 100,000 acres conceded in the present concession, Rhoades suggested that if Gulf Oil secured the remaining islands within a modified concession, outside competition could be eliminated. Holmes and Rhoades were determined that APOC should not enter the contest. Already it had a small presence in Bahrain, evident from the Political Agent's remarks in his 1927 report, where he cited the fact that during that year the Anglo-Persian Oil Company had increased its oil imports into Bahrain by 50-per-cent, by virtue of the company's agents giving credit to purchasers.

The consensus was that at least £2000 would be needed to secure the Additional Area. It would be necessary also to find out what terms Shaikh Hamad required for this privilege, but Holmes did not doubt that at least the same terms as in the first concession would be requested and that there was not the "smallest chance" of negotiating a reduction in the oil royalty. In short, the recommendation to Wallace was that if Gulf Oil wished to explore *any* of the Bahrain islands for oil, it should be all or nothing.

In reply, Wallace wished to know exactly how the Additional Area was being defined. From the new map which Rhoades had supplied in late February, Gulf Oil had calculated that the present designated area on Bahrain island was 131,500 acres, pretty close to the island's total area. This calculation did not include the smaller islands of Sitra, Nabiih Saleh, Muharraq, nor any of the other islands in the archipelago. Having studied the map, Wallace asked Holmes to

> "… straighten us out definitely with respect to the extent of the territory over which the sovereignty of the Shaikh of Bahrain actually extends. … I note also that our competitors are keenly alive to the fact that some new move is on foot as evidence by the presence of these American geologists, but I feel safe in relying upon your ingenuity and skill to keep our competitors from ascertaining the facts until we ourselves are ready to make the disclosures. We are also glad to know that the ruling authorities in that part of the world are interested in the prospective development of some of their resources by joint British and American capital" [Wallace, letter to Holmes, 30 March 1928].

The Syndicate directors made their intentions clear. They hoped to obtain an additional area, at Gulf Oil's expense, thus securing the rights which existed under the present concession and the possibility of increasing the oil area. If Gulf Oil exercised its option, no addition to the purchase price was payable, but the Syndicate expected that as the two concessions virtually formed one property, oil obtained within it in excess of a total of 750 tons per day should count to the Syndicate for a royalty of one shilling per ton.

The evidence suggests that by this time Wallace (Gulf Oil) and Janson (Eastern and General Syndicate) may not have trusted each other entirely. Certainly, Wallace did not plan to make a deal with the Syndicate regarding the Additional Area. An added complication was that Holmes and the Syndicate directors did not always enjoy a harmonious relationship, and Gulf Oil, recognizing Holmes' position as middle-man and his conflicting loyalties, was not always certain of the Major's motives.

It was during this episode that a difference of opinion between Holmes and Rhoades reached Wallace's desk in New York. The Major's written complaint was that Rhoades had refused to give him a copy of the geological map of Bahrain. Rhoades had argued that such an action would have been contrary to Gulf Oil company policy, particularly given the competitive commercial climate which prevailed in Bahrain at that time. The dispute was no short verbal exchange, but one which was being carefully documented, even though the two disagreeing parties were in Bahrain.

Unabashed, Holmes had bypassed Rhoades and informed Wallace that his purpose for having the map was to place himself in a better position with the Political Resident of the Persian Gulf (PRPG) if, and when, it was decided to negotiate for the additional 100,000 acres. This would include harbour rights at considerably lower rental and less cost than if it were taken up as being valuable as oil bearing territory. Given his experience during the Al-Hasa concession episode, Holmes, somewhat unconvincingly, suggested that if any documents were shown to the PRPG "they will not outrage this confidence". With equal lack of conviction, he added that there had been no friction between himself and Rhoades, "we are both too old for that nonsense" [Holmes, letter to Wallace, 20 April 1928].

A month later Wallace resolved the map dispute. While supporting Rhoades for upholding company policy, he conceded that a map would be helpful in "all common interest". Since it might be construed later that the precise areas which Gulf Oil wished to select and include in the mining licence had been defined, and to avert the interest of competitive eyes, blueprints of each of the topographical and geological maps of Bahrain were sent to Holmes, but without any demarcation lines.

Wallace suggested that after the required 100,000 acres had been located under the terms of the mining licence, it would be possible to show this area without drawing any lines on the map whatsoever. [14] Holmes was instructed not to give any indications as to where the boundaries might be, but instead to take a rectangular sheet of paper and calculate the dimensions of the rectangle to cover 100,000 acres on the same scale as that of the map. Then, the sheet of paper should be moved back and forth over the map to indicate to the Ruler of Bahrain the approximate area of the main island remaining, after the selection of the 100,000 acres.

Wallace emphasized that the selection of the final area for a mining licence had to be made with care and, if possible, only after prospecting drilling, which might present a geological condition underground totally unsuspected from the surface geology. The fact remained that exploring for oil on the island of Bahrain was a pure gamble: the indications were that a test hole would be justified, but in view of his long experience Wallace would not say (on behalf of Gulf Oil) whether or not oil would be found.

Six days after Wallace had written this letter to Holmes, Frank Leovy, Vice-President of Gulf Oil, wrote to the Syndicate's New York representative on the 28th of May outlining the understanding and agreements for the proposed further Bahrain concession which would incorporate the Additional Area. If all went well, the documents would be signed on the 28th of November that year. On the 4th of June, Holmes met with the British Political Agent and the Financial Adviser to the State of Bahrain, and left the conference with the impression that agreement had been reached. Ten days later, the draft concession, as considered agreed, was returned amended with provision for the yearly payments to be doubled.

Upon sighting the revised document, "which did not show one single point yielded to me", Holmes withdrew his consent to the alterations and requested that the application should be submitted to the Political Resident and the Colonial Office. Holmes' protest did not augur well for the concession documents under preparation. This setback, however, was soon to be eclipsed by an even greater problem.

Implications of the Red Line Agreement

Whilst Gulf Oil and the Syndicate attended to Holmes' protest, the Red Line Agreement was formalized at the end of July 1928. As explained earlier, this understanding required the companies participating in the agreement not to conduct independent operations in the territories which comprised most of the former Ottoman Empire, the object being to stop these shareholding companies competing against each other.

Since Gulf Oil Corporation was a party to this agreement, albeit indirectly, the company now faced a difficult decision. Since Kuwait was *not* within the defined Red Line territory, Gulf Oil was eligible to pursue its interest in the Kuwait concession, and retain its 20-per-cent shareholding in the Near East Development Corporation, which was a 23.75-per-cent shareholder in the Turkish Petroleum Company, the successor to the Iraq Petroleum Company. The net effect of this complicated shareholding arrangement was that Gulf Oil Corporation held only 4.75-per-cent of TPC shares. Nevertheless, Gulf Oil was bound by the Red Line agreement. Since Bahrain was within the "Red Line" territory, Gulf Oil could *not* continue its involvement with the Bahrain concession, *and* retain its shareholding in the Near East Development Corporation.

At the end of September, Holmes, Ward and Wallace met at a "crisis" meeting in New York to discuss where everyone now stood with regards to the Bahrain concession. Ward recalls that Holmes suffered a great blow when he was informed of the Red Line implications as he had counted on an early start to drilling a test oil well in Bahrain. Already, during August, Holmes had reminded the Syndicate that if it required an additional extension to the Exploration Licence, [15] due to expire on the 2nd of December, application should be made by letter now. He had recommended that also on the 2nd of December 1928 a Prospecting Licence should be asked for, his point being that such a licence was an advancement, and all the Arab rulers were watching Gulf Oil's progress. Now he learned that the Red Line Agreement had thrown his plan into complete disarray.

Holmes and Ward recalled their earlier communication with Captain C. Stuart Morgan in 1926 when he had shown a personal interest in the prospect of oil exploration in Bahrain, although his principals had not been "inclined to immediate interest". Perhaps Morgan's interest could be rekindled and his assistance secured to find a compromise solution since he was well-connected with the Standard Oil group of companies, and was Secretary of the Near East Development Corporation (NEDC). This could be expedited as it was learned that Morgan was about to leave New York for London and the Middle East on a scheduled business trip. An extra item was prepared for his briefcase, a letter signed by the Standard Oil Company (New Jersey) and Gulf Oil Corporation, both NEDC shareholders, urging that the Turkish Petroleum Company directors agree to joint operations in Bahrain.

As planned, Morgan tabled the letter at the next Turkish Petroleum Company board meeting on the 30th of October 1928, through the American Group's attorney and representative on the TPC Board. In effect, Morgan virtually offered Bahrain "across the counter" to their Anglo-Dutch-French associates for $50,000, this figure being the approximate expenditure incurred on the Bahrain survey by Rhoades and his assistants. It is reputed that a strange scene then occurred:

> "One of the Dutch representatives, leisurely replenishing his pipe from his oilskin pouch declared the general territory was 'not unknown' to his company, that they had had their geologists out there long before and they had encountered considerable difficulties and unpromising political conditions but *no oil*"
> [Moore, 1948, p. 32].

Stuart Morgan reminded the board that as a means of protecting its flanks, since a vast new oil belt seemed likely to be developed, the Turkish Petroleum Company could take over Bahrain or leave it with Gulf and keep a "rain check" on it. The Dutchman ended the debate. "No, I'm afraid there is no oil in Arabia" [Moore, 1948, p. 32]. Several years later, Morgan remarked: "People are on occasion not as stupid as they sound. Perhaps this was one of those occasions - to discourage oil tourists" [Moore, 1948, p. 32].

Gulf Oil's dilemma remained unresolved: did it keep its 4.75-per-cent share in Turkish Petroleum Company, pursue the Kuwait concession and divest itself of concession interests in Bahrain; or, did it withdraw from the Red Line Agreement so that it could develop its Bahrain plans? Unexpectedly, the answer was presented by Andrew W. Mellon, Secretary of the Treasury of the United States (later United States Ambassador to the United Kingdom), whose family controlled Gulf Oil. He was of the opinion that the company should not renege on its commitment and abandon the Red Line Agreement. This decision left William T. Wallace with no choice. Gulf Oil stayed with its American associates including Standard Oil Company (New Jersey), who had, as of the 31st of July 1928, joined the Anglo-Dutch shareholders of the Turkish Petroleum Company partnership. [16] The Eastern and General Syndicate had been severely constrained by circumstances beyond its control.

The American Capital Debate

On the 19th of June 1928, the Secretary of State for the British Colonies, L. S. Amery, wrote a confidential memo to the Political Resident in the Persian Gulf concerning any organization awarded the Bahrain concession. He had drafted a clause to ensure that any enterprise involved with the development of the Bahrain concession should remain a British company registered in Great Britain or a British colony. The chairman and managing director (if any) and a majority of the other directors should be British subjects, and that

> "neither the Company nor the premises, liberties, powers, and privileges, hereby granted and demised, nor any land occupied for any of the purchases of this lease, shall at any time be, or become directly, or indirectly controlled or managed by a foreigner or foreigners or any foreign corporation or corporations, and the local General Manager of the Company, and as large a percentage of the local staff employed by them as circumstances may permit, shall at all times be British subjects" [Amery, letter to the PRPG, 19 June 1928].

Four months later, on the 17th of October, Frank Holmes and Edmund Janson called at the Colonial Office in London and informed the Secretary of State that the Syndicate was negotiating with American and Canadian interests, their objective being to find sufficient capital to finance prospecting operations in Bahrain and ultimate commercial development. They also told him of the problems facing Gulf Oil following the Red Line Agreement, although at that time Gulf had not yet decided what action to take as a shareholder of the Near East Development Corporation. Nonetheless, the Secretary of State enquired, if negotiations with the American group were to be successful, would most of the capital be American?

Janson confirmed that this was likely to be the case. He pointed out, however, that the Syndicate had sought British capital without success. Having spent £60,000 on exploratory work in Bahrain, the Syndicate could not afford to finance prospecting and oilfield development operations. As a result it had been left with no choice but to look across the Atlantic Ocean for funds. Janson conceded also that if United States capital were forthcoming, the real control would pass into American hands and that the Eastern and General Syndicate would be kept alive in some way but only as nominal concessionaries. The Secretary of State reminded the Syndicate that under the terms of the Mining

Lease [17], the Bahrain concession could not be conveyed to a third party without the consent of the Ruler of Bahrain and that the British Government had been assured in 1925 that control would remain entirely British.

On the 8th of November, O. G. R. Williams of the Colonial Office sent a letter recording this meeting to the Board of Trade, copied to the India Office and Foreign Office. By this time a new dimension had been added to the deliberations. Gulf Oil had decided to stay within the Red Line Agreement. Whether or not Williams knew of this is unclear, but nevertheless he noted that as the Syndicate had not submitted its proposals, he assumed that it was still hopeful of obtaining British capital. On that basis, he intended to inform the Syndicate that if it would undertake to ensure that the Company would at all times be, and remain, a British company, then the British Government would be prepared to recommend to the Ruler of Bahrain that he should grant the renewal of the Exploration Licence, due to expire within a month on the 2nd of December, for a further year.

Meanwhile, on the American side of the Atlantic Ocean, Ward had been busy stimulating the interest of Standard Oil Company of California (Socal), not a party to the Red Line Agreement, with the intention that Socal might agree to take over Eastern Gulf Oil's option on the Bahrain concession. On the 23rd of November, Thomas Ward cabled the Syndicate's London office requesting approval to assign the Bahrain contracts to Socal. During the following day, unaware of a looming political crisis, Ward informed the Syndicate:

"No time has been lost at this end in preparing for an emergency, as you will have noted from our cablegram yesterday. Mr Wallace and his associates are extending to us the most sincere co-operation and I shall meet representatives of the Standard Oil Company of California, in Chicago, within the next ten days. The matter has been discussed by telegraph with the President of the Standard Oil Company, Mr K. R. Kingsbury, and he has expressed his interest" [Ward, 1965, p. 111].

225 Bush Street, San Francisco, the corporate headquarters of the Standard Oil Company of California until the company became known as the Chevron Corporation in 1984 (*Chevron archive*).

Already, Dr Volney Lewis (the Gulf Oil geologist) was on his way back to New York from Colombia to meet with his counterpart at Standard Oil Company of California, in the event that the Syndicate approved the assignment of the Bahrain contracts to Socal. At the same time, Sir John Cadman, Chairman of the Anglo-Persian Oil Company, was travelling to New York to attend the annual convention of the American Petroleum Institute. The plan was to discuss the Bahrain situation with Cadman upon his arrival. The meeting took place, following which the Chairman of APOC flatly rejected the proposal. [18]

A week later, Ward, barely disguising his frustration with what he described as the Syndicate's lack of constructive action, cabled London:

Western Union cable heading (*Chevron archive*).

"DO YOU OR DO YOU NOT AUTHORIZE ME GIVE YOUR CONSENT IN WRITING TO PROPOSED ASSIGNMENT BY GULF TO STANDARD CALIFORNIA OF NOVEMBER THIRTIETH NINETEEN TWENTY SEVEN OPTION CONTRACT AND MAY TWENTY EIGHTH NINETEEN TWENTY EIGHT LETTER AGREEMENT STOP IMPERATIVE YOU REPLY THIS POINT IMMEDIATELY STOP DO YOU UNDERSTAND CLEARLY THAT IN EVENT STANDARD CALIFORNIA THUS BEING SUBSTITUTED FOR GULF TERMS AND CONDITIONS ... WILL NOT BE CHANGED BUT ... WILL BE PERFORMED STRICTLY AND NOMINEE OF STANDARD IN SUCH CASE WILL BE BRITISH CORPORATION AS PROVIDED IN CONTRACT AND SUCH BRITISH NOMINEE WILL BE COMPANY TO WHICH EASGENSYND WILL TRANSFER CONCESSIONS IF OPTIONS EXERCISED" [Ward, 1965, pp. 113-14].

November ended and December dawned. The President of Standard Oil Company of California, Kenneth R. Kingsbury, then in Chicago, continued to liaise with Michael E. Lombardi, Socal's Vice-President, who travelled to New York with authority to pursue the negotiations for the transfer of the option agreement from Eastern Gulf Oil

Company to Socal. On the 13th of December, Lombardi left New York for Socal's head office in San Francisco to make the necessary arrangements. These included the immediate organization of a Canadian company, so that the option on the Bahrain Concession could be exercised less than three weeks later on the 1st of January 1929.

The Catch

In London, the Syndicate was under renewed and increased pressure. Its request for a year's renewal of the Exploration Licence (the First Schedule of the 1925 Bahrain Island Concession) had been answered by the Colonial Office on the 23rd of November with an unacceptable proposal. Carrying out his intentions, already indicated two weeks earlier to the Board of Trade, Williams communicated to the Syndicate that His Majesty's Government was prepared to recommend the renewal if the Syndicate would insist that its agreement with The Ruler of Bahrain include a clause drafted by the Colonial Office requiring that:

> "The Syndicate shall at all times be and remain a British Company, registered in Great Britain or a British Colony, and having its principal place of business within His Majesty's dominions. ... Neither the Syndicate nor the premises, liberties, powers, and privileges, granted ... by the Shaikh [Ruler] of Bahrein, nor any land occupied for any of the purposes of the Exploration Licence ... shall at any time be or become directly or indirectly controlled or managed by a foreigner or foreigners or by any foreign corporation or corporations" [Williams, memo to the Eastern and General Syndicate, 23 November 1928].

On the 19th of December, the Syndicate replied, quoting Article XIII of the Third Schedule (Mining Lease) of the 1925 Bahrain concession:

> "The rights conveyed by this Lease shall not be conveyed to a third party without the consent of the Sheikh acting under the advice of the Resident in the Persian Gulf. Such consent shall not be unreasonably withheld."

Emphasizing that the *nationality of the assignee* was not specified, the Syndicate's interpretation was that:

> "the new condition you seek to impose upon us prior to recommending the granting of a renewal of the Bahrein Exploration Licence is not acceptable to the Gulf Company, and ... is not one which we are in a position to impose, the only stipulation to which we have a right to adhere being that the *country of registration of the ultimate company* shall be British or Canadian" [Adams, letter to the Colonial Office, 19 December 1928].

The Syndicate added that if some satisfactory arrangement could not be worked out with Gulf Oil, with whom it had entered into negotiations in good faith and on the terms of the Bahrain concession agreement as it existed at the time, then Gulf Oil would almost certainly resist the new condition and claim a breach of contract. Further, this would cause the liquidation of the Syndicate, with the loss of all the capital so far invested in the Arabian oil propositions.

Two days later, on the 21st of December, in Pittsburgh, Pennsylvania, F. A. Leovy (Eastern Gulf Oil Company) signed the company's agreement to assign its option on the Bahrain concession to Standard Oil Company of California in return for $157,149 expenses involved. It is unclear whether or not Gulf Oil knew, when the documents were signed, of the Syndicate's recent dialogue with the Colonial Office, and its latest communication anticipating a breach of contract and the possibility of Eastern and General Syndicate being forced into liquidation. Perhaps suspicious, but also familiar with the Syndicate's pattern of intermittent communication, Ward reported to William T. Wallace on the same day:

"I cannot understand the delay in the discussions with the Colonial Office, inasmuch as I was advised by cable on November 2nd, that Mr Janson and Major Holmes had discussed the extension of the Bahrain Island exploration licenses with the Government in London. I wrote London on December 1st, to the effect that arrangements were proceeding for exercising the option as per the Bahrain contract, on January 1st, and it remained for the Eastern and General Syndicate Limited to secure formal written and effective assurances of the British Colonial Office and/or of any other proper and necessary Governmental officials that it and/or they will not raise any objection to the assignment of the Bahrain Concession. I am writing again today to confirm the above cables and at the same time am stating that the Bahrain Oil Company Limited is being organized under the laws of the Dominion of Canada to be in readiness for the exercising of the option on January 1st 1929" [Ward, 1965, p. 126].

Six days later, still nothing from the Syndicate suggested that the plans now in their final stages in America might be at risk. On the 27th of December 1928, W. H. Berg authorized Ward to exercise, on behalf of Standard Oil Company of California, the option dated the 30th of November 1927 and letter option dated the 28th of May 1928 covering Bahrain island, entered into between Eastern and General Syndicate Limited and Eastern Gulf Oil Company, and to designate Bahrein Petroleum Company Limited of Canada as Standard Oil Company of California's nominee. Meanwhile, in the city, county and state of New York, Frank Feuille, Attorney-in-Fact for Standard Oil Company of California (Socal), signed the duplicate copies of the assignment on the same day. During the next day, the 28th of December, Socal exercised the two options relating to the original 100,000 acres on the Bahrain concession and the negotiations for a concession to the "Additional Area". In so doing, it confirmed the nomination of the

Bahrein Petroleum Company as that to which the Concession and Option covering any further concession relating to the "Additional Area" should be transferred.

As the year closed, Gulf Oil Corporation had honoured its obligations to the Red Line Agreement and, in good faith, had assigned its Bahrain commitments to the Standard Oil Company of California. Socal had conceived the Bahrain Petroleum Company, although it had not been born. Financial pressures on the Eastern and General Syndicate seemed to have been relieved. Commercial interest in Bahrain's oil concession now seemed assured. Implicitly, American capital had been invested in the operation. Yet the practical reality was about to become all too evident. Effectively, Socal's plans had been halted, and its hopes thwarted.

CHAPTER FOUR

From Deadlock to Development

Birth of Bapco

Francis B. Loomis, a former Under-Secretary in the United States State Department, had spent much time in recent years arranging the details of Standard Oil Company of California's foreign operations. Together with Judge Frank Feuille of New York, he had determined that a wholly-owned subsidiary company of Socal could be registered in Canada, so accommodating the British nationality requirement. Thus, on the 11th of January 1929, the Charter of the Bahrein Petroleum Company Limited (Bapco) was sealed in Ottawa, Province of Ontario, Dominion of Canada by Thomas Mulvey, Under-Secretary of State for Canada. Bapco's registered office was to be the Trusts Building, 48 Sparks Street, Ottawa, Canada, but practical day-to-day administration was to be conducted from 225 Bush Street, San Francisco, California, Socal's head office in America. It was hoped that this arrangement would enable Standard Oil Company of California to make immediate plans for drilling in Bahrain. This was not to be, however, since the re-emergence of Anglo-American rivalry for the control of crude oil production in the Middle East had a direct impact on Bahrain. The nationality of Bapco, of the capital financing it and of the personnel operating it, represented three interrelated political issues which fuelled an intense international controversy for eighteen months.

The Catch Revealed

On the 2nd of January 1929, Thomas Ward had received a letter from the Syndicate dated the 20th of December 1928, together with a copy of the Syndicate's 19th of December letter to the Colonial Office [1] which answered the Colonial Office's proposal of the 23rd of November. The main points were that agreement to the assignment of

Office of the Under-Secretary of State for Canada, Ottawa, January 1929 – now the Victoria Museum (*National Archives of Canada*).

the original option to a third party "shall not be unreasonably withheld" and that the *country* in which the nominated company (Bapco) was to be registered should be British. Ward learned also that should the matter not be resolved, Gulf Oil might claim a breach of contract and the Syndicate could be forced into liquidation.

After piecing the sequence of events together, it became obvious that when writing to Ward on the 20th of December, Syndicate secretary H. T. Adams was aware that *one day later,* in good faith, Eastern Gulf Oil Company was to assign both options on the Bahrain concession to Standard Oil Company of California. In other words, although the principle of a British registered company had been established, the practice of it being a wholly-owned American subsidiary, *de facto* financed with American capital, had not been approved by the Colonial Office. Based on Adams' letter, it appeared that both Eastern Gulf Oil and Standard Oil Company of California now faced a major problem. Who would control Bapco? Adams' letter to Ward stated that during the negotiations preceding the first assignment of the original option from the Syndicate to Eastern Gulf Oil more than a year previously (30th of November 1927), "we knew of no reason why the Colonial Office would, or could, veto such transfer" [Ward, 1965, p. 140].

Adams had drawn several conclusions. Aware of the Anglo-Persian Oil Company's opposition to such an arrangement, he assumed that when particulars of the Syndicate's interests were given to the Turkish Petroleum Company, the Anglo-Persian Company (which managed TPC) placed all the facts before the Colonial Office and drew attention to the probability that the properties would cease to be under direct British or Canadian control. It seemed apparent that some representations had been made to the Colonial Office and that APOC already knew of the Syndicate's arrangements with Eastern Gulf Oil Company. Hence, a new condition to prevent direct or indirect foreign control had arisen which, it was reasonable to suppose, resulted directly from negotiations with the Turkish Petroleum Company to which the Syndicate and Eastern Gulf Oil had not been privy.

The Syndicate's reply to the Colonial Office, dated the 19th of December 1928, stated that the nationality condition it sought to impose upon the Syndicate and Gulf Oil, prior to the granting of a renewal of the Bahrain (Concession) Exploration Licence, was not acceptable to the Gulf Company. The only stipulation the Syndicate and Gulf were obliged to adhere to was that the new company's *country* of registration should be British or Canadian, and this was being honoured.

However, unknown to Ward and Wallace, another meeting had taken place on the 28th of December 1928. Adams, as Secretary of the Eastern and General Syndicate Limited, and G. S. Pott, solicitor to the Syndicate, had been summoned to call upon Messrs Williams, Bushe, Hall and Bigg at the Colonial Office in London. The object of the meeting, "without prejudice", was to enable the Colonial Office to find out the Syndicate's precise position in relation to Eastern Gulf Oil Company. Also, it wished to know exactly what options the Syndicate had granted to Eastern Gulf Oil in respect of the concession which had been granted to the Syndicate by the Ruler of Bahrain on the 2nd of December 1925. The Colonial Office delegates indicated that this should not be misconstrued to mean that they wished to force the Syndicate into liquidation. On the contrary, they would do anything they could to "avoid such a consequence". However, *prima facie* it appeared to the Colonial Office that granting the Syndicate an extension to the Exploration Licence would not work to its advantage, if permission for the Syndicate to assign its rights to Gulf Oil "was not forthcoming". The Colonial Office suggested that "it might, therefore, be more in the interests of the Syndicate for the extension to be refused" [Williams, 12 January 1929, Enclosure to the Memorandum No. 8].

The Colonial Office also informed the Syndicate that it could not be assumed that the British Government would be prepared to instruct the Political Resident in the Persian Gulf to advise the Ruler of Bahrain to consent to the transfer of the Syndicate's rights under the Mining Lease (Third Schedule of the Bahrein Island Concession, 2nd of December 1925) to Eastern Gulf Oil Company's nominee. Furthermore, even if the British Government did co-operate, the Syndicate should not assume that the Ruler of Bahrain would agree to the proposed transfer. There is ambiguity as to whether or not the Colonial Office knew of Socal's involvement, since the document recording this meeting does not specifically mention Socal by name.

Although the Colonial Office was perfectly aware of the Syndicate's negotiations with Gulf Oil, it would appear that it did not know that the Syndicate had assigned its option to Eastern Gulf Oil on the 30th of November 1927, and that Eastern Gulf Oil had reassigned that option to Standard Oil Company of California on the 21st of December 1928, together with the additional letter option of the 28th of May 1928. Further, on the day previous to the very meeting now in hand, W. H. Berg of Standard Oil Company of California had *authorized* Ward in New York to exercise the two recently acquired options and to designate the Bahrein Petroleum Company Limited of Canada as Socal's nominee. Therefore, allowing for a transatlantic time difference, almost at the same time that the Colonial Office and the Syndicate were conferring in London on the 28th of December, Socal *exercised* the two options relating to the original 100,000 acres on the Bahrain concession (Exploration Licence) and the negotiations for a further concession to cover the "Additional Area".

The Syndicate was now in a tight fix both financially and legally. The added difficulty it faced was that the Exploration Licence, provided for in Schedule I of the concession, had already expired on the 2nd of December 1928. Since it had not been renewed, and if no retroactive renewal were to be granted, and if no Prospecting Licence had been applied for upon its expiry, provided for in Schedule II of the concession, the agreement would automatically cease.

Since it is unlikely that the 28th of December rendezvous at the Colonial Office had been a hurriedly-convened meeting, or that the agenda was unknown beforehand, a curious question remains unanswered. Why did the Syndicate directors in London allow legally binding documents to be signed in America the day previously, knowing that the contractual commitments could not be fulfilled to the satisfaction of all parties concerned, including the Syndicate itself? It was a matter of little time before Eastern Gulf Oil and the Standard Oil Company of California would learn that the former was *not* relieved of its option obligations and that the latter was immobilized regarding its newly acquired options.

Deadlock Assessment

Pott, the Syndicate's lawyer, deduced that in the event of the Syndicate failing to secure an extension to the expired Exploration Licence, then the Syndicate would have no rights under their concession upon which Eastern Gulf Oil Company could exercise an option. In those circumstances, Eastern Gulf Oil Company would have no further claim on the Syndicate. However, in the event of the Syndicate being able to secure an extension to its Exploration Licence, upon terms agreed by the Colonial Office, this would prohibit the Syndicate from transferring *any* of the rights which it possessed under its concession, either to Eastern Gulf Oil Company *or* to a nominee of that company. [2]

Pott was aware of what this implied. First, the Syndicate would be in breach of contract if it accepted the Colonial Office's terms for extending the Exploration Licence, and thus it had declined to do so. Second, Eastern Gulf Oil Company would have a rightful claim against the Syndicate, if the Syndicate were to accept the Colonial Office's terms. In any case, if such a new concession contained a "British control" clause, as proposed by the Colonial Office, Eastern Gulf Oil Company would derive no benefit at all.

The Colonial Office had maintained its position and declined to reply to the Syndicate's 19th of December letter until the 1st of January 1929, since by then it would know whether or not Eastern Gulf Oil Company intended to exercise its option. Meanwhile, the Colonial Office suggested that the Syndicate might perhaps give further consideration to the question of its position *vis-à-vis* the Eastern Gulf Oil Company and decide whether, in fact, it desired to press its application for an extension of its Exploration Licence, in which case some time must necessarily elapse before a decision could be reached in the matter. [3]

When following up the 28th of December meeting, the British stance was expressed

Thomas E. Ward (*Caltex*).

clearly in the Secretary of State for India's memorandum, enclosed in Williams' communication to the Board of Trade. [4] The Colonial Office sought the Board's opinion as to whether or not it would be desirable to allow American interests, as represented by the Eastern Gulf Oil Company, to obtain a foothold in Bahrain. If this were so, the Secretary of State wished to know whether the Board of Trade considered it preferable to attempt to induce the Turkish Petroleum Company to become interested in Bahrain, if, as it seemed, 100-per-cent British capital could not be attracted to the undertaking.

Since, on the previous day, the Bapco Charter had been signed in Canada incorporating the company as a wholly-owned subsidiary of Standard Oil Company of California, implicitly, American interests *had* obtained a foothold in Bahrain. In a masterly understatement commenting on the situation, Ward wrote to William T. Wallace three days later: "It is very difficult to construe the intention of the Eastern and General Syndicate Limited in the absence of more specific advices" [Ward, letter to Wallace, 15 January 1929].

By the third week of January, the Syndicate's financial circumstances were becoming increasingly apparent. It had not acknowledged payment for the Bahrain rental for which Eastern Gulf Oil was pressing for a receipt so that in turn it could settle its accounts with Socal. No accounts had been submitted since the 7th of November 1928. Queries on both the Bahrain and Kuwait concessions remained outstanding, and Ward's personal expenses, amounting to $5000, had not been paid.

As January ended, and as a preliminary to tackling the problems now facing the Syndicate, Professor Madgwick had met E. W. Janson (a Syndicate director) and the solicitor G. S. Pott in London. Carrying copies of documents hitherto unsighted in America, Madgwick sailed for New York and a meeting with Ward and Wallace. Having studied the new material, Ward informed Adams, the Syndicate's secretary in London, "you will understand that without the information which you now place at our disposal, I could not, nor could Mr William T. Wallace, nor any of his associates, fully appreciate the position of your negotiations with the Colonial Office" [Ward, letter to Adams, 31 January 1929].

Having considered the Secretary of State's views, [5] Wallace conferred with Francis B. Loomis at Standard Oil Company of California. It seemed advisable that a representative should go to London to participate in the discussions with the Colonial Office. Having read the new evidence too, Judge Frank Feuille noted that the condition imposed by the Colonial Office prevented the Eastern and General Syndicate Limited from meeting its obligation under the option agreement with the Gulf Company, and that the Syndicate's rights under the concession had lapsed because it had not obtained either an extension of the Exploration Licence or the Prospecting Licence on the 2nd of December last.

"If Mr Adams' statement is correct, we have a remarkable case in which the Colonial Office had forced a British company under British control to lose a valid concession which it had obtained from the Shaikh of Bahrein Island. The fact that the Syndicate was prevented from transferring its concession to a foreign company would not, in the least, affect the Syndicate's right to hold the concession for itself" [Feuille, letter to W. F. Vane, 31 January 1929].

Feuille drew two conclusions. One, the Eastern and General Syndicate was acting in good faith, although perhaps naively. Two, the pressure against the assignment of the concession to American interests was not coming from the Colonial Office at all, but from the Anglo-Persian Oil Company.

Meanwhile, James M. Greer, Eastern Gulf Oil Company's lawyer, informed Feuille that his company intended to take the matter up with the United States' State Department, and Socal's co-operation was sought in making this appeal to the American Government. Kenneth Kingsbury, Socal's President, agreed to co-operate. As a preliminary, Francis B. Loomis (Socal) wrote to Paul T. Culbertson, US Department of State:

US Department of State logo (*Chevron archive*).

"I think you will be interested in seeing this ironclad proposal emanating from the Colonial Office. … The Colonial Office is evidently not willing that an American company should engage in any activities in the matter of oil prospecting or production in the Island of Bahrein, which is under their political control. It appears from information and correspondence at hand that the Anglo-Persian Company is doubtless behind the action of the Colonial Office in this matter" [Loomis, letter to Culbertson, 6 February 1929].

A week later, Judge Edward C. Finney, First Assistant Secretary, Department of the Interior, was presented with a more precise legal question: "It would seem that Great Britain is not acting in a manner that would entitle her to reciprocal advantages and treatment as provided for in our Leasing Law" [Loomis, letter to Finney, 13 February 1929]. Two days later, Ward informed the Syndicate that the existing contracts between the Syndicate and Gulf Oil were at variance with Gulf Oil's understanding of international agreements between the American and British governments regarding oil matters. He added:

Radiogram heading (*Chevron archive*).

"UNTIL UNITED STATES GOVERNMENT STATES SUCH CONDITIONS JUSTIFIED GULF DOES NOT PROPOSE SERIOUSLY CONSIDER THEM STOP … GULF HOPES HAVE REPRESENTATIVE THERE DURING COURSE NEXT MONTH" [Ward, radiogram to the Eastern and General Syndicate, 15 February 1929].

Crisis Meetings

By the end of February, arrangements had been made for urgent meetings in Washington and London. On the 26th of February, Holmes left the Middle East for London. On the 20th of March, Francis B. Loomis (Socal) and William T. Wallace (Gulf Oil) met with Secretary of State Frank B. Kellogg. Within eight days, Kellogg had cabled the US Chargé d'Affaires in London requesting him to discuss the case informally, at an early date, with the appropriate authorities of the British Government and point out that existing United States legislation was extremely liberal regarding the operations of petroleum concessions held by foreign-controlled companies. He also requested the Chargé d'Affaires to obtain a statement of the British Government's policy respecting the holding and operating by foreigners of petroleum concessions in territories such as Bahrain. [6]

During the 29th of March, Major Harry G. Davis, who had been selected by Gulf Oil to represent its case in London, left New York to attend to his brief, namely to co-operate with the Syndicate. On the 3rd of April, Ray Atherton, the US Chargé d'Affaires, visited the Foreign Office in London to pursue Kellogg's requests. [7] Meanwhile, departments within the British Government had been preparing their case.

On the 2nd of March, H. W. Cole, Director of the Mines Department, had expressed his views in a memorandum to the Colonial Office: while it might be very desirable to maintain British control over oil concessions in the region in question, it might be preferable that the oil, if it existed in commercial quantities, *should* be produced by a foreign controlled company, rather than not at all. Cole suggested that the Secretary of State "may think fit" to consult the Foreign Office regarding the possible effects of excluding American companies and thus the resultant attitude of those companies and of the United States Government towards British oil companies. In any case:

> "It would seem doubtful whether any considerable production is to be anticipated
> in Bahrein, and from the point of view of oil supplies in this region, having regard
> to the Persian output and strong probability of a large output in Iraq, there is no
> urgent necessity to take steps to promote oil production in Bahrein" [Cole, memo
> from the Mines Department to the Colonial Office, 2 March 1929].

Based on Cole's understanding of the Red Line Agreement, it would be impossible for the Bahrain concession to be held solely by the Anglo-Persian Oil Company or the Gulf Oil Corporation's subsidiary company, Eastern Gulf Oil, or any other oil company associated with the Turkish Petroleum Company. Nonetheless, the Secretary of Mines agreed that it would be preferable for the Turkish Petroleum Company, rather than an American company, to become interested in Bahrain.

On the 20th of March, the same day that Francis Loomis and William Wallace had met Frank Kellogg in Washington, O. G. R. Williams of the Colonial Office wrote to the Admiralty in London conveying the Secretary of State's views. If the British Government were to withdraw its opposition to the proposal that Eastern Gulf Oil Company should be allowed to obtain a controlling interest in the Bahrain concession granted to the Eastern and General Syndicate, the likely outcome would be that oil, if it existed in Bahrain, would be exploited by the Turkish Petroleum Company or a subsidiary company formed for that purpose. If, on the other hand, the British Government maintained its opposition to the proposal, the likely outcome would be that Bahrain's oilfields would remain undeveloped, since the Syndicate had failed to obtain 100-per-cent British capital, and even if the Anglo-Persian Oil Company were prepared to provide capital, it could not be utilized for the development of Bahrain's oilfields because of its relationship with the Turkish Petroleum Company and the Red Line Agreement. The Secretary of State was under the impression that the Eastern and General Syndicate had not exercised its option on the Bahrain concession; however, as a caveat in case his assumption was wrong, which it was, he suggested that it would be advisable for the British Government to abandon its opposition to the introduction of American capital to operate the Bahrain concession and to concentrate instead upon obtaining as much British control as possible. [8]

Now, having modified his views, L. S. Amery, the Secretary of State, advised the Admiralty, the Foreign Office, the India Office and the Petroleum Department that the final attitude adopted by the British Government should be worked out *before* conversations were opened with Eastern Gulf Oil and the Syndicate. He suggested that it would be quite reasonable to insist that the company to be formed to take over the concession should be registered in the United Kingdom, that its chairman should at all times be a British subject, and that the majority, if not all, of the local staff should be British too.

The Control Controversy

Within three weeks, the Americans stated their position. On the 8th of April 1929, the Colonial Office learned from the Eastern and General Syndicate that Francis B. Loomis of the Standard Oil Company of California and Major Harry G. Davis, representing the Eastern Gulf Oil Company, had arrived in London with the object of effecting the arrangements the Syndicate had with them relating to the Bahrain oil concession and other oil interests in Arabia.

The Syndicate, on behalf of Loomis and Davis, informed the Colonial Office that the British nationality provisions were quite unacceptable to the American corporations. The Eastern Gulf Oil Company, although under no obligation to do so, by offering the Bahrain concession to the Turkish Petroleum Company, had given fair opportunity for adequate British representation. That offer having been declined, there could be no valid reason or right for the exclusion of American interests in the manner contemplated by the British nationality provisions, the result (so the Syndicate believed) of an afterthought.

The Colonial Office was also assaulted with the news that the Standard Oil Company of California had already formed a corporation, not in the United Kingdom but under

Canadian laws, known as the Bahrein Petroleum Company, for the sole purpose of taking over the *Bahrein Island Concession*. To that end, Standard Oil Company of California now insisted that the Syndicate should settle the disputed issue of nationality forthwith so that the new company could commence development operations. [9]

During the morning of the 9th of April, Syndicate director, Edmund Janson, and Frank Holmes presented themselves at the Colonial Office to start discussions. It became clear that the British Government was not prepared to move with any speed, in any direction. For ten days Francis B. Loomis had been working on the Bahrain matter in conjunction with Gulf Oil's representative, Harry G. Davis. When Loomis reported to Michael E. Lombardi at Head Office in San Francisco, he told him that everything had been done to expedite the matter, but it had got "into the treadmill of departmental routine". The Foreign Office had referred to the India Office, its group of Persian advisers and its Mesopotamian experts, and it would take the matter some time to get around the various officials. Loomis had gathered from the American Chargé d'Affaires and the Syndicate that, in official circles, it was felt that the Colonial Office had gone too far and made a quite unwarranted demand, namely that Bapco should be British in every respect. Loomis had gleaned indications from other sources which seemed to confirm the truth of his understanding. In conclusion, he thought that he could not justify staying in London longer, partly because: "I also have the feeling that the Gulf representative [Harry Davis] ought to bear the burden of the work because after all, it is their fight" [Loomis, letter to Lombardi, 15 April 1929].

On the 10th of May, Janson and Frank Holmes met J. H. Hall, Acting Principal of the Colonial Office, who told them that the Foreign, India and Colonial Offices were due to have a tripartite conference within a week. Whilst Hall was not optimistic that a decision would be made quickly, he suggested a way of speeding things up. The original Bahrain concession and the concession for the Additional Area in Bahrain (including the surrounding waters) could be reoffered by Eastern Gulf Oil Company to the Turkish Petroleum Company, but under conditions which would cause TPC to retransfer the concessions back to the Eastern Gulf Oil Company, should the proposal be taken up. It would be better still if the Standard Oil Company of California performed the deed. It would ease the situation further if Janson were to visit Sir John Cadman of the Anglo-Persian Oil Company that day, inform him of the plan and explain that it was the wish of the Colonial Office that Sir John should use his influence and endeavour to induce the Turkish Petroleum Company to receive this reoffer favourably.

After a telephone call, Janson arrived at Britannia House to outline the scheme. Cadman's reaction was still unco-operative. "As the Colonial Office had got itself into this difficulty, he saw no reason why it should not extricate itself from that difficulty" [Memorandum of a Conference at the Colonial Office, 10 May 1929]. Sir John was persuaded, however, to think the idea over. For the next twelve days, Janson telephoned Britannia House. Finally, on the 22nd of May 1929, Sir John gave in and asked where the Eastern Gulf Oil Company stood so that he might lend his support to the proposed negotiations with the Turkish Petroleum Company. Thus, the proposed scheme was outlined again. Eastern Gulf Oil Company would reoffer the concessions to the Turkish Petroleum Company, in exactly the same manner as it had done on the 27th of July 1928, just prior to the Red Line Agreement being effected at the end of that month. Then the negotiations between the two companies would be carried on as if there had been no refusal and no other difficulties had arisen since that date. In this case, the Turkish Petroleum Company would be quite within its constitutional rights to arrange with Eastern Gulf Oil Company the creation of a new company formed to work the concessions, with Gulf Oil Corporation (a TPC shareholder) having full technical control. It was further suggested that a British liaison officer could be appointed, selected by the newly-created nominee company, who would act as an agent between the nominee company and the Ruler of Bahrain. [10]

Meanwhile, on the 12th of May, the Viceroy, Foreign and Political Department, had sent a telegram to the Secretary of State for India contending that existing treaties guarded sufficiently against external encroachments on the British position in Bahrain. In another telegram later that day, he added: "Any increase in American influence, which already is very strong, is to be deprecated".

Four days later, on the 16th of May, J. C. Walton of the India Office wrote to the Colonial Office, enclosing the Viceroy's telegrams but explaining that as there was no clause in the 1925 Bahrain concession document which provided for the continuance of British control,

> "there is no sure ground on which to withhold consent to the participation of American interests. … Lord Peel sees no alternative, in the circumstances, but to concur in the course suggested … that the United States Embassy should be informed that His Majesty's Government are prepared in principle to consent to the participation of American interests."

Conciliation

By the 27th of May, Harry G. Davis, Gulf Oil's representative in London, had become distinctly agitated. William T. Wallace in New York heard that Davis had come to the uncomfortable conclusion that both the Bahrain and Kuwait negotiations were approaching disaster. Davis suggested to Wallace that he should withhold all action until he had read Davis' letter being written that day. A recent meeting with Edmund Janson had led Davis to conclude that there was some "muddy water" which very much needed clearing. Although, with certain reservations, the Colonial Office intended to act favourably regarding Bahrain matters and the Foreign Office would communicate the decisions to the American Embassy during the early part of that week, Davis believed that several points needed clarification if not correction.

In the first place Davis could not understand what had led the British Colonial Office to "concoct" a scheme whereby the Standard Oil Company of California's position on the Bahrain islands should be transferred to the Turkish Petroleum Company, and the

Eastern Gulf Oil Company's position in Kuwait should also be transferred to the Turkish Petroleum Company, all this to be followed by negotiations, on the part of both entities, with the Turkish Petroleum Company for a retransfer back, through nominees. [11]

In the second place, why had Sir John Cadman become involved in the whole affair? "It was here that Mr Janson shot a bolt", for it transpired that the whole idea had been his and not Hall's at the Colonial Office, although the latter had agreed to subscribe to it so that the support of Sir John could be secured on the pretext that the Colonial Office was bringing pressure to bear. Davis, who surmised that the whole scheme had been dreamt up by Janson to get the Syndicate "off the hook", told Janson that the proposition sounded very complicated. Davis "could see no object in the gyrations, except to cause a great deal of delay and complication in the present status of the case". During their heated exchange, Davis had accused Janson of not taking sufficient action to resolve the months-old impasse. Janson should get "some definite expression right away from the British Colonial Office" as to how matters could be solved. However, on that matter Davis was philosophical, "because I am afraid that without constant stirring up, one of his [Janson's] main objects in life is to follow the line of least resistance" [Davis, letter to Wallace, 27 May 1929].

Davis also contended that the Syndicate "had failed to strike while the iron was hot", knowing that it was obliged to Eastern Gulf Oil to bring this transaction to a final and satisfactory close. Everyone wished to know where they stood, and, naturally, Eastern Gulf Oil preferred not to be forced yet again to request the State Department to protect its interests. Another aspect of the whole affair which bothered Davis was the sudden about-face of the British Colonial Office, since it had originally proposed the Turkish Petroleum Company transaction, yet had suddenly veered round, and decided to send its reply to the State Department through the American Embassy in London. The present situation was most unsettling, as Davis revealed:

"I have wondered whether or not you have figured out what the memorandum prepared by Major Holmes and censored by Mr Janson, regarding their conference held at the British Colonial Office of May 10th 1929, actually means. … Before concluding, I want to convey to you an outstanding feature, and that is that Mr Janson has the idea that the terms offered by the British Colonial Office in regard to the Bahrein situation should be swallowed, hook, bait and sinker; that no action should be taken in regard to Koweit until the Bahrein situation is behind us, and that when this point is reached, the Eastern and General Syndicate Ltd should open up negotiations *anew* with the British Colonial Office, and *ask* the Colonial Office under what terms and conditions they would agree to for the actual consummation of the Koweit concession. I believe that Mr Janson's suggestion, if carried out, would be disastrous, and the Koweit concession would then become a dream, or rather, a nightmare, because such a procedure would be tantamount to a confession that Gulf had acquired no rights whatsoever through the Eastern and General Syndicate Ltd by virtue of the agreement entered into, November 30th 1927" [Davis, letter to Wallace, 27 May 1929].

Two days later, Lord Monteagle, signing for the Secretary of State, Foreign Office, London, wrote to Ray Atherton, Chargé d'Affaires at the United States Embassy in London: the British Government was prepared in principle to consent to the participation of United States' interests in this concession (Bahrain), subject to it being satisfied with the conditions on which United States capital would participate and in particular the nationality of the operating company, of its chairman and directors, and of the personnel who would be employed in the islands. [12] During the following day, this remarkable decision was confirmed in writing by the Secretary of State to the Syndicate. Almost immediately, G. Howland Shaw, Chief of the Division of Near Eastern Affairs in the Department of State, received this news in Washington DC which he promptly despatched to Wallace in New York.

On the 3rd of June, Harry G. Davis wrote to Wallace again, but this time on a very different subject. The political situation in the United Kingdom had changed following the recent general election. The Conservative Party had been voted out of office and replaced by the Labour Party, which was known to have a different foreign policy. During the pre-election campaign, Philip Snowden, Labour Chancellor of the Exchequer in Ramsay MacDonald's 1923-4 Cabinet, had declared that Britain must not monopolize the world's oil supply, for jealousy on this score would eventually lead to war between Great Britain and the United States. Snowden had advocated an "Open Door" policy for all countries to obtain their oil supplies. [13] On the same day, Davis wrote to the Eastern and General Syndicate, c/o the National Bank of Commerce in New York, Crosby Square, London EC3, insisting that it should refrain from any commitment with the Colonial Office connected with the nationality debate and the concession negotiations. [14]

Battle-lines Emerge

If the 3rd of June 1929 had been a watershed, the day on which the British Colonial Office agreed, in principle, to the participation of United States interests in the Bahrain Concession, deadlock still ensued. The battle-lines were clear. Primarily, the British wished to impose political influence and control on Socal. The Americans wanted to be free of such restrictions so that they could exercise their options on the Bahrain concession effectively, without the imposition of British control on their operations.

For almost a year, until May 1930, company directors, lawyers, government officials and representatives of the Ruler of Bahrain were engaged in seemingly endless conferences and cable communication. Hours of discussion generated copious reports which were, without doubt, impressive for their comprehensive content. The line-up to devise a strategy and secure a solution was as daunting as the paperwork. Judge Frank Feuille and Francis B. Loomis argued Socal's case, with Messrs Pillsbury, Madison and Sutro and Messrs Freshfields, Leese and Munns acting as the company's attorneys and solicitors in San Francisco and London respectively. William T. Wallace, James M. Greer (company lawyer) and Harry G. Davis acted on behalf of Gulf Oil, while H. T. Adams (company secretary), G. S. Pott (solicitor), E. W. Janson (director) and Major Holmes took on the Syndicate's cause in London, with Thomas Ward representing its interests in New York.

The renewal, or non-renewal, of the 1925 Bahrain Exploration Licence, which was the grantee's first step toward acquiring a Mining Lease with terms set forth in the Third Schedule to the concession agreement, was an equally delicate and daunting matter. The first of many ponderous interdepartmental conferences called by the British Government's negotiators, was convened on the 7th of June. Messrs Williams, Bushe, Hall and Bigg (representing the Colonial Office), Lord Monteagle, G. Jebb and G. H. Thomson (for the Foreign Office), J. G. Laithwaite (India Office) and H. R. W. Giffard (Petroleum Department) stated what was generally known already: it could be concluded from the available evidence that the Exploration Licence granted to the Syndicate had expired on the 1st of December 1928. Further, that since it had not been extended or renewed and that as the Syndicate had not applied either before or upon that date for the grant of a prospecting licence, it was not now entitled to claim such a licence and its concession had lapsed. [15]

In this case, the British Government could impose upon the Syndicate any conditions it chose before agreeing to recommend to the Ruler of Bahrain that he might wish to consider renewing the Syndicate's Exploration Licence for a further period of only one year, until the 2nd of December 1929. However, Bushe, now a well-versed campaigner in this affair, interjected that *if* the Syndicate refused to accept these conditions and, in accordance with its legal right under the terms of the Bahrain concession, decided to place the matter before arbitrators, *then* the position of the British Government was not "unassailable".

The India Office was not persuaded to this view. Laithwaite, on its behalf, maintained that the conditions should be as stringent as possible so that the operation of the concession might remain under British control. However, the Colonial Office and Foreign Office delegates were a little more realistic. Following the involvement of the American Government, through its embassy in London, it was no longer feasible to do more than to provide for the retention of a British *element* in the operations. The Foreign Office considered that if the "British Control Clause" were to be abolished, then the British case would be weakened. However, if the case were to be taken to arbitration by the Syndicate, then onerous terms for the admission of American capital should not be insisted upon, since the Americans were hardly likely to agree to any restriction which would render their financial control inoperative.

The outcome was that the British proposed four conditions. First, the company to be formed to take over the concession should be and remain British, registered in Great Britain. Second, the chairman and managing director (if any) should be a British subject at all times. Third, that between 49.5 and 33.3-per-cent of the directors should at all times be British subjects. Finally, the local general manager of the company and all the local staff were to be British subjects or subjects of the Ruler of Bahrain – with certain exceptions.

The Americans did not agree. Nothing happened for more than a month while both "camps" considered their positions. On the 17th of July, Gulf Oil's lawyer James M. Greer updated Socal's counsel, Judge Frank Feuille. Action by the American Embassy during the previous afternoon had resulted in the Colonial Office calling a meeting with Janson for the 19th of July. After the meeting, Major Harry Davis (representing Gulf Oil in London) cabled New York. Apparently, during the two-hour conference, Holmes and Janson had been informed that the "Additional Area" being negotiated for the Bahrain concession would be conceded, *if* the four British conditions were met. [16] The official record suggests that the Syndicate representatives did not try to dispute the right of the British Government, or that of the Ruler of

Judge Frank Feuille (*Chevron archive*).

Bahrain, to impose conditions upon the participation of American capital in the concession operations or upon the transfer to American oil interests of the rights conferred by the concession. "It was therefore unnecessary to make use of the various arguments referred to in recent correspondence" [Record of a discussion, 19 July 1929].

Holmes and Janson agreed that the first condition was not unreasonable, even though the Bahrein Petroleum Company had been registered already in Canada. However, they disagreed with the suggestion that, with certain exceptions, the local general manager and the whole of the local staff should be British, their argument being that as American oil magnates had a very low opinion of British directing staff, most likely they would employ American subjects as their local technical employees. [17] Thus, the Syndicate wished to replace the inflexible expression "the whole of the local employees" with the phrase "as many of the local staff as is consistent with the efficient carrying out of the undertaking".

The Colonial Office opposed this suggestion and instructed Janson and Holmes to do their utmost to persuade their American friends to accept the proposed conditions as they stood, since the British Government was unlikely to consent to their being modified any further.

Diplomatic Deliberations

After hearing the account of the meeting, James Greer and Judge Feuille deemed that, generally speaking, such demands were not acceptable to them. They felt that Janson and Holmes should invite further discussions with the Colonial Office in the hope that the demands might be softened before being reduced to writing. Their contention was that since the Americans were furnishing the capital, logically they should remain in control of its use and expenditure. They wished the company to be registered in Canada and not the United Kingdom, so that its directors could meet more easily. Furthermore, "while we do not absolutely refuse to consider a British subject as a member of the Board, we would *never* consent to more than one such member out of a total of five". Greer emphasized that, as was well known, one of the largest oil concerns in America was dominated by British and Dutch capital, with most of the official positions, if not the dominating ones, held by British subjects "without hindrance or obstruction upon the part of the American government". Therefore, "we are justly entitled to the same treatment from the British Government" [Greer, telegram to Feuille, 20 July 1929].

Two days later, on the 22nd of July, Davis reported back to New York that the American Ambassador was convinced that Gulf Oil's claims were just but considered that Janson and Holmes were quite unable to handle the Colonial Office. Both he and the Chargé d'Affaires were of the view that the British Foreign Office was not prejudiced against the Americans and would undoubtedly be very relieved to find a formula to satisfy Gulf Oil. [18] After further consultation with the Colonial Office on the 31st of July, Janson and Holmes informed Davis that the British Government was "inclined to consider favourably" four of the five counter-proposals, Janson having told the meeting that Gulf Oil had objected to Bapco being formed in Great Britain. The reason they gave was that taxation imposed on companies registered in Great Britain was heavier than that imposed on Canadian companies. Not only was favourable taxation a consideration for Gulf Oil, but also convenient travel between America and Canada for the purpose of board meetings. However, a reduction in a company's tax liability, to the detriment of the British Chancellor of the Exchequer's revenues, was hardly likely to gain Colonial Office approval.

Since the Americans were not prepared to accept that the chairman and/or managing director of the new company should be British, they proposed that the chairman of the group in America should also officiate as chairman of the new company, and that the managing director (if any) should also be associated already with the American Group. The Americans were prepared to consider one British director. The local manager in charge in Bahrain could be British, providing that he acted as a liaison officer, engaged in all the negotiations with the Ruler of Bahrain and the Bahrainis themselves. He would, however, be subject to the authority of the local manager who would be a technical official in charge of the field operations of the company.

Sir John Shuckburgh (Colonial Office), however, expressed doubt as to whether or not the Syndicate's American associates fully appreciated the position. Without the good will and support of the British Government, they could not obtain or work the concession at all. According to Shuckburgh, the British Government had laid down certain entirely reasonable conditions as the price of that good will. He contended that the appointment of a British local manager was as much in the interest of the company as it was of the British Government, although "the latter attached the greatest possible importance to this condition" [Record of a discussion, 31 July 1929].

At the conclusion of his homily, Shuckburgh asked Janson to which of the four conditions the Americans attached the greatest importance. Janson answered D, namely the British requirement that the local general manager of the company and all the local staff were to be British or subjects of the Ruler of Bahrain – with certain exceptions. Janson was of the opinion that if this could be settled to the satisfaction of both parties, then the remaining three conditions could be "disposed of with less difficulty". In short, Janson reported that the Americans regarded the control of the local operations as the vital consideration in the successful production of oil and they were not prepared to delegate this to a British subject or to someone unsuitably qualified. The local manager must, in their opinion, be an official in whom they had full confidence, and they were not aware of any British subject who would satisfy this requirement.

In an effort to break the deadlock, Shuckburgh suggested that Major Holmes might be appointed as the company's first local general manager, since his New Zealand nationality would fulfil the British requirement. Not surprisingly, Holmes agreed to accept the post, if it were to be offered to him by the Americans. The Syndicate assumed that there would be no difficulty and agreed to a reworded variant of the controversial condition:

> "There shall at all times be a local General Manager who shall be approved by His Majesty's Government. He shall be the chief Representative of the company in Bahrein and shall be responsible for all relations with the local authorities and population. ... For the first five years after the inception of the enterprise the local General Manager shall be Major Holmes" [Record of a discussion, 31 July 1929].

The Syndicate had assumed wrongly. When Harry G. Davis heard of this, he thought it most unlikely that the American group would agree to Major Holmes being designated "local general manager" of the company. The term "general manager" was reserved in America for the official in charge of technical operations. What about some such title as "Director"? Nonetheless, Davis thought that the principle of Shuckburgh's idea would be acceptable to Gulf Oil and Socal, except that they would almost certainly insist that some provision be made to ensure that Holmes would have no authority to intervene in the technical operations of the company. [19]

Janson, Holmes and Shuckburgh reconvened on the 8th of August for another brain storming session. A suitable designation remained elusive. New ideas such as "special delegate", "special agent", "foreign representative" or "special representative" were all considered. For several more weeks correspondence went to and from London and New York, each side shifting position here and giving a little there, but no final agreement being reached. More vocabulary was discussed in London in anticipation of the time when the Eastern and General Syndicate *might* apply to the Ruler of Bahrain to renew the concession. While all was being said, done and written, Wedgwood Benn (Secretary of State at the India Office) felt that since it was considered to be "highly desirable that the principal representative of the company should be a British subject", it was equally desirable to make clear the special treaty relationship between the British Government and the Ruler of Bahrain which had existed for many years. [20] Thus, a suitable new amendment which reflected this point was proposed, the idea being that it should be included in the paragraphs following the defined conditions.

Revisions Revealed and Reviewed

The 16th of September 1929 was a day of revelation. In Downing Street, London, at the direction of Lord Passfield, the Secretary of State (Colonial Office) outlined the four conditions, now revised, which when accepted would enable the British Government to advise the Ruler of Bahrain to extend, if he wished, the period of the Exploration Licence granted to the Eastern and General Syndicate on the 2nd of December 1925 and to agree to the assignment of the Syndicate's rights under that agreement to a new company controlled by American interests. The letter to the Syndicate emphasized that the British Government attached great importance to the conditions being accepted and was not prepared to agree to their modification. The recent suggestion that reference should be made to the special treaty relations existing between the British Government and the Ruler of Bahrain had been inserted. Also inserted was a request for assurance that neither the Syndicate nor the company to which it was proposed to assign the Syndicate's option on the concession would take "any steps which would prejudice the position of the proposed sites for a landing ground and sea-plane station in Bahrein" [Williams, letter from the Colonial Office to the Eastern and General Syndicate, 16 September 1929]. In essence, the revised conditions were those as discussed, but with one or two minor adjustments.

The first condition concerned the formation of a British company, registered in Canada, but with a requirement to maintain a registered office in Great Britain with a British subject in charge of it at all times.

The second condition proposed five directors, one of which should be a British subject at all times, chosen by the British Government, but whose salary would be paid by the company.

The third condition had been and was to remain the most controversial. According to the Colonial Office, the company would maintain in Bahrain a Chief Local Representative whose appointment would be approved by the British Government. He would be the *sole* representative of the company empowered to deal directly with the local authorities in Bahrain, including the Ruler, his family and subjects. He would communicate through the resident Political Agent; and furthermore, for the first five years of the company's operations in Bahrain, this person should be Major Frank Holmes.

The fourth brief condition had been modified to require that the company would employ as many British subjects as possible, rather than only British subjects. [21]

The planked surface of Wall Street was a scene of near-panic as hundreds of bewildered investors milled about after the stock market crash on Black Thursday, 24th of October 1929. A record of 16 million shares changed hands that day, and the decline in stock value by the end of the year was estimated at $15,000 million (*Popperfoto*).

On the 20th of September, four days after the Colonial Office had revealed its revised conditions, Frank Holmes wrote to William Wallace in America saying that the Colonial Office was pressing very strongly that Janson and he should visit New York without delay to discuss the revisions, because only after face-to-face consultation would the Colonial Office consider any alterations or adjustments, should they be necessary. To this end, Holmes, Janson and Davis embarked on the *Aquitania* and sailed for New York on the 12th of October. Six days later they met William Wallace, James M. Greer, Judge Feuille, Francis B. Loomis and Thomas Ward. The eight men discussed strategy. Janson should go to Pittsburgh on Tuesday, the 22nd of October, to meet the Gulf Oil directors and discuss the proposals. Thomas Ward recalled that it was a memorable day for other reasons too. A "tremendous crash in security values" occurred on the New York Stock Exchange [Ward, 1965, p. 170]. Three days after the Pittsburgh meeting, Janson and Holmes set sail for London armed with Gulf Oil's reply to the Colonial Office's proposals.

A long, explicit letter, written on behalf of Eastern Gulf Oil company and Standard Oil Company of California to the directors of the Eastern and General Syndicate, stated that

"at the outset, we wish to express our sincere appreciation of the consideration given this matter by the Colonial Office. … It is a definite fixed policy of both the Standard Oil company of California and of the Gulf Oil Corporation that neither they, nor any of their subsidiaries, officials and employees, engaged in operations in foreign fields, involve themselves in the political activities of the respective foreign countries. We enter such fields on a purely business and commercial basis" [Leovy, letter from Eastern Gulf Oil Company to the Eastern and General Syndicate, 24 October 1929].

The Americans wished to meet the Colonial Office conditions outlined in their 16th of September letter, but cautioned that sound management and control of the business organization demanded clarification of certain points. Gulf and Socal trusted that the Colonial Office would acknowledge that the requests were just and reasonable.

Although the mechanics had to be worked out, the Americans had no problem with the first condition. They accepted the second too, but with one subtle exception. Rather than the British director being *chosen* by the British Government, the Americans wished him to be *acceptable* to the British Government, so that they could, according to their counter-proposal, have a say in the selection too. The third condition, however, was a different matter. This was quite unacceptable to the Americans because:

"It is not good business practice to agree to the appointment of a certain individual for a fixed number of years for any given important post in the operations of any company. It is difficult to foresee the conditions which may arise in the course of a man's employment. … The company feels justified in requiring greater latitude in this connection. The foregoing is not to be construed as questioning Major Holmes for the designation of Chief Local Representative. Both our American companies, in fact, have the highest and most sincere regard for Major Holmes and are convinced that he is the best man available for the post on account of his proved abilities. … However, we do not feel that we should be called upon to now agree definitely on one man, whoever he may be, to fill that position for a fixed number of years regardless of the contingencies that may arise" [Leovy, letter from Eastern Gulf Oil to the Eastern and General Syndicate, 24 October 1929].

The Americans suggested that the Colonial Office's proposed third condition could be retained, but with the proviso that this arrangement between the company and Major Holmes was valid only so long as it continued to be mutually satisfactory to them during such five years, or less. [22]
Finally, the fourth condition had been modified to the Americans' satisfaction.

Easing towards Agreement

At the Gulf Oil board meeting, held on the 22nd of October, it had been paragraph 4 (not to be confused with condition [d]) of the Colonial Office's 16th of September letter which had brought matters to a head. This stated that if the Syndicate accepted the revisions, then the Colonial Office would advise the British Government to advise the Political Resident in the Persian Gulf, to advise the Political Agent in Bahrain, to recommend to the Ruler of Bahrain that he might wish to consider extending the period of the Exploration Licence which he granted to the Syndicate on the 2nd of December 1925. The Americans wished to know, for how long?

Already a year had been lost because of the numerous discussions regarding the conditions. The Gulf Oil board felt that it should be made clear that the one-year extension of the Exploration Licence only ran to the 2nd of December 1930. Careful thought had to be given to starting up operations in a region as remote as the Bahrain islands. Time was required to undertake reconnaissance work, to order and ship materials, assemble equipment and start operations. The present extension left little flexibility.

Following their return to London, Janson and Holmes presented Leovy's letter to the Colonial Office. Williams and Hall, on its behalf, thought the contents were satisfactory. But, as Wallace remarked later to Feuille, "apparently the Colonial Office officials still deem it necessary to refer the matter back to the India Office and the Admiralty" [Wallace, letter to Feuille, 13 November 1929].

Talks began regarding the possible appointment of Major Holmes in Bahrain. After conferring across the Atlantic Ocean, Francis B. Loomis in San Francisco (for Socal) and Janson (for the Syndicate in London) agreed that Major Holmes should be engaged jointly by the three interested companies in the Bahrain concession, namely the Eastern Gulf company, the Standard Oil company of California and the Eastern and General Syndicate Limited. This was to be effective on the 1st of January 1930. It was agreed also that his principal immediate task would be to secure, through negotiation with the Ruler of Bahrain, the "Additional Area" of the islands of Bahrain for inclusion in an additional concession. [23]

By late November, Gulf Oil and Socal had received notice that the Colonial Office had drafted the form of a transfer of the Bahrain concession to the Bahrein Petroleum Company. Whilst the Americans were delighted, Wallace was cautious as to how this would progress, since "our experience with the English lawyers is that their language is unnecessarily involved and differs from the direct, positive language which we are so accustomed to using in this country in legal documents" [Wallace, letter to Feuille, 23 November 1929].

Five days later, Feuille received a ten page cable from London containing the draft agreement which, in turn, was copied to Socal in San Francisco. On the 4th of December, M. E. Lombardi (Socal) sent a night letter to his colleague Francis Loomis, then staying in Chicago. This confirmed that Socal agreed to the proposed draft but did not wish to accept the assignment unless and until Socal had definite assurance that the term of the Exploration Licence had been properly extended by the Ruler of Bahrain with the necessary approvals to the 2nd of December 1930. Or, failing this, that the term of the Prospecting Licence should not begin to run until assignment of the concession to Socal was effective and the company was in a position to send its men to Bahrain. [24]

The 18th of December was a memorable day. An elated Harry Davis cabled Gulf Oil to say that deadlock between the Colonial Office and the India Office had been broken. Meanwhile, the lawyers continued to work on the document, fine-tuning each party's requirements. On New Year's Eve, as the Colonial Office letter of approval was being waited upon, Feuille heard from Wallace that the Colonial Office was now sticking on another point: it considered a board of seven members, rather than five, would prejudice British interests. Feuille could not see why.

Three days later Gulf Oil was presented with two more obstacles. The appointment of the British director had to be made in consultation with the British Government, and regarding Major Holmes, the British Government wished, in principle, to have the right to his appointment, although consent would not be "unreasonably withheld". If the Eastern Gulf Oil company would agree to these two requirements, then the British Government was prepared to advise the Ruler of Bahrain to recommend that the Exploration Licence should be extended to the 2nd of December 1930, at a cost of a further 10,000 rupees. The communication also stated categorically that the British Government was not prepared to contemplate a partial or limited assignment of the concession. Any assignment made at that time had to be final and complete and in such a form that the assignees were "definitely substituted" for the Syndicate regarding the "enjoyment of all benefits and the performance of all obligations under the Concession" [Williams, letter from the Colonial Office to the Eastern and General Syndicate, 3 January 1930]. The extension of the concession to include the "Additional Area" was another matter.

Disquiet was expressed in America. Davis, Wallace, Loomis, Greer and Feuille conferred. In a confidential cable to Davis, Wallace doubted the Colonial Office's good faith. Feuille "participated in that doubt"; and so, reluctantly, the United States Department of State was approached again. Wallace, when writing to Loomis, confided that he was convinced "some of our present difficulties with the Colonial Office and the evident concern of that Office with respect to Holmes, have arisen out of numerous private discussions which took place between Holmes and Hall (Colonial Office)" [Wallace, letter to Loomis, 14 January 1930].

From this moment, Wallace was no longer prepared to consider what he described as "tacit understandings". In future, he would accept only official written communications from the British Government. On the 18th of January, Feuille admitted to Loomis that he too was experiencing a lack of confidence in the Colonial Office, the Eastern and General Syndicate and Major Holmes. It was possible, he thought, that Hall (Colonial Office) was a personal friend of Major Holmes and, for that reason, the former was aiding the latter in obtaining conditions from Gulf that would guarantee Holmes' employment for five years. On the other hand, the Colonial Office might have been "inspired by the fact that the British government is a competitor in the oil industry, through the Anglo Persian and probably the Dutch Shell, in seeking to impose conditions regarding Major Holmes upon us" [Feuille, letter to Loomis, 18 January 1930].

Indenture of Assent

The matter dragged on for ten days. Writing on the 29th of January from his room at the Hyde Park Hotel, London, Harry G. Davis informed the Syndicate that his principals were now prepared to accept the four conditions set out in the Colonial Office letter dated the 16th of September 1929 and the letter of modification dated the 3rd of January 1930, although a small adjustment to the latter was still required.

A week later, Janson, Holmes and Pott met Williams, Bushe and Hall at their now familiar rendezvous, the Colonial Office. The points raised in Davis' 29th of January letter were discussed, and although each side stuck to its well-worn grooves, it seemed as if little steam remained in the system to fuel further discussion on the revised conditions. Instead, the precise nature of the formal instrument governing the assignment of the Syndicate's rights under the Bahrain agreement became the new *cause célèbre*, an issue, which perhaps predictably, generated many sharp legal exchanges during the coming weeks.

On the 17th of February a posse of lawyers arrived in Downing Street to keep their appointment with the Colonial Office. Discussion focused on the form of assent to be given by the Ruler of Bahrain to the Eastern and General Syndicate assigning the concession dated the 2nd of December 1925 to the Bahrein Petroleum Company Limited. Its precise nature included the right of forfeiture by the Ruler of Bahrain on the breach of the conditions, subject to the right to arbitrate any question in dispute. Whether the Ruler of Bahrain was aware of this new development will, no doubt, remain a moot point. As Ballantyne (Freshfields' solicitor) reported to the Americans later, the lawyers' objection to start with was that this request appeared to involve modifying the Concession Agreement of the 2nd of December 1925. As a point of principle, Ballantyne argued that the 1925 Agreement should be inviolate. The fact that it was proposed to alter it, with the British Government's sanction, filled the lawyers with "grave apprehensions as to the future" [Freshfields, Leese and Munns, 17 February 1930].

Hall maintained that the penalty of forfeiture had been clear in the Colonial Office's letters of the 16th of September 1929 and the 3rd of January 1930. Ballantyne insisted that such an inference was ambiguous. Bushe and Hall argued that the agreement must be a "tripartite deed, entered into by the Ruler of Bahrain, the Eastern and General Syndicate Limited and the Bahrein Petroleum Company Limited, whereby the latter covenanted directly with the Ruler of Bahrain to observe and perform the Conditions on pain of forfeiture of the Concession or any Mining Lease thereunder, subject always to arbitration" [Freshfields, Leese and Munns, 17 February 1930]. Ballantyne maintained that the matter could be permitted "to rest on the correspondence", to which the

Colonial Office replied: "this could not be allowed for a moment". In that case, Ballantyne requested proof of this restriction in the form of the Municipal Laws of Bahrain.

The meeting dispersed with no solution. On the 28th of February 1930, Frank Feuille reported to Loomis that upon reflection he considered that Gulf Oil and Socal had the right to expect the British Government to be guided not only by the principles of equity which prevailed in the United States and England, but by the principles of equity recognized in international law. [25]

Within two more weeks, the substantial weight of legal pressure brought matters under control. On the 13th of March 1930, Lord Passfield (Colonial Office) sent a draft Agreement, to be formalized between the Ruler of Bahrain and the Eastern and General Syndicate, to the Political Resident Persian Gulf. At last, after more than a year's debate, the period of the Exploration Licence under the Concession Agreement dated the 2nd of December 1925 was to be extended to the 2nd of December 1930, according to the conditions which had been agreed. Communications continued back and forth until the 12th of June 1930, when Shaikh Hamad bin Isa Al-Khalifa, witnessed by the Political Agent, C. G. Prior, signed an Indenture with Major Holmes on behalf of the Eastern and General Syndicate. In this document, the Acting Ruler assented to the transfer by the Syndicate to the Bahrein Petroleum Company Limited of the Syndicate's rights under the Bahrain concession, subject to certain modifications. The deadlock had been broken.

CHAPTER FIVE

The Pioneers

Socal on Guard

In March 1930, there was an added warmth to London's spring air. Now that the deadlock had been broken, the British Government was showing signs of softening its stance on the nationality issue. The Secretary of State (Colonial Office) suggested that it might be advisable for the British Government to abandon its opposition to the introduction of American capital to operate in Bahrain and to concentrate instead upon obtaining as much British control as possible. Upon hearing this welcome news, Socal made immediate plans to send a survey party to Bahrain. By the end of March 1930, William Taylor (General Superintendent of the Producing Department) and Fred A. Davies (one of the company's geologists, who later became Chairman of Aramco), were selected to travel to Bahrain and to review the area which had been surveyed two years earlier for Gulf Oil Corporation by Professor Madgwick and Ralph O. Rhoades. Meanwhile, lawyers, company directors, diplomats and government departments continued to work out the wording of the official document which would provide for the assignation of the Bahrain concession from the Eastern and General Syndicate Limited to the Standard Oil Company of California's wholly-owned subsidiary, Bahrein Petroleum Company Limited.

In early April 1930, Davies and Taylor set sail from New York on the SS *Ile de France*. From London, they continued their journey to Turkey, via Paris and Vienna, arriving in Constantinople on the 26th of the month. After crossing the Bosphorus, the boundary between Europe and Asia which is only a mile wide at its narrowest point, the travellers stopped at Haydarpasa, the western terminus of the Asiatic portion of the proposed Berlin-to-Baghdad railway. From Haydarpasa to Baghdad trains followed the same route through the Taurus Mountains of southern Turkey as that taken by Alexander the

Carlton Hotel, Baghdad, letterhead 1930 (*Chevron archive*).

Great during his fourth-century expedition. The train continued from Aleppo along the Turkey-Syria boundary to the railhead at Nisibin. There a convoy of Ford touring cars and trucks operated by the Iraq railway met Davies and Taylor so that they could motor to Mosul, beside the Tigris River, and then pass the ruined biblical city of Nineveh on their way to Kirkuk. At Kirkuk they boarded another train for the last leg of the journey to Baghdad. Once installed there at the Carlton Hotel, New Street, Taylor composed his weekly report for Head Office in San Francisco:

> "We had good dirt roads all the way. Yesterday a.m. [27th] we left Mosul for Kirkuk and it rained on us all day long, with the result that instead of five hours it took us eleven, but the train waited on us at Kirkuk so we got here this a.m. at 6 o'clock. … We were met at the train by the Eastern and Gen. Synd. man, who informs us that Maj. Holmes will be along in a few days or will communicate with us. Weather very nice" [Taylor, letter to McLaughlin, 28 April 1930].

Taking full advantage of their transit stop, Davies and Taylor began to enquire about the Iraq Petroleum Company. [1] When writing his next letter to "Mac", Taylor remarked, "as we do not know the policy of our Co. in this country we are at a loss to know whether they are at all interested, but from the info. so far at hand, I'd say there is plenty of oil here" [Taylor, letter to McLaughlin, 30 April 1930]. One week later, they returned to Baghdad from a tour of the IPC's camps "where we were treated royally and told nothing". It so happened that Harry Minger, one of the Iraq Petroleum Company's senior foremen, was in town. Being one of Fred Davies' cousins also, Minger accepted an invitation to join the two Socal men for lunch. Apart from exchanging personal pleasantries, Davies and Taylor hoped that Minger would provide them with some useful information. When reporting this meeting to McLaughlin, Taylor added that:

> "we are sending under separate cover to Clark Gester 2 copies of the 'conversation'

under which the Irak Pet. Co. operates here, also two copies of a report of the wells, which is the best we can get without compromising the men in the field. … We leave here tomorrow for Bahrein via Basra. Temperature here 109°F [43°C] yesterday, and we are promised some *warm* weather as we proceed toward Bahrein" [Taylor, letter to McLaughlin, 7 May 1930].

Why Davies and Taylor had been so interested in the IPC remains unclear. Curiosity is one explanation, although their energetic zeal would seem to belie this notion. A directive from San Francisco would seem more plausible, yet the fact that these two Socal representatives claimed to have no idea of their company's possible or actual interest in Iraq, somehow weakens that argument. Whatever the truth of the matter, the surviving correspondence gives no hint that either Davies or Taylor suspected that Holmes, who had been in Bahrain for a month, was already more than one move ahead of them.

Open-minded as to what they would find at their eventual destination, Davies and Taylor set off by rail again, travelling south past the ancient cities of Babylon and Ur, finally arriving at Basrah, head of navigation for the Arabian Gulf steamers and reputedly the home port of Sindbad the Sailor. There they embarked on a steamer for the eighty-hour voyage to Bahrain, calling at Muhammarah and Bushire on the Persian coast and at Kuwait on the Arabian coast. On the 14th of May, on schedule, they arrived

Basrah, circa 1930s (*The Hulton Picture Company*).

at an anchorage about six miles (9.65 km) off the islands of Bahrain. Holmes went out to meet them in a motor launch, taxied them ashore and established them in a comfortable building with plenty of breeze. No fieldwork was attempted during the first few days. Instead, the Socal men familiarized themselves with the local conditions. Within three days, they composed a long epistle to their colleagues, G. C. Gester and W. I. McLaughlin:

"Dear Clark and Mac:
"…You would probably be interested in our impression of Major Holmes from the few days we have been with him. He is a rather stout man about 55 years of age and rather bald. … From what we have seen and heard, he appears to be on extremely friendly terms with the influential Arabs and we believe he could go farther toward obtaining concessions from them both in Arabia and Iraq than any other European or American here. During the latter part of the war he was attached to the Admiralty and while there, had access to their confidential petroleum maps. From his study of those maps, as well as personal observations in the field, he knows of many seepages along both the Persian [Arabian] Gulf and Red Sea coasts of Arabia, and believes there are many areas there worthy of exploration" [Davies and Taylor, 17 May 1930].

The main substance of the letter, however, contained some striking disclosures:

"When we left Baghdad we had no idea that Major Holmes was familiar with conditions in Iraq or was paying any attention to them, but on talking with him we discovered that he is very much interested, has been following it closely, and is well acquainted and has considerable influence with the Government officials."

Davies and Taylor added that Holmes perceived "the need for haste in connection with the Iraq work", yet he considered a delay in the attempt to obtain the Kuwait concession would be helpful rather than detrimental. How he had concluded this is not explained. Davies and Taylor had learned too that Holmes had sought William T. Wallace's permission to expedite Gulf Oil's work in Iraq, prior to securing a concession in Kuwait. When Wallace refused,

"Major Holmes was desirous of bringing the Iraq situation to the attention of Mr F. B. Loomis when they were together in London but was prevented from doing so by the attitude of the Gulf representative.[2] in London. The reference to "north and south Iraq" in our cable of the fifteenth referred especially to the following three areas, in all of which our present information indicates [the] chances for commercial oil production are good enough to warrant careful study and consideration."

The three areas referred to were that portion of the present Iraq Petroleum Company concession which would become available to other companies as soon as the IPC had selected the acreage it required; an area near Basrah, which although in Iraq was not included in the concession given to the IPC; and the Neutral Zone between Kuwait and Saudi Arabia which would not be included in any Kuwait concession, but was an area which Holmes felt would be just as good, if not better than Kuwait.

"Major Holmes is very anxious that knowledge of his passing the above information on to us be kept from the Gulf [Oil Corporation]. He feels that Mr Wallace's desire to dictate the order and manner in which his, Major Holmes', work out here shall be conducted is hampering him a great deal and may possibly result in his losing out completely on some favorable area, such as Iraq. He feels that our Company has as much a call on his time as has the Gulf but it appears that Mr Wallace does not feel that way inasmuch as he considers the immediate securing of the Koweit concession above everything else and apparently [is] paying no attention to the necessities imposed by local conditions" [Davies and Taylor, 17 May 1930].

That Gulf Oil wished to pursue the Kuwait concession was nothing new, since this preference, and the Mellon family's wish that Gulf Oil should remain within the *Red Line Agreement*, had been the main factors which had caused Gulf Oil to relinquish its interests in Bahrain. The worrying feature was Holmes' perpetual manoeuvring between one company and another, an activity which, if not checked, might jeopardize Socal's position in Bahrain. Davies and Taylor also learned of grievances within Bahrain's pearl diving industry which involved another competitor. Two pearl merchants with whom they dined, told them that in the course of their business operations they consumed at least 100,000 cases of gasoline a year, in addition to large quantities of kerosene. This represented about half of the Bahrain pearling industry's requirements. However, the merchants "heartily disliked" the Anglo-Persian Oil Company, from which they had to buy all their supply; consequently, they would welcome any competition [Davies and Taylor, 17 May 1930]. With this new intelligence and the benefit of recent experience, Socal could not afford to lower its guard.

Socal Surveys the Bahrain Scene

Once they began to explore, Taylor and Davies found that most of Bahrain island was low-lying desert, "sprinkled here and there with fertile spots encircling small villages. These fertile regions are irrigated from springs and wells, and grow dates, melons and a few vegetables". By the 24th of May, Fred Davies had managed to achieve a little fieldwork, confining it mostly to the northern end of Bahrain island. From what he had seen so far, he and Taylor found that Ralph Rhoades' 1928 report was closely correct.

"We drove out to the old seepage one morning and, while we were only there a very short time, I believe there is faulting present at that point. At least one of the [water] wells which have been dug on and near the seepage gives clear evidence of a fault in its wall and other erratic dips in the immediate vicinity seem to bear it out" [Davies, letter to Gester, 24 May 1930].

Six days later, Taylor's weekly report recorded that:

"Fred has been roaming around all last week, each day from dawn till 1 o'clock, checking up on the previous geologist's report. ... He seems to think that it is not too hot, but he is giving Clark [Gester] a long letter today about a lot of things, so you'll know from that. Think we will have no trouble getting enough water from dug wells, as there is a large basin close by ... one well has water up to eight feet [2.4 metres] of the surface. The water is slightly salty but it seems better than the water we drink here in town and would certainly be all right for drilling and maybe for boilers also. From present indications, we believe that Cable Tools will be the method used but are reserving final recommendation until Fred has had a chance to see a section exposed" [Taylor, letter to McLaughlin, 30 May 1930].

Davies and Taylor planned to choose the location for the camp and the oil well on the 1st of June. They expected that the oil well site would "fall close to the southeast corner of the mountain Jebel Dukhan (440 feet high, so not so much of a mountain). ... In an entirely new country where nothing is known concerning the behaviour of the structures, I presume the best shot is right up close to the top" [Davies, letter to Gester, 31 May 1930].

In their official reports during their stay in Bahrain, the Socal men recorded that the British and American residents of the island played host to them with "colonial spirit". Unofficially, this was not the whole picture, for they had experienced "a great deal" of jealousy from the British residents. However, when everything was signed up, Taylor wrote to McLaughlin that they could tell them all "to go to hell, if they get tough" [Taylor, letter to McLaughlin, 7 June 1930].

Disturbed by Taylor's letter, M. E. Lombardi in Socal's Head Office remarked to his colleague Francis Loomis that no doubt he too had gathered that the geologists' visit had not turned out entirely satisfactorily, since they seemed to be facing many restrictions which they had not anticipated. Whilst the primary object of the visit was to determine the advisability of drilling on

Fred Davies (left) and Bill Taylor (*Chevron archive*).

Bahrain island and to gather as much practical information as possible about the country, labour conditions, material supply and so on, Taylor and Davies were also required to assess the general situation in Bahrain. On the basis of this information, Socal would decide whether or not it wished to acquire more territory than simply the Bahrain islands.

By now Lombardi shared the general concern regarding Major Holmes' allegiances to Gulf Oil, the Eastern and General Syndicate and Socal. "While employment under such conditions is fundamentally unsound because of the possibility of conflicting interests among the employers, nevertheless, it is an existing fact and, consequently, I would be very interested in obtaining Mr Wallace's views as to whether we have any call on Major Holmes' services (and, if so, to what extent)" [Lombardi, letter to Loomis, 11 June 1930].

Despite this anxiety, Lombardi hoped that the Exploration Licence extension would be signed on that or the following day (the 11th or 12th of June), whereupon it was expected that Holmes would apply for the Additional Area. Taylor thought the 100,000 acres would cover all the oil in the islands, "but the additional acreage may keep someone else from causing trouble if we get lucky and find oil. ... I have never seen anything quite like the suspicion in which the Anglo-Persian Oil Co. is held". To make the situation even more complex, the Bahrainis believed that Socal was in partnership with APOC, "and it took a lot of explaining to convince their difficulty and am not sure they are yet convinced" [Taylor, letter to Stoner, 11

Michael E. Lombardi (*Chevron archive*).

June 1930]. An added frustration was that things did not move as quickly in Bahrain as Davies and Taylor had expected. In the same letter, Taylor added that

> "these darned British agents are sure up stage. We go to tea and all that sort of thing, but they remain very aloof. ... This is some place ... where you have to wait a week for mail and a month for the stock reports. As for the stock market, I'm out, but Fred is never sure whether he is out or in, hence the worry. ... The application for the additional acreage on these islands will be made on the 18th and the Lord only knows how long it will take to get it."

The strained relationship between the Americans and the British in Bahrain was never far below the surface. "Our Company sure seems to be a secondary or *less* in the consideration of all our dealings here. … Am writing this by lantern light so shall not go into details. … Fred runs a typewriter and has given Clark a very full report on all conditions outside of operating" [Taylor, letter to McLaughlin, 14 June 1930].

Nevertheless, Davies finished checking Gulf Oil's map as far as he could. The Exploration Licence extension, retroactive to the 2nd of December 1928 and valid until the 2nd of December 1930, was signed on the 12th of June as had been anticipated. [3] Major Holmes was expected to file his application for the Additional Area on the 18th of June. By the end of the month, Davies and Taylor had left Bahrain for Baghdad. But, as both had intimated, other minds had focused on the Additional Area application, including that of Lieutenant-Colonel H. V. Biscoe, Political Resident in the Persian Gulf.

On the 30th of June, Biscoe lobbied the Secretary of State (Colonial Office) and informed Lord Passfield that he considered Holmes' application, on behalf of the Eastern and General Syndicate and through the Political Agent, for permission to enter into negotiations with the Ruler of Bahrain for an Oil concession to cover the remaining areas of the Bahrain archipelago, "altogether premature" [Biscoe, letter from PRPG to CO, 30 June 1930]. He argued that if the British Government waited until the results of preliminary drilling were known, then it would "possess a good bargaining counter" for subsequent negotiations with the Bahrein Petroleum Company Limited, particularly with the idea of seeking a higher annual premium. The Ruler of Bahrain, however, intervened. He was anxious to conclude the agreement as this would generate an immediate payment of 10,000 rupees to his Government.

Concession Agreements - One Door Opens as Another Closes

The concession agreement, embodied in the 1925 document and the subject of so much intense international negotiation for the previous few years, was assigned, at last, on the 1st of August 1930 from the Eastern and General Syndicate Limited to the Bahrein Petroleum Company Limited (Bapco). On the same day, as required by the agreement, Judge Frank Feuille paid the syndicate $50,000, a further similar sum being due a year later. On the 18th of August 1930, as provided by the 27th of December 1928 contract, Bapco reimbursed Eastern Gulf Oil Company's expenses, a sum of $57,449.23.

As required also under the conditions governing the assignment of the concession from the Eastern and General Syndicate to the Bahrein Petroleum Company Limited, Bapco designated a London representative. They appointed D. Duncan Smith, a partner of Freshfields, Leese and Munns, a firm of London solicitors well-versed in the Bahrain concession paperwork and liaison with Socal's attorneys, Messrs Pillsbury, Madison and Sutro in San Francisco. [4]

Meanwhile, on the 8th of August, William Wallace had written to M. E. Lombardi enclosing a copy of a private and confidential letter received from Edmund Janson (Eastern and General Syndicate director). From this he had concluded that the additional concession on the Bahrain islands had received Colonial Office approval, a fact which he found to be "quite gratifying". Wallace's expectations had been raised by

Janson's late July meeting at the Colonial Office, during which the latter appreciated that there was not enough space on Bahrain island for a separate group to operate successfully and, although the Colonial Office was against monopolies, it quite realized that it was not fair to expect the Bahrein Petroleum Company to prove the ground only to let other people reap the benefit of their work. Consequently, the Colonial Office was in favour of granting the Exploration Licence extension. However, following a now predictable pattern, the matter was referred to the India Office.

The summer passed, and on the 11th of September 1930 Frank Holmes sent a copy of the 1st of August 1930 Deed of Assignment (of the Bahrain Oil concession to the Bahrein Petroleum Company Limited) to the Political Agent in Bahrain, asking that it be registered appropriately. On the same day, Holmes applied for a Prospecting Licence,

"… for and on behalf of the Bahrein Petroleum Company Limited as assignee of the Bahrein Oil concession, upon terms as laid down in the 2nd Schedule of the Bahrein Oil concession dated the 2nd of December 1925. The tenure of the Exploration Licence of the Bahrein Oil concession comes to an end on the 2nd of December 1930 and it is requested that the Prospecting Licence [drilling licence] become operative from that date. The period for which the Prospecting Licence is requested is of two years' duration dating from the 2nd of December 1930. The annual payment of Rupees 10,000 as provided by the concession, will be made on or before 2nd of December 1930, and is to be applied in payment for the first year of the Prospecting Licence" [Holmes, letter to the Political Agent, 11 September 1930].

Everything went smoothly until a month later, when F. C. Starling of the Petroleum Department, London, wrote to the Colonial Office. The outlook for finalizing the details of the Additional Area concession application suddenly clouded. Starling claimed that the Eastern and General Syndicate had not put forward any specific reason for "demanding" an extension of its concession. He presumed that their object was mainly one of acquiring control over as large an area as possible before drilling actually commenced. He argued that, from the Ruler of Bahrain's point of view, the concession was quite favourable, particularly if operations were carried out on behalf of either the Gulf Oil Company or the Standard Oil Company of California in an "efficient manner", as no doubt would be the case [Starling, letter to the Colonial Office, 11 October 1930].

The Petroleum Department view was that the area the Syndicate was entitled to take up under the lease already, about 156 square miles (404 sq. km), was large enough for it to acquire control of a substantial oilfield, if oil were to be discovered under the terms of the Prospecting Licence. Starling believed that there was always the danger that a company, forced to demarcate the area of its concession before sufficient drilling had been done for the geological structure to be thoroughly understood, might be in an unfortunate position if, subsequently, a good oil-producing structure were to be discovered close to the boundary of the concession. The result could be that companies granted subsequent concessions were well-placed to drain oil from territory held by the original concessionaire. Although it was expected that companies would assume risk in such situations, Starling suggested that in the case of Bahrain it might be extremely useful to the British Government in future negotiations if a certain amount of territory were to be withheld.

"In practice, the investigations of the Anglo-Persian Oil Company have indicated that the prospects of finding oil (in the Arabian peninsula region) are not very favourable, and it therefore seems advisable to encourage the Bahrein Petroleum Company to undertake as much active prospecting as they are prepared to do. ... Reviewing the question generally both from the point of view of Bahrein and of the British Empire, the Secretary of Mines is of the opinion that no further concession should be granted at present" [Starling, letter to the Colonial Office, 11 October 1930].

The memorandum was copied to the Admiralty and the Air Ministry, as well as the Foreign, India and War Offices. Unfortunately, there is a gap in surviving documents which might otherwise provide stimulating reading on the American reaction to this communication. A decision was waited upon for several weeks. On the 3rd of December 1930 the Colonial Office informed the Eastern and General Syndicate that, after careful consideration but with regret, the British Government had decided not to advise the Ruler of Bahrain to consider the "grant of an extended concession until the mining licence under the existing concession had been taken up and the 100,000 acres had been demarcated" [Flood, letter from the Colonial Office to the Eastern and General Syndicate, 3 December 1930]. In that event, the Government would be willing, of course, to advise the Ruler of Bahrain to reconsider the matter.

Socal Takes Stock

While the British Government was setting up its "bargaining counter" upon which it would negotiate applications for an Additional Area concession and a Mining Lease, [5] Socal's immediate preoccupation was the outcome of Fred Davies' survey of Bahrain. On the 26th of November he released his long-awaited report:

"We had in our possession copies of a geological report and map on Bahrain Islands, prepared for the Gulf Oil Company by R. O. Rhoades in April 1928. This report and map were the results of a very thorough study lasting over a much longer period of time than we were prepared to spend there. The principal object of our visit, geologically, was to check over this former work sufficiently to ascertain its general correctness. Since Rhoades' work was found to be very detailed and essentially correct, ... I shall not attempt to make a new structural map of the island

Fred Davies (left) resting during his fieldwork in Bahrain, circa 1930 (*Bapco archive*).

nor to repeat the facts of a general nature discussed in his report. Consequently, my report should be considered somewhat as a supplement to his, and for a complete picture, the two reports should be filed and referred to together" [Davies, *Supplementary Report*, 26 November 1930].

Davies had noticed that during the six-year period when water wells had been drilled on the islands, the head had dropped at the rate of a foot a year. He was uncertain whether this large decrease was entirely due to a fall in pressure in the reservoir horizon or whether it might have been caused by insufficient and inaccurate information as to the condition of the wells at completion. Davies knew that sixty-seven wells had been drilled with initial flows ranging from 400 to 20,000 barrels per day. He may not have known that a number of the wells had been drilled very cheaply, at prices which did not allow for a proper lining to be installed, thus immense quantities of water had been lost through the ground, as well as into the sea.[6] During their study, Davies and Taylor had found too that

"with a little care and search, a car could be driven close to almost any point it was desired to visit, except on Jebel Dukhan itself and in some soft sand areas in the southern part of the central depression. A proposed location was selected at the eastern base of Jebel Dukhan ... and it is believed no especial difficulty will be experienced in constructing a road direct to it down the east side of the hill, if such a course proves preferable to following the present road down the west side" [Davies, *Supplementary Report*, p. 5].

After analysing the stratigraphy in considerable detail, Davies ventured to draw some long-awaited, and hopefully favourable, conclusions:

"Practically nothing is known concerning the character of the section to be expected in a well drilled on Bahrain. ...

"The only surface evidence of oil or gas on Bahrain is the bitumen deposit which lies about 3 miles [4.8 km] south and a little east of Jebel Dukhan. All the asphalt at present exposed is in the edges of one pit 15 to 18 feet deep [4.57 to 5.49 metres] and is all in a local accumulation of surface sand. … There is no liquid petroleum in evidence anywhere but in the walls of the pit, the saturation is such that the rock is rubbery under the hammer and freshly broken surfaces present the black, glistening appearance of semi-fluid asphalt" [Davies, *Supplementary Report*, pp. 10-11, 23-4].

Although Davies had detected what he called a large definitely closed domal structure (anticline), he devoted many pages of his report to a comparative analysis of such other oil-bearing structures as were known at that time in the Middle East. Whilst cautioning that "little of a definite nature is known of the section which may be expected in a well drilled on the island" [p. 26], he boldly committed himself by stating that "Bahrain could be considered a much more likely-appearing wildcat venture *if* the structure was of the steep-dip variety,[7] similar to the present producing fields of the Near East" [p. 28].

Short of "plane-tabling" the Bahrain structure and drilling, Davies could not answer that question. Based on the information which he did have at his disposal, Davies knew that drilling a well in Bahrain was unlikely to be a first-class wildcat operation, but he knew too that this proposition was not an ordinary one in which more information could be obtained easily, if at all, before a final decision was made as to whether or not drilling should go ahead. Davies admitted that the lack of information relating to the character of the formations which *might* be expected in Bahrain was, to put it mildly, "especially unfortunate". However, he concluded that as Bahrain was a closed structure with some surface evidence of oil, it had a better than even chance of being underlain by a considerable thickness of limestones (as reservoir rocks) and marls (as a seal). It would seem, therefore, that the logical thing to do would be to go ahead and drill a test well.

The cairn of stones which Fred A. Davies built in 1930 to mark the position of Bahrain's first oil well (*Bapco archive*).

"I recommend that such a well be drilled, located 500 feet [152.4 metres] south and 350 feet [106.68 metres] east of point 28-Q on Rhoades' map. This point is marked in the field by a cairn of rocks surmounted by a flag" [Davies, *Supplementary Report*, p. 29].

As it turned out, Davies had pin-pointed the position for Bahrain's first oil well to within 50 feet (15.24 metres) of the apex of the domed structure. It was a remarkable achievement.

Jebel Camp Takes Shape

On the 18th of March 1931, an enthusiastic Frank Holmes wrote to his friend Ward in New York:

"With regard to our oil business, the Standard Oil Company of California have now made arrangements to send six men out to arrive at Bahrein in the middle of May. The first work is to arrange for their water supply for their drilling rig. This entails sinking a water well and piping it at a distance of six miles [9.66 km]. They expect to have all the preliminary work completed and to commence real drilling in September or October of this year" [Holmes, letter to Ward, 18 March 1931].

Everyone looked to successful results. Holmes cabled Lombardi in San Francisco informing him that a crane had just been erected at the end of Bahrain pier, capable of lifting 5 tons (5080 kg). This would be necessary for unloading the drilling equipment now on order. On the 6th of April 1931, Francis Loomis informed Holmes that he had pleasure in introducing to him Mr E. A. Skinner, now proceeding to Bahrain to take charge of Socal's affairs on the island. Since Skinner had had a good deal of experience in foreign countries, Loomis thought that Holmes would find him willing to co-operate with him in the "east" and maintain pleasant and comfortable relations with the Ruler and the local population of Bahrain. He emphasized that Skinner understood "quite thoroughly our desire to keep aloof from anything in the way of political entanglements" [Loomis, letter to Holmes, 6 April 1931].

In May 1931, Edward Allen Skinner and his team arrived in Bahrain. An estimated 1000 tons (1,016,047 kg) of equipment followed on successive steamers. C. G. Prior, the Political Agent, recorded in his annual report for the year that "the organisation of the Oil Company is remarkable. Of the immense quantities of stores required, nothing was mislaid and everything arrived on time" [Prior, *ARPGPR*, 1932, p. 47]. He had noted too that all the requirements from machinery to lead pencils had been imported from the United States, this influx having had the marked effect of introducing American-made goods into the suuq. Jebel Camp was fitted with refrigerators, electric light, heaters, full length baths, hot and cold showers and water-borne sanitation, all considered by the Americans at that time to be "*sine qua non* on location, though they would be considered luxuries by most officials in India" [Prior, *ARPGPR*, 1932, p. 47].

A panorama of the oil pioneers' camp below Jebel Ad-Dukhan, Bahrain island, in the mid 1930s (*Steineke Collection, Aramco archive*).

Left to right: Mrs and Mr Ed Skinner, Michael Lombardi, Major Frank Holmes and Edmund W. Janson visiting an oil well, early 1930s (*Chevron archive*).

Mrs Skinner later recalled the Camp:

"I had never seen a Nissen hut [8] before, but it made a splendid home. The rooms were large and it was very pleasantly furnished. None of us realized that it would ever be cold here, and what a shock we had that first winter. We hadn't brought many blankets, but Jack Schloesslin built stoves from 12 inch pipe for all the huts and I think they must have saved our lives." [Bapco archive].

In contrast, during the summer, these tunnel-shaped corrugated metal structures provided no protection from the intense heat of the sun. Unless insulated, the Nissen huts became like hot ovens. As a result the walls were plastered on the outside with a thick coat of mud, utilizing a mixture of local clay and chopped straw which had been puddled by human feet in a shallow basin. This solution was an adaptation of the material used locally for roofing houses. [9]

Drilling Begins at Well Number One

Life in the new camp had taken shape. No doubt speaking for everyone with hopes for the drillers' success, including the Ruler of Bahrain, Ward had commented:

Mr C. C. Grenier, Customs and Port Officer for Bahrain, celebrating the "spudding in" of the first oil drilling bit in Bahrain, on the 16th of October 1931 (*Bapco archive*).

"The depression [10] has hit the Gulf Company like all the others, and the stock is selling for one fifth of the price of two years ago. As far as I am concerned, I am worth about twenty cents on the dollar and business is correspondingly bad. I am going to have faith in you … that the deal comes off for we sure need the money" [Ward, letter to Holmes, 8 October 1931].

Eight days later, on the 16th of October 1931, Well Number One was "spudded in". His Excellency Shaikh Hamad bin Isa Al-Khalifa, Deputy Ruler of Bahrain, worked the drill for its first few blows. As the rig quivered under the impact, His Excellency glanced at it and is reported to have said jokingly, "The machine is drunk!" [Prior, *ARPGPR*, 1932, p. 47]. Later drillers swore that Davies had chosen the only place on Bahrain island that never had a breeze. Conditions were hot and humid. But if humour was sometimes jaded because of the weather, spirits were high in anticipation of finding oil. In November, Holmes reported that the standard rig was operating and "so we are on the way to prove if there be oil or not. I have much faith we will have luck and strike a good flow" [Holmes, letter to Ward, 2 November 1931].

By the spring of 1932, the drilling rig had became something of a tourist attraction. Socal director, M. E. Lombardi, was very aware of the potential hazards of this interest. Attentive competitors were seen as a real concern. Wishing to protect the interests of the Ruler of Bahrain, the Bahrein Petroleum Company and the Standard Oil Company of California, yet appear courteous to well-wishers, Lombardi suggested:

"I know nothing whatever of the psychology of the Arabians and any suggestion that I make is purely on the basis of our desires, which may be impossible of accomplishment under foreign conditions. However, it occurred to me that possibly we might avoid the appearance of inhospitality and unnecessary secretiveness and accomplish our object, if we require of all visitors a pass issued by the

Shaikh himself [The Ruler], thus making the Shaikh, or one of his representatives, the judge in case of visitors from companies other than our own. This is just a suggestion" [Lombardi, letter to Holmes, 12 March 1932].

Less than a month later, an enthusiastic Holmes wrote to his friend Thomas Ward that the well was doing splendidly. Already it was down to 1764 feet (537.66 metres) and had been in oil bearing formation for 200 feet (61 metres) of shale and limestone.

"We hope to run a test towards the end of this month. ... We have a fair crowd here, all efficient and good chaps. Skinner, the head, is an exceptionally good chap. ... Be sure you keep this to yourself, as San Francisco would not be pleased if they heard that I had let anything out" [Holmes, letter to Ward, 3 April 1932].

Ward replied that

"practically all our [11] hopes are now centred in the success of Bahrein. Every other castle ... did not last. Since the crash of 1929 they have toppled to the

Jebel Ad-Dukhan No. 1, Bahrain's discovery well, was the scene of a ceremonial spudding-in, attended by the Deputy Ruler, His Highness Shaikh Hamad bin Isa Al-Khalifa on the 16th of October 1931 (*Bapco archive*).

ground or disappeared in the air. Some of our old friends at the Gulf Company have disappeared in the course of economic events and you are to be congratulated in being in the thick of the fight where something is actually doing. ... I look forward to your news and trust that the production from Well Number 1 will exceed the figure you mentioned" [Ward, letter to Holmes, 9 May 1932].

Meanwhile, Lombardi having reconsidered Socal's interest in the Bahrain concession, proposed possible changes concerning the company's application for a concession to cover the remaining territory of the islands of Bahrain after the 100,000 acres had been selected. Holmes advised against any alteration. He believed that it would concern the Ruler of Bahrain, since he was looking for oil revenue to relieve the financial distress of the Depression, which was being felt in Bahrain too. In Holmes' view, a delay in producing oil would have undesirable repercussions and injure the oil companies' prestige. [12]

In early April, Holmes left Bahrain for a short trip and happened to encounter the Bahrain Government's financial adviser on the boat, who later discussed Lombardi's proposed changes with the Political Resident. Predictably, the British Government was not in favour of any

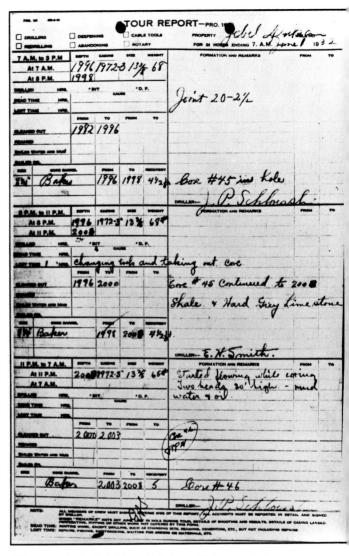

The entry in Jack Schloesslin's log which recorded the first flow of oil in Bahrain on the 1st of June 1932 (*Bapco archive*).

modifications. Nevertheless, in his capacity as Bapco's Chief Representative, Holmes applied to the Political Agent in Bahrain on the 22nd of April to extend the Prospecting Licence for two years from the 2nd of December 1932. Correspondence continued between the Ruler of Bahrain's advisers, Bapco and Socal. As the first day of June dawned, nothing further had been decided regarding the Additional Area concession application.

Discovery

At 6 o'clock in the morning on the 1st of June 1932, "the drill pierced a layer of blue shale. The men smelled oil and heard an ominous rumbling. Very cautiously, they drilled another eight feet" [Bapco archive]. Oil flowed from a depth of 2008 feet (612 metres) from the Mauddud zone. Edward Skinner recalled: "The well came in like a lamb. It was a driller's dream" [Bapco archive].

The next day, Lombardi liaised with Francis Loomis:

"As you are aware, we have encountered a very good show of oil in Bahrein, ... the last cable we had indicating that the well was capable of making upward of 2000 barrels a day of 34 gravity oil. [13] ... We are willing and anxious to proceed with the prospecting by starting another well as soon as we can get the necessary material on the ground, but if we get oil in the second well you can readily see that we cannot test it without producing over the limit of 100 tons [14] and from then on we are

Bahrain's first oil well (*Joe C. Torres*).

practically stopped from prospecting unless we can get the contract changed [15] as we have tried to do" [Lombardi, letter to Loomis, 2 June 1932].

It seemed to Lombardi that the "most pressing thing to try to accomplish"was the revision or clarification of Clause IX of the Prospecting Licence, the second schedule of the 1925 Bahrain concession which stated

"the right to win up to 100 tons of oil free of payment and further quantities of oil on payment of the royalty per ton provided in the Mining Lease, but on condition that the Company shall apply for a Mining Lease in respect of each area in which work is proceeding as soon as more than 100 tons of oil are won from one single bore hole within it."

On this issue, Holmes advised Socal that the Ruler of Bahrain, the Political Agent

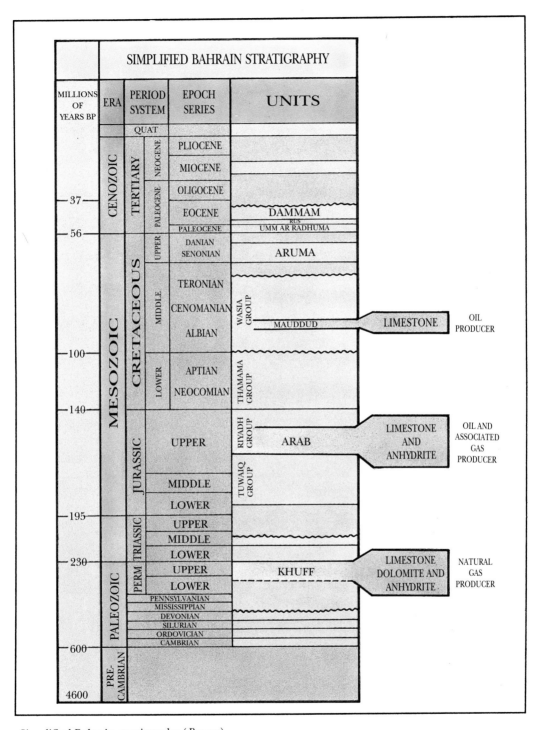

Simplified Bahrain stratigraphy (*Banoco*).

and the Financial Adviser in Bahrain should be "lined up", before the Colonial Office was approached again. However, Lombardi could not see anything in the correspondence which prevented Socal from discussing the present situation with the Eastern and General Syndicate in London directly, so long as "we keep away from the Colonial Office" [Lombardi, letter to Loomis, 2 June 1932]. He ended his assessment saying:

> "Our interests, as I see them now, lie (1) in securing delay for further prospecting before we have to select acreage or apply for a mining lease, and (2) in securing [an] option on further territory both on the islands and on the mainland."

With both matters still unresolved, the rig was dismantled and re-erected two miles (3.2 km) away on a new site to the north; drilling started on the 1st of August. About the same time, a second rig, which had arrived recently from America, was erected on a third site about 8000 feet (2438 metres) north-east from the second well. As work progressed throughout the summer, Lombardi became increasingly anxious:

> "The situation, as I see it, is something like this. We have asked Major Holmes many times to get permission from Ibn Saud for our men to visit the mainland of Arabia … and so far Major Holmes has done nothing positive in that direction. …

> "Therefore, we have broached the subject through Major Holmes to the Eastern and General Syndicate and are waiting for them … to take some action, which so far has not been forthcoming. To that extent we have a right to be somewhat concerned at the situation because, on account of our showing of oil at Bahrein, other oil companies may very well have had their interest in … the neighbouring Arabian coast aroused, and Major Holmes and the Syndicate have done nothing positive to protect our interests there" [Lombardi, letter to Loomis, 3 August 1932].

On the other hand, Lombardi recognized that Socal still had a lot of work to do in order that the company's position could be consolidated in Bahrain, not least of which was the "reform" of Clause IX of the Prospecting Licence. Nevertheless, "taking it all in all, it seems to me we are in quite a delicate position". During the time between the dictation and the typing of Lombardi's letter, he had received what he described as an interesting communication from Mr Skinner "which throws some new light on the Bahrein situation". The resulting postscript noted:

> "Possibly Major Holmes has been more successful than we had realized … and this may influence our future course."

As the autumn leaves fell in London, so did Socal's hopes of immediate solutions to the problems which prevented the company from consolidating its position in Bahrain.

In late October the Colonial Office indicated that it was not prepared, at that time, to advise the Ruler of Bahrain to give the Bahrein Petroleum Company a further extension of the Prospecting Licence to the 2nd of December 1934. But it was willing to reconsider it during the next year. Amendments to Clause IX of the Prospecting Licence had been delayed. The Additional Area concession application had stalled. During the last two months of the year, while the attorneys in San Francisco and the solicitors in London put their legal minds together once more, operations continued in Bahrain. Just before the year ended, Well Number Two "came in" with a rush on the 25th of December 1932. Holmes sent a message to Charles Belgrave and his wife to come at once to the oilfield.

> "When Marjorie and I reached the well, which was in the foothills near Jebel Dukhan, we saw great ponds of black oil and black rivulets flowing down the wadis. Oil, and what looked to us like smoke, but which was in fact gas, spouted gustily from the drilling rig and all the machinery. The men who were working were dripping with oil. … It was not a pretty sight but it was an exciting moment for me. I could see, without any doubt, that there was an oil field in Bahrain. It was a great day for Major Holmes, who now saw the visible proof of what he had always believed" [Belgrave, 1972, p. 83].

Time for Transition – A New Way of Life

Irrespective of nationality or skills, the realities of working in the new oilfield were tough. For many Bahrainis it was a time of transition, adapting to a way of life that was new but also provided job security and a guaranteed income. For many Americans, the oil industry in Bahrain provided an alternative to unemployment. Most pioneers realized that life in Bahrain would be demanding, but many also knew that for them it was "make or break". Although Socal had three big refineries at El Paso, El Segundo and Richmond, chances of being promoted to senior positions seemed a long way ahead for many young men. Somehow, the changing political situation in Europe and the remoteness of Bahrain, were not sufficient to discourage ambitious pioneers and their families who saw Bahrain as a grand opportunity.

Initially, Bapco worked a seven day week which meant that local employees returned to their families perhaps once every two months. Eventually, one day of rest was allowed after thirteen days of work and the Bahrainis returned home once every two weeks on a Thursday afternoon. At about five o'clock on the following day, Friday, Bapco trailer lorries travelled along the few main roads which existed at the time and collected the men for the following weeks' work. One former Bapco employee recalls that the buses were only available for the "big people" and the labourers were not allowed to travel in the pickups either.

For those who lived on Muharraq island, particularly in the east-coast village of Hidd, the journey to Jebel Ad-Dukhan took about seven hours, much the same time as it takes now to fly from Bahrain to London. Those employees who lived in Hidd had to walk for an hour to Muharraq town, then, because there was no causeway at that time, they took a boat across the channel from Muharraq island to Manama, the capital on the northern shore of Bahrain island. The boat waited for ten or twenty passengers before it set off. The return fare was the equivalent of one fil. [16] If anybody came and offered a rupee (100 fils), then the boatman would go. If there was no wind, or if the wind was blowing from the Manama side, they took the sail down and rowed or poled the boat across the channel, a journey of an hour or more depending on the tide and the current. At low tide, the boat had to zig-zag through the deep channel. Even when the tide was high, strong currents could prolong the journey considerably.

Once at Manama pier, there was another wait for a taxi or a truck, a unit with a flat trailer and no side walls. Initially, the roads were beaten tracks which became rough and muddy during the winter. Then it was necessary to lay three or four layers of *mangour* (palm frond matting) on the ground to compact the earth and prevent the wheels sticking in the mud. If a driver turned sharply, the passengers could be thrown over the edge of the open trailer, something which did happen on three or four occasions, with fatal results. There were no mudguards on the wheels, and so once the practice was adopted of rolling the roads with a mixture of sand and tar, oil from the road surface splashed onto the passengers. Employees soon learned to cover their clothes with sacks

The Jebel Camp (*Joe C. Torres*).

to prevent them from being spoiled. Later, long passenger trailers capable of carrying 100 people were introduced. Although they were covered, thus cleaner and safer, they did not have windows.

Home, for the majority who worked at Jebel Ad-Dukhan, was a *barasti* house made from palm leaves which had been transported to the desert. There were three such camps – one each for Bahrainis, Iraqis, and Indians, who totalled about 350 employees. The Nissen huts were occupied by western expatriates. During the summer, many men slept in the open air where the occasional breeze made conditions more bearable. However, the companionship of other desert inhabitants, particularly scorpions, spiders and centipedes, often made outdoor sleeping a painful trade-off for the humidity inside the *barasti* houses. Once the desert had been sprayed with oil, these creatures became less frequent camp visitors. Bapco did not provide the *barasti* camps with food, only water, which was pumped along a pipeline from Zallaq to tanks on top of the Jebel Ad-Dukhan. As there were no water coolers, water was stored in tins which had contained kerosene, imported from APOC's refinery at Abadan. When empty, the tins were washed out with hot water, filled with fresh water and then covered with sacks and put in the wind to cool.

Provisions were delivered to the three camps by villagers on donkeys. Fish, mutton, chicken and eggs came from Askar and Zallaq. Sugar, rice and oil were brought from Manama. Dates came from the

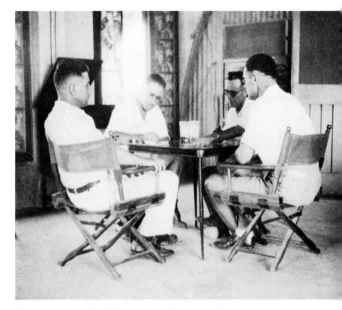

Recreation at the Jebel camp (*Bapco archive*).

villages of Zallaq, Askar, Jau, Rafa', Dar Khulaib and Al-Makhiya, together with vegetables, such as ladies' fingers (okra). Often the men organized themselves into small groups, say of five men, all of whom came from different villages. To start with, it was agreed who would bring the food for the week and who would cook each day. Later, some groups hired a cook each to live with them. Charcoal fires or pressure primus stoves fuelled with kerosene (paraffin) were used for cooking meals, which often comprised dhal, prawn biriani or fish curry. Wheat was unavailable, as it was too expensive at the equivalent of 100 fils a kilo. Instead, the Bahrainis took flour to the Jebel site and made chapattis. Sometimes they made ghee, but since some men did not like it, it was not a popular dish. Basam, another bread variant, was made frequently, since it lasted for a month without spoiling.

At the end of each fortnight, after pay-day, each group calculated how much had been spent on food, divided this between the number of people in the group, so that each one paid his contribution towards the provisions and the cook's salary (about one rupee or 100 fils). During cool evenings, or in the winter, heating was supplied by charcoal braziers. Hurricane lamps provided light, since candles were impractical either because they blew out or melted in the summer. To relax, the men sometimes danced among themselves, sang, played a gramophone, cards or dominoes, read, or just slept.

Inevitably, the stress of long working hours in the intense heat of the desert created misunderstandings which often focused on language and cultural differences. The "goonie" episode was one example. One day a driller asked a Bahraini driver to find him a "goonie". After the driver had produced three or four empty jute sacks, the driller became agitated, climbed down from the rig, pushed the driver over and kicked him. After a few minutes, Mr Skinner arrived on the scene to find out what was going on. When asked what had happened, the driver reported the incident as it had occurred, saying that he didn't understand what he had done wrong. The driller interjected that instead of bringing a piece of carpentry apparatus so that he could cut a right-angle, as a joke the driver had brought the bags instead. Skinner immediately realized that the driver had been right and that he had not been making fun of the driller. Although in Arabic the carpentry instrument is called *zawiia*, meaning a corner or angle, it is also known colloquially as a goonie. In Farsi, the driver's language, goonie also means an empty sack.

Another episode occurred one day, when the head carpenter asked one of the Arabic-speaking Bahraini drivers to go to Hut One and bring a bottle of water to him. At that time ice was imported from Persia wrapped in aluminium sheets and stored in wooden boxes. These ice-boxes were used to chill fish and meat, as well as bottles of water. As requested, the driver lifted the lid off a box to remove a bottle of cool water. An American driller, who had been watching him, caught the driver's hand, hit him and asked why the Bahraini was opening the box. When the two men returned to the site, the head carpenter asked: "Where is the water?" When the driver explained what had happened, the two Americans started a fight. Although the carpenter was perhaps sixty years old and thirty years older than the driller, his body was strong. The Bahraini driver remembers that when the older man hit the driller, "Ohhh! Afterwards, they shook hands together and so my hit was freed [avenged]."

Differences of opinion down at the Jebel Camp were not the only ones which involved the oil pioneers. Bapco's General Manager, Russell Brown, and Charles Belgrave who assisted the Ruler of Bahrain in administrative matters, were often at loggerheads. Most of the disputes concerned the former's resistance to interference over Bapco's pay scales and the latter's reluctance to encourage the training of Bahrainis. Many Bahrainis, including several Bapco retirees who remember those pioneering days well, resented this attitude and criticized Belgrave for being a "typical imperialist". His immortal, if apocryphal statement, that "Bapco has the cash and Bahrain has the class", is considered among many Bahrainis to have been greatly responsible for fuelling a major misconception for several decades. Many people believe that Bapco and modern Bahrain developed separately. In reality, they grew together.

CHAPTER SIX

Dawn of the Oil Era

Confrontation

As oilfield exploration continued to advance in Bahrain, attorneys and solicitors continued to find solutions to the problems which surrounded the Bahrain concession agreements. The first success occurred on the 15th of February 1933, when an Indenture between the Ruler of Bahrain and the Bahrein Petroleum Company Limited was signed. In effect, this was a supplemental contract to the 1925 *Bahrein Island Concession* and the much sought-after modification to Clause IX of the Prospecting Licence (Schedule 2 of the concession). Bapco now had the right to produce up to 100 tons of oil, free of royalty payment, and further quantities of oil on the payment of a royalty of 3 rupees 8 annas per ton of net crude oil produced and saved. Thus this revised clause now made it clear that the company would not be compelled to commit itself to a Mining Lease before the expiration of the Prospecting Licence.[1]

However, international awareness of Socal's activity in Bahrain had added to other remaining difficulties, not least of which was the continuing and expensive presence of the Anglo-Persian Oil Company's products on the islands. Throughout 1932, the local population had complained bitterly about the "unduly high rates" charged by APOC for petrol. When news reached Bahrain that the Persian Government had cancelled the APOC agreement, a general fear that the nation's supply of petrol and kerosene (paraffin) would be cut off aggravated matters further. The immediate repercussion was that in half a day the price of paraffin rose to more than double the market price. In order to combat the profiteering, the Bahrain Government arranged for the Municipality to buy a large quantity of paraffin and sell it at the ordinary price. When this happened, the price reverted to normal.[2]

Meanwhile, the Eastern and General Syndicate's activities were being monitored, particularly by Lombardi who had been in Kuwait for a while. On the 17th of March 1933, his forceful cable to W. H. Berg in San Francisco left no doubt as to how he saw the present situation:

"E. W. Janson, Maj. Holmes propose breaking relations with S. O. Cal. and strongly opposing S. O. Cal. in Arabia and withdrawing support Bahrein Pet. Co. because S. O. Cal. approached Ibn Saud for concession through others. STOP. I understood Loomis notified Eastern and Gen. Syndicate Ltd last year London S. O. Cal. would proceed through others after two months. STOP. Maj. Holmes says notice did not refer to concession but only geological work for the benefit of Bahrein Pet. Co. STOP. ... Situation serious. Have seen Eastern and General Syndicate telegrams to Gulf Oil Co. Wire Cairo" [Lombardi, cable to Berg, 17 March 1933].

Reacting to Lombardi's agitated cable, Loomis telegraphed Ballantyne, the London solicitor: "I distinctly told Janson last summer that we proposed establishing contact with Ibn Saud if they [the Syndicate] did not do so before the end of October. STOP. Janson agreed this fair" [Loomis, telegram to Ballantyne, 18 March 1933]. For the next two days Ballantyne considered the recent flurry of cables. Even then, certain questions remained unanswered. What exactly does the Syndicate mean? With whom will they break, and on what grounds? Assuming that the second question applied to Socal and not Bapco, it appeared to Ballantyne that Socal might find itself in competition with the Syndicate. Nothing, however, should interfere with or embarrass the Bahrein Petroleum Company's interest in Bahrain, including the application for the Additional Area concession. [3]

On the 21st of March, Francis Loomis replied to Ballantyne's latest advice, "referring to our exchange of cables yesterday ... it seems to me that Holmes was talking in a very loose and unwarranted manner, and his remarks have the flavor of what is known in racetrack parlance as a 'bluff'. So, I do not think we need to take the matter very seriously." The following day, Ballantyne confirmed this view: "My legal opinion confirmed today by Chancery counsel. ... I think E. W. Janson, Major Holmes legally and morally inde-

Telegram heading (*Chevron archive*).

fensible. Cannot help feeling they are bluffing" [Ballantyne, telegram to Vane, 22 March 1933]. Within a few days, Lombardi had calmed down too. From Cairo he cabled W. H. Berg once more:

> "Dear Bill – You must have been quite upset by my rather wild telegram from Kuwait – but you couldn't have been as upset as I was when I sent it" [Lombardi, cable to Berg, 28 March 1933].

Nevertheless, the whole question of the company's relationship with the Eastern and General Syndicate was being reviewed as a matter of urgency. A lengthy memorandum, written by W. F. Vane to W. H. Berg on the 20th of March, summarized the status of Socal's "Arabian venture" and considered whether the quoted excerpts from relating correspondence placed Socal under any legal or moral obligation to consider the syndicate in further negotia-

W. H. Berg (*Chevron archive*).

tions in the region. Within a few days, Ballantyne advised Socal to postpone any action until full discussion had taken place. Certainly, if Holmes persisted in leaving Bahrain and neglecting his responsibilities to Bapco, then arrangements should be made with the Colonial Office to appoint a successor and dismiss Holmes. Whatever happened, Ballantyne stressed that it was vitally important to maintain the prestige of both Bapco and Socal in the Near East. [4]

Just over a week later, on the 31st of March 1933, Ballantyne added that "no doubt the Colonial Office and the Anglo-Persian Oil Company have knowledge of Mr Lombardi's cable to Mr Berg and are therefore aware of a split in the camp". He added: "For the Company [Bapco] successfully to be set at naught, at the outset of its career in the Near East, by a man like Holmes, would do us incalculable harm." In Ballantyne's view, the present situation demonstrated that Holmes and his associates were not the kind of ally one can "lie with comfortably: they have tried before this to play the fool with the Company and I do hope that we shall find a way this time to teach them that such a game is likely to be expensive for them" [Ballantyne, letter to Loomis, 31 March 1933].

Suspicions continued to grow. To what extent the Colonial Office and the Anglo-Persian Oil Company had deciphered cables between Bapco and Socal, if at all, remained a matter of speculation. In San Francisco it was felt that the opposing parties were unlikely to have enough samples of Socal's code to achieve that purpose. Nevertheless, when Loomis wrote to Ballantyne in early April, he ended by saying:

"Should the Syndicate go over surreptitiously to the Anglo-Persian, it would be an almost unbelievable piece of treachery to the Gulf [Oil Corporation]. Should anything significant or serious develop in connection with this notion, I hope you will inform me promptly by cable" [Loomis, cable to Ballantyne, 10 April 1933].

Prospecting Licence Extension and Additional Area Applications Falter

While London and San Francisco had been debating the Eastern and General Syndicate's activities, Holmes had returned to Bahrain. On the 4th of April 1933, as Chief Representative of the Bahrein Petroleum Company, he had applied on Bapco's behalf for a one-year extension of the Prospecting Licence, to be effective from the 2nd of December 1933. In support of the application, Holmes pointed out that Bapco had had a modern and powerful deep-boring drilling plant in constant use since the middle of October 1931. Moreover, the company had added a second drilling machine of the same character.

Three weeks later, the Political Resident Persian Gulf despatched a telegram to the Secretary of State for India saying that he believed privately that, in return for granting a one-year extension of the Prospecting Licence, the Bahrain Government had a fee of 60,000 rupees in mind, payable over a maximum of five years from royalties once they exceeded one lakh (100,000 rupees) a year. He also understood that the Ruler of Bahrain might be willing to grant the Additional Area to the company, if it offered really

"Bait Skinner", Mr and Mrs E. A. Skinner's home between 1932 and 1937. Prior to their occupation, the house accommodated visiting pearl buyers from France (*Bapco archive*).

good terms. Having discovered this intelligence, the PRPG urged London to impress upon the British Government how important it was to do nothing concerning oil interests which might give the Ruler of Bahrain the idea, rightly or wrongly, that Bahrain was being sacrificed to other interests. [5]

On the 8th of May, Holmes and Skinner sent a joint-cable to W. H. Berg in San Francisco. The Anglo-Persian Oil Company had acted. Just as feared, it had offered the Ruler of Bahrain 230,000 rupees cash to be non-recoverable in exchange for a concession to cover the Additional Area, on the same terms and conditions as the original concession which had been signed on the 2nd of December 1925. [6] Four days later, Ballantyne expressed surprise to Holmes at the Anglo-Persian Oil Company's curious tactics. If APOC was trying to convince the Ruler of Bahrain that the Additional Area was worth between £70,000 and £80,000, why was that company offering only £17,000? Ballantyne suggested to Holmes that APOC was only making mischief. "If so, there must be some reason for this as they have no occasion to shew ill will to the Bahrein Petroleum Company" [Ballantyne, letter to Holmes, 12 May 1933]. Ballantyne added that no doubt a letter of explanation was on its way which would "make interesting reading". He had noted too that in Holmes' application for an extension to the Prospecting Licence, he had inserted another curiosity. The inference was that the Licence "may" be extended rather than "shall" be extended. Not wishing to start another legal argument, Ballantyne merely drew this matter to Holmes' attention. However, in truth, the battle-lines were drawn once more. Berg questioned APOC's sincerity. Socal and the lawyers were of the opinion that Holmes should be paid off.

As Socal and the lawyers contemplated this turn of events, drilling problems were encountered in Bahrain during the summer of 1933 which raised doubts about the timing of the Additional Area concession application. The difficulty concerned technical problems experienced with Well Number 3. Socal had hoped that drilling from this well would provide a clearer indication as to the exact position and the extent of the oil-bearing structure encountered in Bahrain. Although expectations of a commercial flow had been realized in part, inconclusive results from Well Number 3 indicated that substantial new prospecting work was needed within the existing concession limits, before the company applied for rights to drill in an Additional Area. Thus, on the 11th of August 1933, Bapco notified the Political Agent in Bahrain that "at the moment", Socal, in other words Bapco, preferred to press for an immediate extension of the Prospecting Licence to the 2nd of December 1934, as per the request made in writing on the 4th of April 1933. [7]

Five days later Holmes visited the Ruler of Bahrain to inform him of this decision which, he explained, had been made with great regret. On the 18th of August 1933, Holmes sent a strictly confidential cable to Bapco's registered office in London in which he stated that the Ruler was much disturbed at the unexpected departure by the Bahrein Petroleum Company from its intention to pursue the acquisition of an Additional Area of Bahrain. As a consequence, the Ruler of Bahrain had had to resist much pressure from the Political Agent regarding Anglo-Persian Oil Company efforts to

Front cover of the *Standard Oil Bulletin*, July 1933 (*Chevron archive*).

secure a competitive footing in Bahrain. Notwithstanding this turn of events, the Ruler of Bahrain had "stood firm" in his belief that Bapco would pursue its offer, as proposed in the letter of the 17th of May. The Ruler feared also that if he were to accept a loan against royalties, as proposed by the Political Agent, then this would dissipate the "last remnant of his liberty to act independently, and domination by the Anglo-Persian Oil Company would follow". As to who intended to supply the funds, Holmes was not certain, but it was felt generally that the finance would be provided by either the British Government or APOC, but more likely the latter. [8] In Holmes' view, there was little doubt that "quite a degree of discrimination" had been exercised against Bapco in favour of the Anglo-Persian Oil Company. Had not the time arrived to hint to the India Office that this situation had not passed unobserved?

Holmes' anxious cable of the 18th of August added that during his difficult meeting two days earlier, The Ruler had told him (through his interpreter Mohammed Yateem) that the British "political people had attacked him with sail and oars in action", meaning that he had been brow-beaten over Bapco's decision to withdraw its offer regarding the Additional Area concession. The Ruler's opinion was that the Anglo-Persian Oil Company Limited wanted to obtain a foothold in Bahrain, if only one square foot, and that APOC and the British Political Agent were "joined internally together". The Ruler, as Holmes interpreted the situation, did not want APOC "at any price" [Holmes, letter to Ballantyne, 16 August 1933].

The Political Resident Persian Gulf was not convinced of Holmes' motives. He believed that the whole deal had been orchestrated by Holmes without reference to either the India Office or the Colonial Office. In particular, the latter disapproved of Holmes' direct representation with the Ruler of Bahrain, believing that his account of his interview with Shaikh Hamad was entirely inaccurate. This, together with the fact that, in the Colonial Office's opinion, he had violated the provisions of the 12th of June 1930 Indenture, made it even more imperative that the whole matter should be taken up by Bapco's office in London.

Holmes Resigns

In composing its case, the Colonial Office had come, as far back as the 1st of July 1932, to the mathematical conclusion that, in one year and eleven months previous to that date, Holmes had spent only six months and fourteen days in Bahrain. Hitherto this had not been important, but now that oil had been struck the Colonial Office took a different view. However, Ballantyne had written to Whitehall stating that Holmes, when acting as Bapco's Chief Local Representative, had been in Bahrain at all the times required by the company under Condition C of the concession assignation document. The Colonial Office disagreed, arguing that Holmes was, after all, the British Government's nominee and, in any case, if he were to be forced to stay in Bahrain for ten months of the year, he would probably resign. In this case, perhaps insistence on such a period of residence might well be the easiest way out of everyone's difficulties.

Major Frank Holmes (*Bapco archive*).

On the 22nd of August 1933, Ballantyne informed Loomis in San Francisco that he had just received a further cable from Holmes stating that the Ruler of Bahrain had refused to grant the requested extension of the Prospecting Licence. The offer for the Additional Area, however, could be made after Socal had selected the acreage under the terms of the existing concession agreement. Two days later, the India Office pressed its demands that the Chief Local Representative "shall be resident continuously within territory". Thus, Ballantyne informed Socal that

"... whilst reserving all company's rights concerning this interpretation, have tentatively arranged with India Office and Syndicate that Major Frank Holmes shall be permitted to retire, with Skinner appointed at least temporarily, thereby avoiding hiatus" [Ballantyne, telegram to Vane, 24 August 1933].

On the 8th of September, Lombardi cabled London. His opinion was that Major Holmes should resign, *but* it should "be definitely and obviously forced" by the India Office and not by Socal. Lombardi did not know of any "overt act" which Major Holmes had committed in his official capacity as Chief Local Representative of the Bahrein Petroleum Company Limited as a result of which his resignation should be demanded. The differences of opinion had arisen over the mainland transactions. Although Lombardi did not see any obvious grounds for complaining,[9] minds in London had been made up. In a cable sent at 5.53 a.m. on the 9th of September 1933, Ballantyne informed Lombardi that Major Holmes must go without delay: "I am satisfied that we have sufficient evidence ... but if further proof is wanted I am sure it can be obtained" [Ballantyne, cable to Lombardi, 9 September 1933].

On the 12th of September, Ballantyne suggested to Lombardi that the first necessity was to secure "Frank's immediate retirement" as representative by asking Holmes to resign by the 21st of September. This method would "obviate undesirable consequences" [Ballantyne, cable to Lombardi, 12 September 1933]. Two days later, Holmes tendered his resignation. Effective from that day, the 14th of September 1933, Edward Skinner was appointed Bapco's Chief Local Representative. While the prudence of continuing such a position had yet to be decided, Ballantyne closed his file for the time being with an intuitive omen: "No doubt the voice of the British Government will play an all important part" [Ballantyne, cable to Lombardi, 15 September 1933].

From Pearls to Petroleum

During 1932 and 1933, as oilfield operations gained momentum, the Bahrein Petroleum Company had begun to play an active part in the development of the islands' economy. The Eastern Bank, having opened its new building on the 25th of June 1932, now provided facilities for Bapco's payroll. The procedure was that the Chief Cash Clerk would go each fortnight to draw cash for the Thursday pay-day. Having taken delivery of the coins, he then sat on the bank floor and bounced each rupee on the ground before he bagged it. Part of the Bapco pay-day equipment comprised six large rocks on the ground outside the company's pay windows. As each employee received his wages in silver rupees, he proceeded to one of the rocks upon which each rupee was bounced to check their purity. Any counterfeit rupee would give a dull thud rather than a sharp, metallic ring.

The quality of the silver coins was not the only aspect of Bapco's wages which concerned Bahrainis. In the late 1930s, Tom Berry recalls one Bapco management meeting at which he took the minutes. A deputation of local merchants had complained to the Government that Bapco was overpaying its employees, as a consequence of which the merchants couldn't find any staff to work for them. The merchants insisted that a minimum wage should be 12 annas per day. Bapco paid one rupee a day, about 16 annas. The company maintained that no Bahraini could keep a family on less than one rupee a day. The merchants argued otherwise. Eventually, a compromise of 14 annas was agreed.

As time passed, such feelings of disparity diminished. Instead, Bapco's credit-worthy reputation permeated throughout the suuqs. Ali bin Ebrahim Abdul Aal remembers that

> "our travel number was about the size of a 100 fil coin which had a hole in it. When we wanted credit in the market, we showed that number and we received credit. If I tell you I had a 100 rupee problem [a 100 rupee debt], you will laugh! But it showed the importance of Bapco at the time. The people who worked for the Company were trusted."

Nevertheless, by the mid-1930s, feelings had become strained in some Bahraini communities, particularly those which had depended upon pearl-diving for their livelihoods. The Depression and the introduction of Japanese cultured pearls had hit their incomes hard, yet many people were either resentful or suspicious of the burgeoning oil industry which was as yet little understood in Bahrain. These emotions were expressed in a narrative poem, a dialogue between oyster beds and oil wells, written in 1353 AH (1934). The author, a learned man, lived in the village of Bani Jamra in the north-west extremity of Bahrain island. In the poem, the author played the role of the judge. Having listened to each side of the case, he made his summation.

"Diving was a source of income which may never occur again. We have seen how people became rich and then later became bankrupt. We hope that the industry will recover one day and those lakhs will return to the people. But today, the oil wells are the winners. Listen to my advice and let everyone of you compromise and operate independently. People will become rich and conditions will be prosperous" [Atiyah bin Ali, 1353 AH].

According to the poem, the oyster beds and the oil wells forgave each other, the angry attitude between them disappeared and, instead of abusing each other, they joked, shook hands, embraced each other and thanked the judge for his counsel.

Much of the problem had arisen because many Bapco employees still had a commitment to the *nowaakhdha* (pearl diving captains) to dive or work on the dhows during the pearl diving season. As was well known, if the divers had a good season, they might be able to repay the loans they had been advanced by the captains. If not, they remained in debt until the next season. Many Bapco employees found themselves locked into an obligation they could not break away from. Gradually, a system was worked out to pay off the *nowaakhdha* so that the men could be cleared of their commitment and work full-time for Bapco. There were still some seasonal workers, but this was discouraged.

Bapco also found itself assisting the Manama Post Office which was run by the India Post Office as a sub-office, consisting of one room in the basement of a house near to the Bab Al-Bahrain. Once the mail-bags arrived aboard a British India freighter at *Mina Manama*, they were delivered to the post office, the contents emptied onto the floor, and then sorted into piles for Manama, Muharraq, Awali and so on. Since the floor level was very low and the post office flooded at high-tide, the timing of this latter operation was critical. One day, Leslie Smith (Bapco's Administration Manager) discovered the cause of damp letters when he went to inspect the company's mail. After consulting with Ward Anderson (Bapco's General Manager), they commissioned the company's carpentry shop to make some racks which were duly delivered to the post office. The postmaster, however, was very reluctant to receive the racks, fearing that he would be reprimanded by his superiors in India for accepting gifts from outsiders. Smith's solution was simple. "You don't have to accept them [the racks]. We are just going to leave them here." And so a dry mail-sorting system was created for Bahrain.

After Imperial Airways established Bahrain as a stop-over on its eastern route in December 1932, airmail became a regular service. Thereafter, about a two-week wait followed until a reply was received. Jim Foley remembers that he had to write out all his material in long-hand for the Indian stenographer (not a secretary) who worked for him. All communications were numbered, including reports. Sometimes they arrived at their destination in the wrong order. Others did not arrive at all, so the missing ones had to be repeated. This usually meant retyping them, since there were no photocopying machines and the carbon copies would have been distributed.

There were no telexes or international telephones at that time, but faster or urgent communication with London, New York and San Francisco could be achieved by

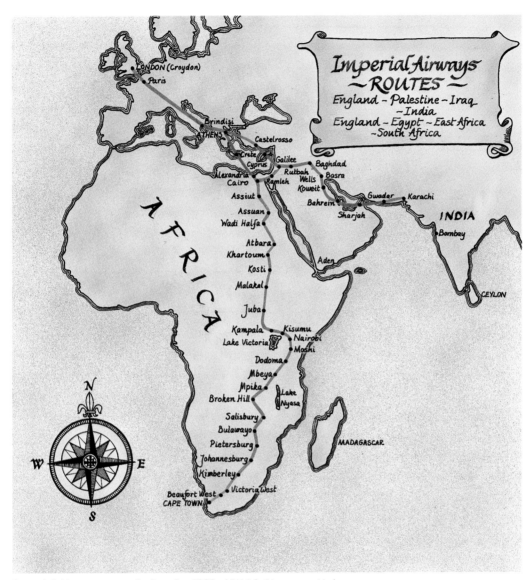

Imperial Airways routes during the 1930s (*British Airways archive*).

sending radiograms from the Cable and Wireless station in Manama. In the 1930s, a five-letter commercial code was developed for both secrecy and economy. In Bahrain, the code books were kept in the General Manager's office. From these, the tedious task of unscrambling the jumbled letters began. For highly confidential documents, there was an additional secret code, a feature which became especially critical during the continuing negotiations over the Bahrain Prospecting Licence, Mining Lease and the Additional Area Concession application.

Hannibal, an Imperial Airways Handley Page 42 G-AAVD Hanno aircraft, at Bahrain Airport in 1935 (*Bapco archive*).

Awali Township is Established

Within two or three years of oil being discovered at Well Number One, it became apparent that the *barasti* and Nissen hut camps at the Jebel would soon outlive their purpose. Plans were made to build a permanent town. A site was chosen on a hillock, a few miles north of Jebel Ad-Dukhan. The first houses to be built made the most of any breezes and so the township became known affectionately as *Mughaidrat*, the place of little dust storms. Later, the Ruler of Bahrain suggested that the name was too difficult for the non-Arabic speaking residents to pronounce. So, a few years after the township had been established, a memorandum was circulated to all employees:

> "In order that our community settlement may be known by a name appropriate to the place it has taken among the institutions of Bahrein, and in harmony with long-established custom here, it has been suggested by His Highness Shaikh Sir Hamad bin Issa al-Khalifah that it be called Awali. Recognizing the spirit of friendly interest which prompted this suggestion, as well as the merit of the suggestion itself, we have accepted the name proposed by His Highness, and request that it be used hereafter in place of other names such as New Camp, Jebel Bapco, Bapco or Bapco City or any other" [Deacon, 23 April 1938].

Since Awali, meaning "high place" in Arabic, had been used to designate the elevated region around Jebel Ad-Dukhan, and the entire Bahrain island was known in classical times as Awal, Deacon added that these associations made the adoption of the town's official name especially appropriate.

As the settlement developed, the bleak site assumed some character. The first *gutch* houses were designed for bachelors. They were constructed in the form of a succession of rooms off a central corridor, with walls 12 feet high (3.65 metres) to allow for maximum circulation of air. The whole structure was surrounded by a verandah to provide shade. Vern Raven recalls that in House 2, for instance, lived six bachelors. Each man had a room, but most pioneers recall their utilitarian character, very much like a hospital room with an iron bedframe and metal cupboard for clothing. A houseboy made tea for the men in the mornings and afternoons, and in the evenings dinner was taken across the road in the mess hall.

Other bachelor accommodation was provided in bunkhouses, eight to ten temporary, long bungalows with about twenty two-man rooms and separate washhouse and toilet facilities between all the houses. Like the *gutch* houses, each room had two beds, two steel cabinets, one table, one table fan, as there was not enough electricity to power a ceiling version, and one chair. After air-conditioning was introduced, the rooms were cooler but since the temperature rarely dropped below 90°F during the day, the men often boarded-up the windows to keep the sunlight out. One British resident found life in the "lower camp" particularly trying.

> "I was saving every penny to get back to England. But if you resigned, you not only had to pay your own fare home, you had to repay your outgoing fare too. … I protested about my living accommodation and they found me a better room in one

of the *gutch* houses. I made friends and after about three months I had long since given up any idea of going home. I think everybody was pretty miserable. We did lose forty-per-cent of staff within a few months, particularly the British."

By late 1938, centralized air-conditioning plants had been installed in Awali. At the time, it was considered a novel system and believed to have been the first of such magnitude to have been built anywhere in the world. The installation comprised a central refrigeration system near to the *gutch* houses on the Awali hillock from which chilled water was piped through all the streets to each house which had a so-

Gutch house verandah, Awali (*Bapco archive*).

called "bird cage". Each of these devices contained small coils through which the chilled water flowed and a fan which blew air over the coils; thus the air in the room was chilled and dehydrated to remove the humidity. The resultant moisture dripped into a pan, was piped outside and dispersed in the garden. Meanwhile, the water in the coils was recirculated to the refrigeration plant to be chilled once again.

The system was designed to keep houses at a temperature of 85°F (30°C), but that didn't always happen, with the result that many residents complained about the efficiency of the air conditioning system. But as one engineer explained

> "these plants can't run seven days a week, twenty-four hours a day, month in and month out during the long spring, summer and autumn season without something happening once in a while. A lot of people thought that 85° was too high a design temperature in the first place and said so. The humidity contributed to the discomfort. ... The lack of air conditioning in the kitchens was a source of complaint in the 1940s and 1950s too. The engineering approach is that the kitchen is a hot place anyway and if you designed in air conditioning for the kitchens, you would have to have twice as big a unit and twice the investment capital" [Extract from interview with Denny Berdine, 16 July 1987].

Although the living accommodation was air-conditioned, initially no such comfort was installed in either the dining hall or the Awali office.

> "It was very miserable working in that office without air conditioning, particularly in the summer months, because everything would stick to your arms. You would have a sweat towel around your neck and your nose would drip. You couldn't sign a piece of paper without getting the signature blurry as we were still using fountain pens - there were no ball points in those days" [Extract from interview with Ray Andresen, 12 August 1987].

After the refinery was built, drinking water for Awali was supplied from steam condensate. Bob Bartlett recalls that this distilled water was blended with a small quantity of well water, just enough to give it taste for drinking purposes. Eventually an ionics water-treating plant was installed in Awali, since the refinery plant could not provide its full requirements. Later, Hussain Yateem, who had acted as interpreter for Bapco on many occasions, asked the company to build a distilled water plant for him in Manama so that he could make life sweeter for the Bahraini tea-drinking population. Since more sugar is required to sweeten brackish water than distilled water, Yateem concluded that he could assist the local tea-drinkers by selling them distilled water so that they could save money by reducing their sugar purchases. Ed Howard remembers that an old boiler, left over from the first drilling operations at the Jebel Camp, was reconstituted and converted into a distilled water plant in Manama for the purpose.

El Segundo being loaded in San Francisco, 1934, for its voyage to Bahrain (*Chevron archive*).

The *El Segundo* Task Force

When the foundations of Awali were being laid in 1934, Bapco was considering the implications of commercial oil production. At that time, in Socal's view, it was quite impossible to predict world political and economic conditions for, say, the next three or four years, and therefore it was difficult to assess Bapco's market for crude oil. Thus, Socal, *de facto* Bapco, was not willing to produce more oil than it might reasonably expect to sell at a fair profit.

One major problem was Bahrain's immense distance from sources of supplies, as well as their irregular delivery. Another factor which slowed down oilfield operations was the inadequate harbour facilities: coral reefs and shallow shore waters prohibited large ocean-going vessels from discharging their cargo onto the two stone piers in Manama. As it was, steamers which drew up to 20 feet (6 metres) had to drop anchor between 2 and 4 miles (3.2 and 6.4 km) offshore and then offload freight and passengers into lighters. Tankers with more than a 20-foot draft (6 metres) would have to anchor at least 14 miles (22.5 km) offshore. In any event, commercial production would require the construction of a submarine pipeline and terminal.

Despite these obstacles, notable speed had been made in prospecting the Bahrain concession. Although in mid-1933 it was still too soon to determine when Bahrain's oilfield would be of sufficient size and productivity to warrant commercial development, the present indications were favourable. On the 13th of December 1933, the *San Francisco Chronicle* ran a news item announcing that in the New Year a Standard Oil tanker would leave San Pedro

"on an expedition unique in industrial history. The party will go to the island of Bahrein ... to build a submarine loading line and terminal to link an island oil supply with a tank fleet" [p. 6].

The *El Segundo* tanker, a ship of 3400 tons, had been selected for the task. The vessel, which was to voyage westwards to avoid Panama and Suez Canal tolls, would carry a party of 45 men, including crew and a deep-sea diver, as well as all the necessary mechanical and technical equipment to build a 3-mile-long (4.8 km), 12-inch-diameter (30.48 cm) pipeline to an offshore loading terminal, storage tanks and a trestle draw-bridge. Few merchant ships had ever been equipped quite like the *El Segundo*. Fuel for the round trip, everything necessary for the subsistence of 45 men for nine months, together with miles of pipe, sheet-steel for tank construction, five 8-ton anchors, six marking buoys, three trucks, a trailer, two tractors, numerous air-compressors, electric-welding generators, a derrick and an unloading platform for landing freight from lighters, and tools, were all on the list. Since the tanker had no general cargo hold, openings were cut through the steel deck so that heavy equipment and supplies could be stored in the cargo-tanks. Cargo-booms were added to the ship's masts, an incongruous sight on board a tanker whose usual cargo was normally taken on and discharged through hoses.

The pioneers had inspired Californian imagination. Then, as now, news editors sought a special angle to catch their readers' eyes:

"Hard-bitten tanker sailors at San Pedro rubbed their eyes twice yesterday when the Standard Oil tanker *El Segundo* put to sea; her decks filled with strange gear and her crew equipped as though she were bound on some trading expedition to the Sahara Desert. ... To properly impress [the Ruler of Bahrain] of the dignity and eclat of the Standard Oil Company, Capt. I. B. Smith, super-diplomat among shipmasters, has been supplied with heavily gold-incrusted uniforms and dress hats equally adorned" [*Los Angeles Times*, 31 December 1933, p. 6, syndicated to the *New York Times*, 14 January 1934].

"That the *El Segundo* sails from the port of Los Angeles gives one a clue – perhaps misleading – as to the linking up of Cleopatra and the Queen of Sheba with the prosaic routine of the oil business. Are we right in discerning the hand of Hollywood in the outfitting of this expedition? ... If so, imagination goes further and presents a picture of Cap'n Smith going ashore on a canopied throne with a pair of water-tenders waving peacock-feather fans before his weather-bronzed face. In any event, the Sheikh [the Ruler of Bahrain] is due for a treat" [*Christian Science Monitor*, 22 January 1934].

Laying up the floating roof on a Sitra oil tank, 1934 (*Bapco archive*).

The *El Segundo* (*Chevron archive*).

Almost two months later, *El Segundo* dropped anchor off Sitra island on the 22nd of February 1934, the anniversary of George Washington's birth. Following its 55-day trip from the United States port of San Pedro, the vessel was used for more than three months as a freighter, storehouse and floating dormitory. From her anchored position 16,000 feet (4880 metres) offshore, the construction work of the marine loading-terminal began. A submarine pipeline was laid and three crude-oil tanks with floating roofs were built on Sitra island, the tank shells having been transported from California and assembled with rivets and welding by a Bahraini workforce. An additional 7500-barrel tank was built for storing ship's ballast water. Four sets of radio telephones for communication between ship and shore and a telephone line connecting the present line on Bahrain island to the facility were installed also.

The terminal included a pumphouse, built of stone and corrugated sheets, equipped with four 225-horsepower gasoline engines connected directly through a vapour-proof partition to four pumps. Eventually, the units pumped oil along the 16,000-foot sea-line to tankers, at any desired rate of flow up to 5700 barrels an hour.

A gathering line was laid across sandy ground which, in places, was marshland too. Trenches were dug by local labourers who had had no previous experience with picks and shovels. Nevertheless, they became able assistants to the team of welders. The construction of a new road to the terminal, including a 1200-foot-long (365.7 metres) trestle draw-bridge across the channel between the islands of Bahrain and Sitra, was all part of the *El Segundo* team's task. Teakwood piles, imported from India, were driven home by an improvised pile-driver which utilized an old drill-stem and a 500-pound (226.8 kg) hammer operated by 25 men.

By mid-spring, media attention had focused on Bahrain. On the 18th of April, the *Pacific Coast Wall Street Journal* announced that some time during July 1934, Standard Oil Company of California would load its first tanker of crude oil. Where this first shipment of Bahrain crude was to be sold had not been determined. But, that notwithstanding, this event "will mark the inauguration of commercial production of Standard's first Near Eastern oil field development venture which has successfully culminated in the proving of a field where the company [Standard Oil Company of California] has 100,000 acres of land on which six wells have been completed and on which eight more now are being drilled".

On the 7th of June 1934, *El Segundo* left Bapco's new marine terminal, bound for Yokohama, Japan, where it docked just over a month later, on the 9th of July. Originally, the oil had been intended for the nearer European markets, but with the widening of the Asiatic markets, particularly in Japan, the decision had been changed. The cargo, Bahrain's first shipment of crude oil, comprised 25,082 barrels of oil.

During the months the construction work had been under way, "an extremely amicable relationship existed between the Company personnel and Sheikh Hamad, founded on mutual respect and appreciation" [*Standard Oil Bulletin*, August 1934, p. 10]. As a token of this friendly feeling, the Ruler gave the ship's party a farewell banquet on the eve of its departure.

"Fifty-two days out from Bahrein Island, practically half-way round the world from San Francisco, the Standard tanker *El Segundo* came into her home port on August 1, completing a round-trip voyage begun just eight months before" [*Standard Oil Bulletin*, August 1934, p. 4].

Covert Competition Threatens Bapco's Search for Markets

Simultaneous with *El Segundo*'s mission during 1934, questions had been asked about the exact clearance which Gulf Oil Corporation had obtained from the Turkish Petroleum Company in 1928 (now the Iraq Petroleum Company) regarding the title of the Bahrain option. Had Gulf Oil been entitled legally to assign the option to Standard Oil Company of California?

As the lawyers tackled their latest challenge, on the 26th of October 1934, William Fraser of the Anglo-Persian Oil Company visited the India Office in Whitehall to discuss the future of the Bahrein Petroleum Company's concession with Messrs Laithwaite and Walton. Fraser revealed that for some months past very tentative and confidential discussions had been proceeding between the Iraq Petroleum Company and the Standard Oil Company of California, precipitated by two factors: "the difficulty which the Bahrein Petroleum Company was experiencing in finding a market for its oil" and the IPC's wish "to eliminate American interests" from an area in which there might be competition and "to meet the threat which a very large production by those interests might constitute" [10] [Memorandum of a Discussion, 26 October 1934].

Fraser believed that Socal's chairman might even travel to London to examine the possibilities of some arrangement with the IPC more closely. To assist any future agreement between the two companies, the latter had accepted the former's offer to allow three representatives of the "principal negotiating groups" to fly to Bahrain and verify the potential productive capacity of the Bahrain oilfield. Indeed, already, they "had been despatched for this purpose by aeroplane". An added complication was that the French interests in the IPC, not directly associated with the present negotiations, had become very suspicious of the proceedings. As a result, "a representative of the French group serving in Iraq had, on his own responsibility, at once proceeded to Bahrein by air independently".

Although Fraser informed Laithwaite and Walton that negotiations were still at a very early stage, he indicated that "no transaction was likely to be entirely satisfactory to the IPC which did not result in the elimination" of Standard Oil Company of California from Al-Hasa and Bahrain as *independent* operators. Furthermore, the IPC was anxious to secure as a minimum, a 50-per-cent interest in the operations of Socal and its subsidiary companies in that area, including the Bahrein Petroleum Company Limited. If a deal were to be struck between Socal and the IPC, the effect would be one of the following: Bapco would receive some share in markets at present closed to it, in return for some understanding regarding the volume of production; or, the buying out of Bapco altogether; or a joint-operation; or, an arrangement under which the IPC granted marketing concessions to Bapco, in return for being able to secure counter-concessions from Socal in Al-Hasa. Laithwaite and Walton found the entire proposition extremely embarrassing. An interim diplomatic solution was agreed. The Political Resident would be briefed confidentially, but that, in the event of the Ruler of Bahrain asking for an explanation of the representatives' visit to his country, a circuitous explanation could be offered along the lines that

> "we understand that certain groups of the IPC interested in marketing in the Far East were contemplating tentative discussions with the Standard Oil which might affect the disposal of the output of oil at Bahrein, and that they had been anxious, with a view to these discussions, to form their own impressions as to potential production in the area covered by the Bahrein Petroleum Company's concession."

Before many more weeks had passed, Wallace Murray, Chief of the Division of Near Eastern Affairs, Department of State in Washington, had heard conflicting rumours "in the wind". Ill-at-ease, he sought clarification from Socal:

> "With reference to our previous correspondence and conversations regarding the concession which your company holds in Bahrein, I think you will be interested to learn that we continue to receive reports of the imminent transfer of the concession to the Anglo-Persian Oil Company. The reports are to the effect that your company is considering selling its rights in Bahrein, and that it is also probably

willing to dispose of its interests in Saudi Arabia to the Anglo-Persian Company. I should be interested in learning anything you can tell me of the accuracy of these reports" [Murray, letter to Loomis, 4 January 1935].

Whatever the plans of APOC or the IPC may have been, the Ruler of Bahrain, Shaikh Hamad bin Isa Al-Khalifa, and the Bahrein Petroleum Company Limited signed an agreement on the 29th of December 1934 which became the basic document to which all amendments and agreements concerning the Bahrain concession were to refer subsequently, until 1970. In essence, the 1934 agreement converted the original Bahrain concession into a Mining Lease, provision for which had been made under the terms of Schedule Three of the 2nd of December 1925 document. As a result, Shaikh Hamad granted Bapco an exclusive right for a period of fifty-five years from the 1st of January 1935 to prospect and drill for, extract, treat, refine, manufacture, transport and deal with petroleum products within the area defined in the First Schedule of the 1934 agreement. With Bapco's position now more securely defined *vis à vis* oil exploration, the next task was to secure markets for the commercial production of Bahrain's oilfield. In February 1935, Socal wrote to the American Ambassador in Turkey:

"Our Company is very likely to become, within a short time, one of the most important producing petroleum companies in the Near East. I base this statement on the fact that the Company has already developed a producing oilfield on the Island of Bahrein … has commenced the shipment of crude oil from that port, and has, a month or so ago, made its first payment of royalty to the Sheikh of Bahrein [the Ruler].

"The Company has a concession in eastern Arabia. … Our geologists have already found there several promising structures, and there is ample reason to believe that a large and valuable oilfield will be developed. …

"We could, very readily, supply Turkey all it needs in the way of fuel oil, lubricants and other petroleum products from our resources. … I feel that we are in a position geographically, and therefore logically, better than any other American company to supply the needs of Turkey. So what I would like to know, if possible, is whether the Turkish Government would consider it worthwhile to discuss some such plan as I have briefly suggested.

"I wish to say in this connection, that we have no affiliation with the Standard Oil Company of New York, or any of the other members of the so-called Standard group. … We have no interests in common, and … if you can, conveniently, give me some enlightenment on the possible views of the head of the Turkish Government … I shall be more than obliged" [Standard Oil Company of California, unsigned copy of a letter to Robert Skinner, 12 February 1935].

Seven months later on the 8th of September, *The New York Times* reported that an "amicable arrangement of the disposition of the oil produced by the Standard Oil Company of California from its concession on the Island of Bahrein ... has just been made in London. ... It was said Walter C. Teagle, President of the Standard Oil Company of New Jersey ... was instrumental in the deal". The article went on to claim that the new arrangement involved the purchase of at least 15,000 to 20,000 barrels of oil daily by three large international companies. Not only was Standard Oil Company of New Jersey named, but also Royal Dutch-Shell and the Anglo-Persian Oil Company. It was reported also that Bahrain's oil production averaged only 2500 barrels a day in the first seven months of 1935, the reason being that Standard Oil Company of California "could not find a market for any greater amount at the prevailing world prices".

Caltex is Born

The press continued its vigil. On the 18th of October 1935, *The Financial Times* reported on a high-powered marketing conference in New York. W. C. Teagle, President of Standard Oil Company (New Jersey), K. R. Kingsbury, President of Standard Oil Company of California, Andrew Agnew, a director of Shell Transport and Trading Company, and R. G. R. van der Woude, President of Shell Union Oil Corporation (the US holding company subsidiary of Royal Dutch-Shell group), had gathered to discuss the marketing of Bahrain's crude oil. If disposition through regular channels could not be agreed, it was understood that Socal would go into the markets alone.

At the beginning of November, Kenneth R. Kingsbury returned to San Francisco, fending off reporters "without comment" [*The Pacific Coast Wall Street Journal*, 1 November 1935]. His answer to those with whom he had been trying to negotiate a marketing deal and who had continued to adopt a "no attitude" stance, was published in the November 1935 issue of the *Standard Oil Bulletin*:

> "On November 14th the Company announced placement of orders for equipment for the construction of a refinery on the Island of Bahrein ... where the Company has developed an oilfield through its subsidiary The Bahrein Petroleum Company. The initial unit of the refinery will have a capacity of 10,000 barrels daily, and will be so built that it will be possible readily to double this capacity to 20,000 barrels daily. The refinery will produce principally gasoline, kerosene, diesel oil and fuel oil. It is expected that it will be possible to begin operations next July."

As the new year dawned, Kingsbury, President of Socal, set off on a visit to Europe. His travelling companion was Captain Torkild Rieber, Chairman of the Texas Corporation. Together, they toured several countries and studied the problems connected with the marketing of Bahrain's crude oil. Already, Bapco's production during the first six months of the year had amounted to 246,000 tons, compared with 42,000 in the same period of 1933. On the 26th of June 1936, Kingsbury and Rieber issued a press statement.

"After several months of negotiations an agreement has been reached between the Standard Oil Company of California and The Texas Corporation, through which the production of crude oil of the Standard Oil Company of California, East of Suez, and the petroleum products from the refinery which is now under construction at Bahrein, will be marketed through the foreign distributing facilities of subsidiaries of The Texas Corporation. The capacity of the refinery at Bahrein, which is nearing completion, is to be expanded, while additional marketing facilities will be erected where necessary. To accomplish the purpose, a new corporation, The California Texas Oil Company, Ltd. has been organized, each of the parent companies having equal representation on the Board of Directors" [Statement for the Press, 26 June 1936, CCL].

Routes to Bahrain undertaken by Bapco and Caltex employees during the 1930s and 1940s (*Jane Stark*)

KEY TO PLACE NAMES

1.	Los Angeles	13.	Paris	24.	Damascus	35.	Penang
2.	San Francisco	14.	Marseilles	25.	Mosul	36.	Singapore
3.	Vancouver	15.	Zurich	26.	Kirkuk	37.	Jakarta
4.	New York	16.	Naples	27.	Lake Habbaniya	38.	Manila
5.	Gander	17.	Brindisi	28.	Baghdad	39.	Shanghai
6.	Azores	18.	Malta	29.	Basra	40.	Kobe
7.	Casablanca	19.	Istanbul	30.	Kuwait	41.	Yokohama
8.	Gibraltar	20.	Alexandria	31.	BAHRAIN	42.	Perth
9.	Shannon	21.	Cairo	32.	Karachi	43.	Adelaide
10.	London	22.	Haifa	33.	Bombay	44.	Tasmania
11.	Southampton	23.	Beirut	34.	Ceylon	45.	Sydney
12.	Le Havre					46.	Hawaii

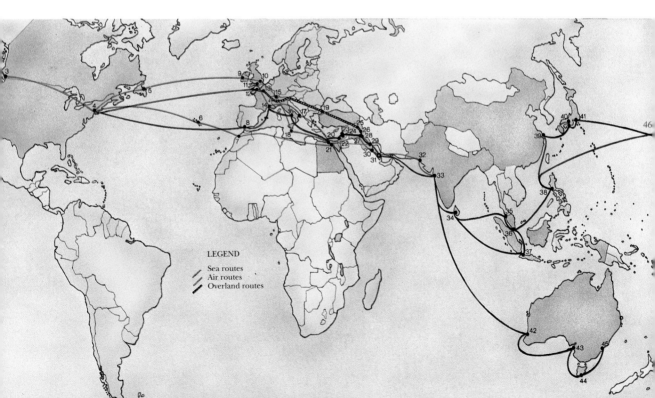

LEGEND
Sea routes
Air routes
Overland routes

Captain Torkild Rieber (*Texaco Inc.*).

Socal was to be represented by J. A. Moffett, who would be Chairman of the Caltex Board, and two Vice-Presidents, R. H. Morrison and Max Thornburg. The Texas Corporation representatives were to be H. M. Herron, who became Caltex's first President, J. V. Murray (a director) and William Kunstadter (Secretary and Treasurer). The administrative offices were located in the Chrysler Building, 43rd Street, New York City. For the time being, the California Texas Oil Company Limited became a subsidiary of the Bahrein Petroleum Company Limited.

Michael Lombardi later revealed that when Kingsbury and Rieber had met to form Caltex, Rieber had insisted that he was not prepared to have a minority interest. It had to be a fifty-fifty arrangement, or no deal. Kingsbury is reputed to have said that this was all right, but Texas Corporation would have to pay $18 million for an even-split. Although Lombardi records that Kingsbury and Rieber liked and trusted each other, and worked together well, this deal is said to have irritated the Texas men for nearly two decades thereafter. No doubt other familiar observers shared a similar view, for when almost two years earlier Fraser of the Anglo-Persian Oil Company had intimated the desirability of eliminating Socal's *independent* interests in Bahrain, it is unlikely that he had considered the possibility of this twist in the "game". Far from American interests being *reduced* in Al-Hasa and Bahrain, they had been considerably *increased*.

The Bahrain Refinery Takes Shape

In October 1935, the ground was broken on the present refinery site for the first time. The first refinery unit, a crude still, was completed as planned on the 12th of July 1936 with a capacity of 10,000 barrels per calendar day. Just over a year later, it was reported that Bahrain had become the twelfth largest oil-producing country in the world. On the 11th of December 1937, His Highness Shaikh Hamad bin Isa Al-Khalifa officially opened the refinery. Another milestone in the development of Bahrain's oil industry had been achieved.

The construction of the refinery had been no easy task. As no heavy-lift equipment was available in Bahrain capable of taking the weight required to lift sections of the fractionating column, it was necessary to ship the column to Bahrain in segments. The entire 92-foot-high (28 metres) and 10-foot-diameter (3 metres) column, having been built by the Chicago Bridge and Iron Company in Birmingham, Alabama, was then cut into 10-foot (3 metre) sections. The sections weighed between 2500 and 3000 pounds (1134 and 1360 kg), the maximum weight which the lifting gear already available in

Bahrain could handle. The sections were then match-marked before shipment so that when they arrived in Bahrain, the match-marks would go together and the column could be rewelded. Ray Andresen recalls that the sections fitted together perfectly, just as they had been designed to be built. The same principle was applied to the shipment and installation of several other major items of equipment.

Not long after the refinery went "on stream", more oil was discovered which required an increased refining capacity. Another crude still, some re-run stills and more treating plants were designed. As more "cracking" capacity was required, Socal contracted Universal Oil Products (UOP) to make a thermal cracking plant with a coking section comprising six large coke chambers. These were huge vessels over 80 feet (24.38 metres) high, about 12 feet (3.66 metres) in diameter and 2 to 2.25 inches (5 to 5.7 cm) thick. To assist with the rigging-up of this expansion, the Pacific Rigging crew from Los Angeles was sent to Bahrain. An immense challenge was finding a way to raise the very heavy coke chambers in one piece. In order to do that, some steam boilers were brought in from the producing field to use with gin polls and sheaves to lift the chambers up onto their foundations. Andresen recalls that it was a Herculean task.

Debugging plant deficiencies was not easy either. One of the worst problems involved the thermal cracking plant coke chambers. The operators could not keep the plant running because the solidified petroleum coke [11] could not be removed fast enough to meet required turnaround cycles since the unit had faulty coke removal equipment. The coke process came from the Texas Company, which had been using it in its refineries in the south and mid-west of America. The Texas Company had also designed the coke drills used for removing the petroleum coke from the chambers.

Petroleum "coke drums" built in the 1930s, being relocated for use as hydrogen storage to serve the Platformer unit under construction in 1956 (*Bapco archive*).

First the vessels were cooled and washed down with water so they could be opened without the operators being burned. Then a cutter was attached which bored down through the middle of the coke to make a hole. The first cutter was removed and another cutter installed to make a bigger hole. After that, a further cutter with balls and chains was attached. They were just like big cannon balls on the ends of chains which were whirled around from under the coke with the idea that they would knock the coke down into a manway. From there the coke would gravitate into trucks so that it could be taken to the coke pile. The main hazard was when one of the balls flew off, the chain broke, an eccentric developed on the shaft which then bent, and then became stuck in the bearings. The only solution was to send men inside the vessel to dig the shaft out by using gas torches to heat the shaft, and then bend it so that it could be removed, a very tedious, hard, dirty and time-consuming operation. Finally, a new drill was designed to an entirely different principle, which allowed the arms to cut through the coke. Instead of the vessel becoming choked so that the drill could not be moved up or down, the new idea allowed the drill to be rotated and the arms folded vertically so that the entire drill could be pulled out of the chamber. When it was put into operation, the coke came out exceedingly fast. The problem had been solved.

Postscript to the Decade

As Bapco celebrated its tenth anniversary in 1939, the foundation of Bahrain's oil industry had been firmly established, despite many setbacks. Now, as the company faced the future, it saw a shadow of a very different kind. Rivalry among competing oil companies had been eclipsed by a new threat, Adolf Hitler's expansionist ambitions.

The Bahrein Petroleum Company's first filling station, opened in Manama in 1938 (*Bapco archive*).

CHAPTER SEVEN

Bapco and World War II

Additional Area Granted

As the war clouds gathered over Europe, the *American Magazine*'s roving reporter discovered Bahrain.

> "On the sand in the middle of Bahrein is a neat, white ultramodern city which cost $25,000,000 to build and which includes 50 oil wells and a $12,000,000 refinery. There are homes, a hospital, clubs, a school, a movie theater, a beauty parlor, and a commissary, all air-conditioned and full of electric gadgets, from kitchen ranges to curling irons. … It is one of the world's great oil fields and, except in the oil trade, a closely guarded secret" [Beatty, January 1939].

Beatty also remarked that within a few years, no longer would there be 4th of July celebrations in Bahrain, or such "terrific complications as exist today when Texans and cockneys, both thinking they speak pure English, fail utterly to understand each other". Instead, the Americans will be satisfied with the profit from oil production. And indeed they were. For, just twenty-three days before Kingsbury and Rieber issued their press statement announcing the formation of Caltex, a *Supplemental Deed* to the 1934 *Mining Lease* was signed on the 3rd of June 1936. This made provision to double the guaranteed minimum annual royalty from 75,000 to 150,000 rupees, a change which recognized the increase in crude production from the day when oil was discovered on the 1st of June 1932. After the creation of Caltex, it became apparent to the British Government that organizational changes, both present and future, together with the uncertainties of the world at that time, posed a serious question. To what extent would the British be able to maintain control over the Bahrain concession and the marketing of the islands' oil products?

K. R. Kingsbury (*Chevron archive*).

Starling, Head of the Petroleum Department, mooted the idea that the British position in the area would be strengthened if Petroleum Concessions Limited (a British company) were to acquire Socal's interests in the Arabian Gulf. The India and Foreign Offices considered the idea. But the former recognized that as the British Government had no legal hold over the Bahrein Petroleum Company Limited, it was doubtful whether any form of words could be devised which would be effective in strengthening British control. Another factor was that discussions which had taken place in 1933, before the signing of the 1934 Mining Lease, had been based on the assumption that the British Government would remain neutral over the Additional Area matter.

Further, the Bahrein Petroleum Company had been informed that it could apply for the unallotted area following the Ruler of Bahrain's request that it should do so, to which the British Government had acceded. If Whitehall were now to advise the Ruler to grant the concession for the Additional Area to Petroleum Concessions Limited, the matter could hardly remain hidden and the British Government's position would be indefensible internationally. [1] Thus, any notion that the British Government should abandon its neutral stance on this issue should be resisted.

The next step towards the consolidation of American oil company interests in Bahrain occurred on the 19th of June 1940 when the second amendment to the 1934 Mining Lease was sealed by the Ruler of Bahrain. Witnessed by H. Weightman, the British Political Agent, and signed on behalf of the Bahrein Petroleum Company Limited by M. Weston Thornburg, Vice-President and Attorney-in-Fact, the document became known officially as the *Deed of Further Modification* of the Lease dated the 29th of December 1934.

The essence of the document was that the mining area in Bahrain had been extended from 100,000 acres to include all land and water areas within the Ruler of Bahrain's domain, including the Additional Area. That notwithstanding, Bapco was not permitted to drill on Umm An-Nassan island, nor within the municipal boundaries of Manama (the capital) and Muharraq (as defined in the Lease), [2] nor within 300 yards (274.32 m) of the Ruler's hunting lodges at Romaitha, Amar and Al Umattala, without permission from the Ruler.

The Standard Oiler

ANNUAL REPORT TO EMPLOYEES ON THE AFFAIRS OF THE COMPANY

MAY 1939

255 Bush Street, San Francisco, the corporate headquarters of the Standard Oil Company of California, as illustrated in the cover of the company's 1939 Annual Report (*Chevron archive*).

Bahrain Enters World War II

In the course of 1939, Standard Oil Company of California had become increasingly aware of German interest in the mainland concessions of Arabia. Although this interest was not new, the Germans had not been able to turn it to material worth because they lacked the necessary finances to carry out an organized exploration and production programme. However, they did have substantial gold holdings in the Selection Trust of London, among other places, and it was thought conceivable that they might convert some of these holdings to pursue oil exploration in the Arabian Gulf region.

These fears were indirectly supported when the Cairo-based *Al-Ahram* newspaper noted in January 1934 that both the Germans and Italians were active in all countries in the Near East, the most recent event being the promotion of the German consulate in Jeddah to the status of a legation. Herr Grobbe, the new German Minister, was leaving for Jeddah to make a big impression among the Arabs, accompanied by agents of the German news agency. The purpose, it was thought, was to open an office for propaganda and to secure a concession for the exploration of oil for Germany in a centre which would be of technical military importance on the Red Sea coast. The Italians in Cairo were of the view also that the Union of Arabs would turn towards the Rome-Berlin Axis [*Al-Ahram*, 24 January 1939].

At a luncheon held in Jeddah on the 22nd of February 1939, W. J. Lenahan of the California Standard Oil Company Limited (Casoc) discussed this report with Dr Grobbe, who claimed that the whole matter was Jewish propaganda and denied that Germany's interest in Saudi Arabia was political. However, that was not the view of Shaikh Hafiz Wahba, who was well-versed in the world situation, particularly from his experience as Saudi Arabia's envoy in London and from having visited America in the summer of 1938. During this trip he had had the opportunity to talk about oil affairs with various company officials, including Fred Davies who was then based in Bapco's New York office.

On the 9th of March 1939, some weeks after Lenahan's encounter with Dr Grobbe, Shaikh Hafiz had occasion to discuss the situation with another Casoc representative, Lloyd N. Hamilton. During their conversation, the Shaikh is reported to have said that, in his view, Germany's interest in Saudi Arabia *was* entirely political. He added that both Germany and Italy "rather anticipate that sooner or later they are going to come to grips with Britain, and if war comes they are hoping to repeat what Britain did with the Arabs in the last War, that is, enlist the help of the Arabs in order to throw off the British yoke and gain their complete freedom and independence. … However, they were getting nowhere", because, in Shaikh Hafiz's opinion, the Arabs fully realized the value of maintaining their friendship with the Allied Powers [Hamilton, memo to Lombardi, 9 March 1939].

However each side felt about the likely outcome of European events, mutual anxiety was not eased when California Standard Oil Company sent Lombardi a coded cable some two weeks later announcing that: "Japanese Minister to Egypt, and party, leaving for Jedda March 22nd, has requested letter of introduction to California Arabian Standard Oil Co.

and Bahrein Petroleum Company, Ltd, so that two of party may visit operations" [California Standard Oil Company Ltd. cable to Lombardi, 22 March 1939].

Max Thornburg came to the conclusion that it would be "difficult" to arrange for the Japanese to visit the Bahrein Petroleum Company. Lombardi, from San Francisco, suggested that the visit should be discouraged without appearing discourteous. On the 9th of April 1939, the London-based *Sunday Times* reported that negotiations for a Treaty of Friendship between Japan and Saudi Arabia were being concluded with King Ibn Saud, so that the Japanese could pursue their hopes of securing a concession to prospect for oil in Arabia.

The Americans, now more anxious, made representation to George S. Messersmith, Assistant Secretary of State in Washington DC. Francis B. Loomis wrote a personal letter from Socal, in which he said

"it seems that after several years of costly effort, oil having been discovered in Arabia on a concession obtained by the Standard Oil Company of California, efforts are being made by representatives of several foreign oil companies to take advantage of the large amount of work and great capital expenditure" [Loomis, letter to Messersmith, 25 April 1939].

On the 3rd of September 1939, Britain declared war on Germany. During the next day, the Ruler of Bahrain declared war on the Axis Powers. Bahrain and the refinery went onto a war basis immediately. British residents were required to surrender their passports to the Political Agent and were not allowed to leave Bahrain. As sabotage was feared, Ray Andresen recalls that the storehouse was manned night and day since, if anyone had removed or damaged the refinery's spare parts, it would have caused the entire operation to shut down. Added protection was provided by the Bahrain Government in the form of a khaki-clad camel corps which policed the oilfield and the refinery:

"We put out water troughs for the camels and had people going round in pickups to make sure that everything was all right. We designed some special reinforced concrete covers for the wells and put in chokes."

When the news came, volunteers started to create a local defence force. John Gornall recorded some years later that

"the force was organised as a company of three platoons, each with two rifle and one light automatic machine gun sections. Discipline was perhaps not that which would have been expected of such a unit in the regular army but, for most part, individuals took their training seriously and on two evenings each week the stamping of feet, the click of rifle bolts and shouted orders coming from the parade ground showed where they were learning the mysteries of drill and the rifle, the intricacies of the light machine guns and the art of throwing hand grenades" [Gornall, 1965, p. 45].

The mounted camel guard on duty at Bahrein refinery, 1938 (*Bapco archive*).

Tom Berry remembers one particular incident during a meeting of volunteers:

> "Mike Bush, a petroleum engineer, issued us all with British army rifles, Lee Enfields, and showed us how to load them. A chap called Chandler said he knew how to use one, and tried. A bullet went bang and ricocheted up through the wall, into the next office and blew a hole in the middle of the table. We all rushed in there, but the Bahraini who had been sitting at the table had disappeared. He came back about ten days later and asked, 'Is the war over yet?'"

The Italians Bomb Bahrain

An unsigned Socal memo, written just before war was declared, recorded that although the Japanese remained very persistent, the great potential danger to Arabia lay in Italy's aspirations. Whoever the author may have been, he would not have been surprised to learn of events in Bahrain on the night of the 19th of October 1940. At about 3.15 a.m. the refinery was bombed by the Italians.

> "Two, or possibly three, machines approached Bahrain from a westerly direction and after circling over the refinery at a height between two thousand and three

thousand feet [609.6 and 914.4 m] dropped salvos of bombs, a number of which failed to explode. Some 84 bombs were dropped, all of them of small calibre. They fell very wide of the target and there were no casualties and no damage was done" [Weightman, *ARPGPR*, 1941, p. 33].

Various accounts followed as to who was responsible. One story is that the Italian priest, Rev. Fr. Irzio Luigi Magliacani, who was living in Bahrain at the time, had something to do with the episode. After the bombing raid, public feeling ran so high that it became impossible to guarantee his safety, and the Political Agent deported him to Bombay on the 13th of November. Not long before the raid, the two large refinery flares which burnt waste gas from the plant had been moved. The rumour was that the Catholic priest had told the Italians to bomb between the two lights. He did not know that one of the flares had been repositioned outside the installation.

Another account suggests that the planes, rumoured to have been led by Count Ciano, Mussolini's nephew, circled the refinery several times apparently trying to locate where the most damaging attack could be made, their information having been received from Italian tanker captains who had loaded cargoes shortly before Italy entered World War II. From the wharf the captains would have seen two flares. One was a low-pressure flare burning near the crude still; the other was further south near the cracking plants. West of that there was the huge coke pile. Ray Andresen recalls that the general considered view was that the position of the flares made the refinery very vulnerable to being bombed.

"They could have used those two flares as a guide to cut the refinery right in half with a bombing raid. So we moved the flare in the north part of the refinery way down beyond the south part and put the south flare underneath the effluent ditch that went out to the Arabian Gulf so that we could burn the gas on the waterway and beyond. In this way, we still had two flares."

Since the information given to the Italian bomber pilots regarding the position of the flares did not coincide with what they found, they had to decide whether to act on the information that had been relayed to them by the tanker captains or use their sights and bomb where they saw the flares. They chose to bomb between the flares, but as one was outside the refinery, no damage was done. Instead, the eighty-four or so shrapnel devices were dumped into the petroleum coke pile. Andresen added that if they had been dropped in the refinery, more than likely the shrapnel would have cut through the piping and the pressure vessels to create a fire ball.

An equally colourful rumour which circulated suggested that Mussolini had wished to get rid of Bilboa, one of the high-ranking officers in the Italian Air Force, who was therefore despatched on this mission and directed to pick up gasoline at various stops going south. The story goes that, as he had heard before he left Italy that at the last stop there wouldn't be any gasoline for him, he loaded less ammunition and more fuel so

Aerial view of Awali town (*Bapco archive*).

that he could continue his flight to Eritrea and safety after having dropping his payload on Bahrain and Dhahran.

Phil McConnell, a veteran pioneer, later wrote that

"in retrospect, it was a comic opera affair, although we didn't think it was then. We believed that no enemy could attack us because we were too far from Europe. I woke in the middle of the night, hearing the roar of the explosions. I immediately concluded that they were bombs and called to my wife to get under the bed and not to turn on the light. In my half-awake state, I struggled into a pair of pants [trousers], then concluded that those were too good to be worn in this situation. I finally dressed myself and dashed down the hall of this big gutch house just as one of those planes roared overhead.

"I saw [Ward] Anderson, the Assistant Manager in charge of refining, just coming out of his house. He jumped into his car and roared off to the refinery. I jumped into mine and roared down the street and out toward the field. There were lights every place, leading to the refinery from all the streets, making it a great beacon of light. As I drove toward the oilfield, suddenly the thought struck me that there might be paratroops out there. I said to myself: 'What do you expect to do when you get out there?' I began to wake up and indicating some common sense, turned the car around and came back just as fast as I had gone out. I drove to the house of Milt Lipp, the Manager. He was out front in his pyjamas. As I roared up and slammed on the brakes, he shouted: 'Get those goddam lights out!' That was the obvious thing to do. The people began to gather at [Milton] Lipp's house and representatives from the British military base down-town arrived. I could talk for half an hour about the absurdity of our conversation. We didn't know what it was all about except we had been bombed" [Extract from McConnell's unpublished notes].

Instructions had been given to the operators to "dump" the units if the refinery was hit. This meant depressurizing the units by opening the safety valves and "dumping" the gas to the flares so that a bad fire would not spread.

"Of course when you do that with cracking plants and other high-pressure units, all the gas goes up in the sky maybe 50 to 100 feet [15.24 to 30.48 metres] and creates luminous flames because they have insufficient oxygen. By the time we reached the refinery, the flares were bursting into the air. We talked to all the operators and found out that as far as we knew, we had not been hit very badly, but we didn't know till next morning when we could take a good look. The planes left thinking they had destroyed the refinery. The BBC counteracted the next day and said the refinery was absolutely all right. So, of course, the more the BBC said that, the more the Italians claimed that they had destroyed the refinery. The propaganda continued back and forth" [Extract from interview with Ray Andresen, 12 August 1987].

Response

The bombing changed the attitude of many of the Americans in the area. McConnell, who was in Bahrain at the time, recalls that "a considerable number said they didn't sign a contract to dodge bombs, so wanted to go home. And many did." Bapco offered American employees the choice of remaining in Bahrain or returning home. One pioneer remarked that "I just didn't see me remaining in Bahrain. ... I didn't feel it was our war and they could get along without me. Anyway, I went home."

Soon orders were received from San Francisco and New York that all American women and children should leave immediately, whether they wished to stay or not. All

British women and children could return home if they wished to do so, but for many wives this proposition was not viable. After the outbreak of war the Political Resident in the Persian Gulf had made a Defence Regulation under the Persian Gulf States Emergency Order in Council 1939, requiring British subjects and British protected persons employed in certain oil companies in the region, including the Bahrein Petroleum Company, to continue in their employment. They could not leave unless consent was received. Persons who disobeyed the Regulation were liable to trial and punishment. This regulation also applied to Bapco's Canadian employees. Therefore, although the wives bound by this directive could have obtained permission to go "home", their husbands could not. In any case, the option offered the British wives less safety since the United Kingdom was virtually under siege at the time.

In the previous July, the Germans had invaded the Channel Islands. Winston Churchill, celebrated for his oratory, had told Britain's lower house of parliament, the House of Commons, that with "rifles and bayonets in their hands", Britain's soldiers were ready to resist invasion. Just a few weeks before the Bahrain refinery had been bombed, the "Battle of Britain" had been fought over England, the London blitz had begun, Japan had aligned itself with the European Axis and, during October 1940, the battle for control of the Atlantic convoy routes had intensified. For those who did choose to leave, the company fitted out a tanker with beds, washrooms and food so that it could take passengers to Bombay. From there, American personnel joined the *President Garfield* and set sail for the USA, via South Africa, arriving in New York on Christmas Eve 1940.

For those who stayed in Bahrain, the priority was to ensure that the refinery remained operational since, at that time, it was (with Iran and Iraq), one of the three major oil producing countries of the Arabian Gulf. Precautions against attack were taken, since there were real fears that another might occur. Within ten days of Bahrain being bombed, the Italians invaded Greece. By the year-end, the first major British land offensive against the Italians had been fought in the Western Desert of North Africa. By early 1941, an advance guard of General Erwin Rommel's Afrika Korps had landed in Tripoli with a panzer division specially trained for desert warfare. No one was certain how German and Italian aspirations might affect the Bahrain and Dhahran oilfields, or the whole of the Arabian peninsula region.

After the raid, the Bahrain refinery was shut down temporarily so that it could be blacked out. The lights were concealed and a new system installed so that the refinery could operate without it being seen at night. Some fake tanks with lime-wash were installed south of the refinery to act as decoys for enemy aircraft. John R. Gilmour, Petroleum Department, London, informed his colleague Lumby at the India Office:

"The active Defence consists of a force of 350 Native Police under control of the British Advisor to the Sheik [the Ruler], who are equipped with motor cycles, camels and ponies and armed with modern rifles and machine guns. There is also a similar force of natives commanded by an ex Iraqi Levy Arab Officer, who will defend the oil fields and the refinery areas. There are also one 3" and one 12-pr. obsolete A.A. guns. There are at present no regular force stationed on the Island, but the question of a regular Garrison is now under discussion between the service chiefs in Cairo" [Gilmour, letter to Lumby, 8 May 1942].

Air-raid shelters were built. Materials were readily available since, just before the war, a large supply of steel plate had been ordered to replace corroded tank floors. Awali was divided into eighteen zones, each with its own shelter. Most were constructed with steel plates covered with sand bags. The swimming pool in the upper camp was covered at the deep end and sandbagged to accommodate between fifty and one hundred people. The caves in the hillside, below House 3 and in the "glen", were used also. Other shelters were made from a combination of *barasti* shelters and drilling pipe.

"Wardens were appointed and Air Raid Instructions issued. In the event of a raid, the warning was to be a long, warbling blast of the siren, signal for all women and children to take shelter and for the men to go to their posts of duty. In the refinery, reinforced concrete shelters were also provided and, in remote areas, ziz-zag breastworks of sand bags were built up in strategic places" [Gornall, 1965, p. 48].

Throughout the winter of 1941 and during the following spring, members of Bapco's management regularly attended meetings of the Iraq and Persian Division of the London Petroleum Board, held in Baghdad. After one such meeting held on the 2nd of May 1942, a scheme for the increased protection of the Bahrain refinery and the Sitra Tank Farm was approved at an estimated cost of £150,000. [3] An officer of the Royal Engineers duly arrived and the scheme was implemented. This included the erection of retaining and blast walls and the sheathing of all storage tanks. Although it was not possible to protect the Tank Farm from direct hits, tanks were protected from possible flying débris by 30 foot (9.144 m) rock walls built around them. [4]

"It had been agreed that the project would be a combined operation. The 10th Army in Iraq was to provide bricks, cement, reinforcing steel and whatever plant it could muster, whilst the Company would be responsible for labour, scaffolding, sand, aggregate and any deficiencies in equipment. In May a small detachment of officers and NCOs arrived in Bahrain and the first shipment of bricks, despatched by dhow from Basra ... was unloaded at Sitra Old Pier.

"The work began on the twentieth May and, for a few weeks, all was confusion. The refinery tank farm, normally quiet and almost deserted, was full of men and bricks and scaffolding. Heaps of sand and aggregate were everywhere and, to crown everything, the Iraqis, who had been brought to Bahrain to act as foremen bricklayers, went on strike. However, order was soon restored and the work went ahead" [Gornall, 1965, pp. 49-50].

During June 1942 a crisis occurred when it was discovered that no more bricks would be sent from Iraq. Instead, a local substitute was used: coral stone from the seabed. Bahraini firms were invited to submit tenders for digging up the coral and bringing it ashore from the sea near Zallaq village on the west coast of Bahrain island.

For some time, Bapco had been anxious to introduce a wireless telephony system between their offices in Bahrain and those of the California Arabian Standard Oil Company on the mainland. As early as the 3rd of November 1938, the India Office had been advised by Bapco's London director H. R. Ballantyne that Casoc wished to install and operate in Bahrain a wireless-telegraphy apparatus which would permit voice and key transmission between the Company's station in Bahrain and the mainland of Saudi Arabia, purely for defence purposes. When the Italians bombed Bahrain, the India Office was still considering the proposal. It was not until the 4th of April 1942 that permission was formally granted, with the stipulation that the facility would operate for the duration of the war only and was subject to withdrawal at any time. Further, Bapco and Casoc were asked to assure the Political Agent that the power of the apparatus would not be greater than that necessary to communicate between Bahrain and Dhahran and that access to it would be confined to carefully selected staff.

The Political Agent insisted on four conditions being observed. Conversations would be confined to matters connected with launches and barges in service between the two companies and no mention would be made of shipping, air or troop movements. Copies of orders issued by the companies and officials entitled to use the facility would be supplied to the British Government, namely the Political Agent and the British Navy. Four specific frequencies would be allowed only, and "silence will be enforced immediately on receipt of orders from the Senior Naval Officer". Eventually, direct radio telephonic communication was established between Bahrain and Dhahran on the 11th of June 1942.

After the Japanese had bombed Pearl Harbor on the 7th of December 1941 and the United States of America had entered the war, communication between Bapco in Bahrain, its offices in London and New York, and Socal's headquarters in San Francisco made it necessary to implement a special code or cipher for inter-company liaison. Permission was granted in early 1943, provided that Bapco supplied the censor with "plain-language" versions of all messages. In addition, all messages containing information about ship movements, especially concerning tankers, had to be sent through the relevant naval authorities or the Political Agent. [5]

Scorched-Earth Policy for Bahrain

Once Rommel's army had established itself in North Africa and intelligence reports suggested that his strategy was to head south for the Arabian oilfields, concern grew that mainland Arabia and Bahrain might be invaded. Details of a scorched-earth policy for Bahrain, to be implemented if the Allies were forced to abandon the islands, are recorded in Foreign Office documents. [6]

During 1942, a notable low-point in World War II for the Allies, the plans included the permanent destruction of all oil wells on Bahrain island and blowing up the Bahrain refinery. In a "Secret Cipher Telegram" despatched on the 16th of April 1942, the Commander-in-Chief Middle East informed the War Office in London and the Tenth Army in Baghdad that he intended to include the Bahrain and Saudi Arabian oilfields and installations within the general schemes for "oil denial". Eight days later, in a *Most Secret* cipher telegram, the Commander-in-Chief informed the War Office *et al*, that he had discussed the oil denial programme with Bapco and Casoc representatives from Bahrain and Saudi Arabia respectively. In principle, it had been agreed to plug forty wells in the outer portion of the Bapco oilfield. In the event of a "grave emergency", the wells could then be immobilized permanently. Two days later, on the 26th of April, the Political Resident Persian Gulf, having consulted with the Petroleum Department (British Government), agreed that the

General Erwin Rommel with his aides during the long-expected offensive in Libya, 27 May 1942 (*Imperial War Museum*).

"mudding-off" of oil wells in Bahrain which were surplus to production requirements was desirable. The Ruler of Bahrain was to be informed confidentially of the company's plans, and, in view of Bahrain's importance from a supply aspect, conditions under which the permanent plugging of the remaining wells should be undertaken were still being considered.

On the 2nd of May 1942, following a conference in Baghdad to discuss the matter, Bapco was prepared to co-operate with the proposal on the understanding that it would receive a formal request from the Political Agent to carry out the programme, that is, the company sought a political and not a military mandate. Four days later, the Commander-in-Chief Middle East informed the War Office that Bapco and Casoc had not agreed to the destruction of oil wells without covering authority from their directors. If this action were to be required, then the Petroleum Department should approach the companies concerned, and not request the Tenth Army to communicate the order to Bapco and Casoc.

Agreement was reached and arrangements were put in hand. At the beginning of June, the War Office was informed that in the event of the situation deteriorating, it was proposed to demolish all Bapco oil wells in Bahrain earlier than originally planned, and to supply the Bahrain refinery with crude oil from Casoc's wells in Saudi Arabia, a plan which would take three days to implement and a twenty-four-hour demolition period. The second stage of the plan would involve the demolition of surplus wells and the permanent plugging and destruction of remaining producing wells. The third stage represented the destruction of all equipment, including the refinery. This would be undertaken only when the emergency became acute and immediate.

On the 26th of July 1942, Ward B. Anderson (then Bapco's General Manager and Chief Local Representative) confirmed the company's compliance in a confidential communication to A. B. Wakefield, the PRPG, in which he noted the company's plans to systematically destroy the thirty-three oil wells still open at that time, should the need arise, and to carry out the other stages of immobilization and destruction if necessary. However, these plans caused the War Office concern. The early destruction of the Bahrain oil wells might create political difficulties, the implication being that the Bahrain refinery was receiving less favourable treatment than Iran's Abadan refinery. When expressing this anxiety, the War Office added:

"It appears to be contrary to security and sound economy to deprive Bahrein refinery of natural sources of supply and substitute a different source more remote and separated by sea. Suggest as possible compromise that as many BAPCO wells should be kept in production as can be permanently junked [destroyed] in about seven days and that the remainder should be permanently sealed early" [War Office, *Most Secret* cipher telegram from the War Office to C-in-C, Middle East, 17 June 1942].

In the end it was agreed that if the situation deteriorated, all remaining Bahrain wells, except twelve, would be plugged and "junked". The twelve would supply 25,000 barrels a day for a limited period. If the need arose, these could be plugged and "junked" during a four-day preparation period for final demolition.

Ingenuity and Improvisation

With the world at war, supplies became scarce. At one point, the situation became very lean indeed. Many of the liberty ships (freighters built in California), though efficient as cargo vessels, were slow and could not outrun the submarine packs. As a result, many of the convoys were sunk en route to Bahrain. When this happened, materials destined for the refinery had to be reordered. In the meantime, as Ray Andresen remembers,

"we improvised … by making parts in the machine shop, saving old pieces that we could reuse again. We kept everything going. It was a hard trial out there in those days. It took so long to get anything from the States. As the quickest time in which we could get supplies was about two months, we had to make sure we had enough in the storehouse when we had shutdowns to tide us over those periods. We had some things with short shelf life and some with long shelf life. Certain things which we needed for maintenance and which wouldn't come up too often, we stocked a minimum amount. Other things which were consumed readily, like chemicals and acids, we had to maintain a constant supply. In a refinery, one of the big items is piping and valves, insulation, tubes and so on. Sulphur and cement came from India. We had to watch the situation all the time to make sure we had enough materials to continue 100-per-cent operations."

When a ship was lost in the Indian Ocean in transit to the Middle East, particularly to Bahrain and Abadan, it usually sank with a consignment of lubricating oil. As a result, a system of recovering lubricating oils and regrading them for lower classification lubrication requirements was implemented.

Ed Howard remembers a few instances which stuck in his mind for many years after the war, because they were unusual. One involved pressure vessels.

"We had lots of machine iron, sheet steel, but we had no means for making a pressure vessel, a cylindrical vessel with a hemispherical cap. The body shop could make what they call a roll-pin head. You took a flat plate, cut it and folded it to make the flange head but we had no way to make the curve. Instead, we took three bolts from the store which had been left over from materials used to manufacture drilling bits. These bolts were 16 inches in diameter [40.6 cm] and 12 feet long [3.66 m]. Once the ends were machined to make them look like rolling pins, they made a frame. Then we took a wind gear from a tractor and covered a small donkey steam engine from the tool shed. We also had a device which would bend steel plate. It looked like the old fashioned clothes wringer with two parallel rollers, one above the other. We did the same thing. We fed in a steel plate, pushed down on it until it began to bend. We worked it back and forth until it eventually made a circle. After that we welded it. It worked like a charm."

Another instance involved the utilization of expended office water chillers, units with a drinking fountain on top and a tiny refrigerator inside. Ed Howard recalls that "in that terrific heat, they would run for six months and then they would give out. There was no way we could repair those compressors. So we decided that we would make a central refrigeration plant and we would pipe distilled water (for drinking) to all the major parts of the refinery." Howard had read about a steam jet chiller technique in one of his engineering journals which, after a complex process, generated chilled water from steam. When the time came to test the system,

"we turned on the steam. I walked around putting my hand on the tank to see if the system was working. My Bahraini crew weren't sure what I was doing but they copied my actions. I would put my head on the tank, feeling it. My Bahraini crew emulated me as usual, until about the third time I tried it, my head man said: 'It's not hot yet, Sahib.' Then I realized they had no idea that this was a new type of chiller. After two hours, we pulled frost on the outside of that tank. They would touch it and lick their fingers and they were just laughing and laughing. They had never seen anything like it!" [Extract from interview with Ed Howard, 28 July 1987].

Bud Machin found himself improvising weld rod coating, important in welding because it keeps oxygen out of the weld so that it does not crack. But first he had to find out how to make it. "I worked with the lab, headed up at that time by Noel Watson, and we tested many solutions and gels with different metals added. We analysed the little bit of weld rod that we had, and we finally came up with a soupy mixture we could put onto the rod and bake it. That got us by until we could get more coated weld rod."

Solving the fire brick problem was an exciting challenge too, the solution of which needed Bahraini know-how. Machin continued the story:

"I had a team of masons. The head Iraqi mason and one of the Arab masons who worked for him lived in one of the villages north of Awali just inside the date palm area. We knew that many years ago they made fire brick in Bahrain, so eventually we found an old man in one of the villages who remembered as a boy going with his grandfather to where they made bricks. He took us to that place. This old man also showed us where to find the mud to make the bricks. After we had experimented with mud samples, we made good brick, but we could never prevent excessive shrinkage or the bricks bending out of shape. I guess if we had ever had to use them, it would have worked, but we never did as our ordered replacement brick arrived on the next boat. I am sorry we never got to test them as it was a lot of fun. Thinking back, I wonder what they did with the fire brick in early Bahrain, where it went and how it was used. But we were too busy making it to think about the early history of it" [Extract from interview with Bud Machin, 29 October 1987].

Another job required the development of repair procedures to keep the rubber loading hoses at the Sitra terminal in safe working condition. That involved experimenting to try and vulcanize raw rubber imported from India in order to strengthen the hose patches sufficiently. The ingredients were liquid rubber, carbon black and heavy linen. Several layers of rubber, separated by linen, were composed to give the patches the necessary strength. Machin remembers that they had to cut out the bad spots, build them up with the raw rubber and carbon black, insert a layer of linen and then some rubber, repeat the process, fill the patch flush, strap it, and finally, surround the hose with tight fitting steam drums. Steam heat and pressure from clamps cured the rubber. "It worked. We kept our cargo loading hoses in safe operating condition for almost eighteen months before replacement hoses were received. The old hoses were being patched up all the time. We got to be pretty expert at it, but we had to learn from scratch. Nobody knew anything about vulcanizing rubber."

The swing-bridge on the causeway linking Muharraq and Bahrain islands, during the 1950s (*Bapco archive*).

Salt in the air corroded very thin meter leads. This problem generated another interesting story. During the war, Caltex in New York issued new specifications to use plastic meter leads and plastic pipe. A lot of plastic pipe was shipped to Bahrain, but it didn't take long to find out that the bugs liked it enough to eat holes through it. So the engineers had to revert to using metal meter leads.

Bapco found itself assisting with the completion of the Muharraq-Manama bridge, construction of which had been interrupted because girder shipments en route from England had been sunk. Military aircraft which landed at Muharraq airfield needed aviation gasoline supplied by the refinery. The new bridge, linking Muharraq island with Bahrain island, was complete except for a couple of main girders. Bill Steel, the State Engineer, visited Bapco and asked the company to help the Bahrain Government to finish the construction of the bridge. It transpired that Bapco could utilize a large pile of worn-out boiler tubes, now rusting in the producing field, to improvise the manufacture of the remaining sections. Steel was contacted and told that Bapco would complete the job, but the bridge would need two welded trusses to be inserted.

The State Engineer is reported to have told Bapco to conform to British standards and use rivets. Bapco's engineer assigned to the project told him that the company had given up riveting tanks years ago. The bridge was completed by Bapco using the old boiler tubes, and welding.

Food – Shortage, Supply and Oversupply

During the war, there were many food shortages. The local population was no longer able to obtain rice from India and Thailand. Instead, wheat was imported from the United States, but it was whole wheat, not wheat flour. The Bahrainis were provided with rations of wheat and ghee every two weeks. Soon, Bapco management found out from other workmen that the Bahrainis could not digest this kind of wheat. So what could be done to solve the problem?

The answer was that Russell Brown, Ward Anderson and Ray Andresen went into the flour milling business. They read up the process in books and encyclopedias. Having found the biggest available emery stone, Andresen informed the chief draughtsman that he wanted to grind flour. The outcome was that a flour mill was built in Awali. "It got to the point that we were running it twenty-four hours a day to keep up with the demand for our cracked flour. We had it going night and day. That was in 1943 but it got our employees through the war. It was an improvised machine, but it worked."

On one occasion Bapco's Swiss chef in Bahrain ordered thirty tuns of cheese from New York, a tun being a large cask or vat. The New York office assumed that there had been a typographical error, corrected the imagined mis-spelling of *tun* to *ton*, and then processed the order. Ed remembers that instead of receiving 30 casks of cheese, "we received 60,000 pounds – 30 tons, 2000 pounds each. What could we do with it? We expanded our cold storage as far as the physical space and the capacity of the compressors allowed. Somebody in the New York office got the 'felt pen' for that mistake." After this, the President of Caltex in New York sent a cable saying that if any more cheese was ordered, the guy responsible would be fired. Ray Andresen commented: "The Kraft Cheese Co. in the United States said Bahrain was the best customer it had. Five tons of cheese at a time, nobody was ordering that much."

Fresh eggs were supplied to the commissary, but they did not remain fresh for long since they spoiled quickly in the heat. Claire Raven remembers that the storekeeper would replace them, but the bad eggs had to be returned in a jar so that he could see how many had to be replaced. However, since you couldn't tell they were bad until they were cracked, it was hard to tell how many eggs were in the jar. "Several methods were devised to determine a spoiled egg, but a spoiled egg is a spoiled egg. The strange thing was that there really wasn't an odour because as the eggs went bad so quickly, a smell did not have a chance to develop."

Claire Raven helped to solve another problem. Although the Awali bakery bread looked perfectly crusty on the outside, it would be soggy and sour inside. The term for this condition is 'rope' or 'ropiness'.

"I had one of my old textbooks with me which described this thing to a tee. Nobody knew what to do about it. It is a yeast or fungus which occurs because the containers are not sterile. They may have been washed clean but this organism wasn't killed and there was no way to get rid of it without using very high temperatures, or acid such as vinegar. So I sent my little book down to Bill Goodyear who was Head of the Commissary. He teased me about this for years, because I solved his problem. Vinegar is one thing which will kill the fungus. It was a big problem for the whole of Awali because the fresh bread was as bad as one day-old bread."

Wartime Communication

A daily activity was to gather before the short-wave radio and hear the BBC news from London. The reception was not always good, but it was always audible, particularly from the loudspeakers. Ed Howard remembers that the whole community would gather to hear the news. "We used to get American newspapers by mail. The most striking part about *Time* was it always came out showing the progress of the Germans moving eastward toward the Middle East. They were trying desperately to get to Iraq and Iran. The Japanese were moving up through India in the other direction. We were right in the middle. Particularly during the middle years of the war, Bahrain was very tense." Many people subscribed to the Sunday edition of the *New York Times* also, though it took as much as two months to arrive.

During the war, cable communication was maintained between Bapco and New York. There was no telephone or air mail service to the United States, although there was irregular sea mail. Letters were censored. All letters were opened and resealed and didn't always come in the order in which they were written and mailed, some taking as much as six weeks to arrive. Numbers were cut or blacked out, including weights and dates, because it was feared they might be code. For instance, when Vern Raven returned to Bahrain, he wrote to his wife informing her of the route he was due to take. "He told me to look at a certain article in the *Reader's Digest* which he thought I would find interesting. It was an article about travel and there was a map of the route by which the author had flown out to the Middle East. This was exactly the route Vern took. All he could say was that he thought I would find this article interesting."

Muriel Smith used to work as a decoder at the Naval Base in Juffair during the war. "Each numbered code represented a letter of the alphabet, which was changed all the time. A lot of it we didn't understand, of course. I'll give you an amusing example. We had a Rear Admiral, Cosmo Graham, stationed there for a while and a message came through for him one day to be coded and sent out. The message was: 'When is my baggage arriving?' The baggage was his wife. We found that out afterwards."

On one occasion, Bapco became involved with wartime communication of an unusual kind. A British patrol boat had run onto a reef just outside *Mina Manama*. The officer complained that the buoy was in the wrong place, but since they had no way to survey the Gulf to find out where that buoy was, Bapco was asked to assist:

"It was the middle of summer. We couldn't do a thing in the daytime then. The water would shimmer so much you couldn't take a sight. So we posted one of our people down at the Manama harbour and one over at the Portuguese Fort. Then John and I [Ed Howard] and one of the boys went out in a boat, the owner of which thought he knew where all the shoals were. Our object was to tie up to the buoy, then fire a flare-pistol into the sky at which point these two people would take sights on us and triangulate our position. The trouble was that the guy who took us out in the boat didn't know as much as he thought he did and the tide went out and we fell on the reef. We had to wait six hours until the tide came in to get off the reef.

"This was the middle of the night. I had people over at the Portuguese Fort watching with an eagle eye for some signal. When the tide came in and we were refloated, we decided we weren't going all the way back to Manama, but would land on the shore by the Portuguese Fort. We signalled with our pistols to let them know that it was all over. We went into the beach right below the Fort, landed the craft and climbed up that steep bank only to find ourselves looking into the muzzles of three rifles held by three Bahraini mounted soldiers of the camel corps. I said: 'They think we are an invasion from the war'. Abdulla chatted with them for a while. Gradually, as we watched them in the moonlight, they relaxed and the guns came down. I said: 'Abdulla, what did you tell those guys?' He replied: 'I told them it was all right, we were just signalling to ships in the sea'. Signalling ships at sea was the worst thing we could have been doing" [Extract from an interview with Ed Howard, 28 July 1987].

Market Sharing

At the beginning of the war, the Bahrain refinery faced an "outlet" problem. Despite the loss of Romanian supplies, there was ample refining capacity in the Middle East, especially as Egypt was meeting an increasing proportion of her needs from indigenous oilfields and refineries. Haifa refinery had been constructed and Abadan refinery had been developed to reach a throughput capacity of thirteen million tons a year. Since the end of 1938, the Bahrain refinery had been capable of handling over one million tons of crude oil per year supplied from its own field, together with approximately 5000 barrels of Arabian light crude oil per day which had been barged from Casoc's Dammam field in Saudi Arabia. After the middle of 1940, particularly after the adoption of a "short-haul" policy, the problem in the Near East and the Arabian Gulf was not one of raising production but sharing limited outlets.

From the 15th September 1941, Bapco curtailed Bahrain's oil production to 15,000 barrels a day, although it was increased to 18,000 barrels per day at the end of the year. It was understood that Casoc had cut crude production in the Al-Hasa field by 25-per-cent. The External Affairs Department of the Indian Government in Delhi reacted sharply to this news:

A section of the Bahrain oilfield and former Jebel Camp, circa 1940 (*Bapco archive*).

"It is improbable that the curtailment of production is due to any lack of market demand, considering the difficulties experienced in building up a petrol reserve in India. It must then be due to a shortage of tankers, and a state of affairs which results in one-third or more of the Bahrein refinery capacity being left unused is unfortunate" [memo to R. T. Peel, India Office, 18 November 1941].

Although the proportion of Bahrain kerosene and petrol imports into India for the five months to August 1941 was higher than for the similar period in 1939, the fact remained that India was short of fuel. Equally, the India/Bahrain/India round trip was considerably shorter than the voyage to Abadan and back. Whilst it was recognized that everyone "must suffer" as a result of the tanker shortage and that APOC had a right to equal consideration, if a saving of sea-time could be effected, it seemed that much could be said for taking the maximum possible lift from Bahrain. However, it was not long before this debate triggered a dispute between the Anglo-Iranian Oil Company (AIOC) and Caltex over their respective markets. By the middle of 1942 an agreement was reached to settle the argument. AIOC's Abadan plant would be held to a production limit representing only 70-per-cent of its potential. Bahrain's output was held to the average of the previous three years, which allowed the refinery to work at 90-per-cent capacity. In the later years of the war, Abadan was able to meet almost all new demands in the Indian Ocean area. In addition, from the spring of 1944, it was able, together with Bahrain, to contribute to the supply of the United States Navy in the South-Western Pacific.

The fluid catalytic cracking unit (FCCU) under construction in November 1944 (*Bapco archive*).

Aviation Gasoline Prompts Diplomatic Dispute

By the middle of 1942, the United States Government had decided to build additional facilities for oil refining in Bahrain which included facilities to manufacture aviation gasoline. It was contemplated that once the plant went into production in July 1944, it would produce 5800 barrels per day of 100-octane aviation fuel. The project, consisting of a Fluid Catalytic Cracking Unit (FCCU), an isomerization plant, an alkylation plant and an acid treating section, would remain the property of Bapco once it was built, although the US Government was to lend US$13.4 million out of a total estimated cost of US$18 million. [7] In the end, Bahrain produced very little aviation gasoline, as the project fell far behind schedule and was not commissioned until almost the end of the war, after which demand for the fuel fell. However, the FCCU's construction was a major achievement. It was a massive project which involved the Defense Plant Corporation, a US Government agency, and the Bechtel Corporation. It also caused a diplomatic argument between the United Kingdom and the United States, sparked off by the need to increase substantially the number of American personnel resident in Bahrain.

On the 8th of February 1943, C. W. Baxter of the Foreign Office was asked by the

American Embassy in London whether the establishment of an American Consulate in Bahrain would be agreeable to the British authorities. Some six weeks later, the American Chargé d'Affaires, H. Freeman Matthews, read that:

"We have given very careful and sympathetic consideration to your letter. ... We have hitherto not permitted any foreign Consuls or other Government Agents to reside in these Arab States [countries of the Arabian peninsula region] and for many reasons we should not wish now to alter the attitude which we have consistently maintained on this point. For instance, it is evident that, once a Consul from one Power was admitted, claims by other Powers on most-favoured-nation grounds or otherwise would almost inevitably follow. ... Further, if a foreign consul were to be appointed, difficulties might arise over the question of jurisdiction ... we should not wish to give any scope on the part of foreign powers for advancing such claims. ... This system, we are convinced, has in the past greatly contributed to the maintenance of peace and good order. It had enabled the British representatives in this area to exercise their influence to the fullest extent. ... We therefore hope that the United States Government will be content not to press for an alteration in the existing arrangements ..." [Peterson, letter to Matthews, 18 March 1943].

The Americans were *not* content with Maurice Peterson's three-page homily. The Secretary of State for Foreign Affairs at the Foreign Office, London, was lobbied. Whilst the nature of the relations between the British authorities and the Ruler of Bahrain was understood and appreciated, the American State Department "calls attention to the fact that American economic interests have for some time been substantial in Bahrein and, as previously explained, will be enlarged in the immediate future. ... It might be well to note here that the American authorities have already been caused no little embarrassment by repeated, severe criticism for not having provided consular facilities for the citizens of the United States living in Bahrein and for the American seamen and merchant vessels stopping there" [US State Department, secret letter to Rt Hon. Anthony Eden, 12 April 1943].

Two weeks later, the PRPG informed the Secretary of State for India that, in his view, the British should press for opposition to the American proposal to establish a consulate in Bahrain on the basis that there was no justification for the post. Moreover, the company was not American but British. "If we permit Consul actually to reside at Bahrein we must do so in certain knowledge that we are undermining our whole position in the Gulf" [PRPG, telegram to the Secretary of State for India, 27 April 1943]. Neither side was prepared to give way. Nearly three months later, the dispute was no nearer resolution. "We cannot say whether the Americans will return to the charge, but for the moment danger of appointment being made is not imminent" [Secretary of State for India, telegram to PRPG, 14 July 1943].

The problem was not resolved until well after the war had ended. Meanwhile, on the 2nd of September 1944, the United States of America opened a Consulate in Dhahran.

Parker T. Hart was appointed Vice-Consul. Assisted by C. J. McIntosh, he paid regular visits to the Bahrein Petroleum Company's premises in Awali, Bahrain having been included informally by the United States Government within the Consulate's jurisdiction.

Other Wartime Developments

In 1938, a programme was implemented to augment Bahrain's crude supply by using barges to shuttle crude oil continuously from Saudi Arabia to Zallaq on the west coast of Bahrain. Later, Bapco obtained two barges from Basrah and tug-boats from Gray Mackenzie, the shipping agents in Bahrain. The barges made daily trips between Dhahran and Zallaq. A small pump station and a storage tank were built at Zallaq using second-hand material.

Although no new operating facilities were installed during 1940, major improvements or additions were made to the power plant, crude distillation units, SO_2 Plant, cracking plant and the pitch pond. Alterations were made to the wharf facilities at Sitra and Zallaq, where the barges from Saudi Arabia berthed. Pipelines were re-routed and a new storage tank was constructed at the Zallaq marine terminal.

By 1942 the *Caltex-2* had come into service, carrying about 10,000 barrels of oil on a draft of about 12 feet (3.66 m). Unfortunately, the 20 mile (32 km) navigable channel between Saudi Arabia and Zallaq had never been charted. The known shallow reef running parallel to the Saudi Arabian coast about a mile offshore was a constant hazard. To assist with the avoidance of an oil spillage which might occur if the tanker grounded, Bill Stine (former Manager, Caltex Marine Services) undertook a detailed hydrographic survey of the area to identify a channel which would permit the tanker to navigate safely.

Zallaq barge terminal which operated between 1938 and 1945 (*Bapco archive*).

Hydrographic charts were made, a marker buoy was placed at the safe crossing point in the offshore reef, and the *Caltex-2* went into operation to make consecutive voyages carrying additional crude oil to Bahrain.

Also during 1942, facilities for receiving up to 35,000 barrels of Arabian light crude oil per day were installed at Zallaq, including a deep-water wharf, submarine line, steam plant, pumping equipment and a new 13,000 barrel tank. An additional crude oil line from the Zallaq terminal to the refinery was installed to cope with the estimated 4.429 million barrels of crude which were barged from Saudi Arabia that year. It was during the months May to July that crude runs averaged only 662,000 barrels per month (approximately 66-per-cent of normal capacity) with the result that several sections of the refinery were shut down completely. However, for the remainder of the year crude runs averaged 97.5-per-cent of normal capacity. A new shipping line was installed

Aerial view of the Bahrain refinery, Sitra tank farm and Sitra wharf, circa 1950s (*Caltex*).

between the refinery and Sitra terminal, to cope with the cargo lifting requirements of the 111 tankers handled that year.

By the end of 1944, as part of the refinery expansion programme, a 12-inch (30.48 cm) crude-oil line running from Zallaq to the refinery was completed. The work of laying a submarine pipeline from the Arabian American Oil Company's facilities in Dhahran [8] began during the year, with the hope that the Bahrain island overland section would be complete by early 1945. Partly because of the aviation gasoline project's manpower needs and partly because of other expansion, very considerable additions to the existing accommodation in Awali and Rafa' were made. One 52-room and nine 24-room air-conditioned bunk houses were constructed, with the idea that the former building would be transformed into the basic unit of a new hospital for eastern expatriates as soon as materials could be obtained.

Anticipating aviation gasoline production, a second-hand drum plant was installed near the Sitra Tank Farm, having been imported from Cairo. Replacement and missing parts were designed and manufactured, including some of the grinders. Eventually, it worked so well that as many as a thousand drums were made in an eight-hour shift. It became quite a showplace where several conveyors reduced the manual work involved and speeded up output. John Creecy who ran the plant remembers that:

> "it was quite a thing. They had a man from Texas Co. come over to start it up. I'd never made a drum in my life. But I tell you, we transformed the drum plant. Even the Bahrainis who worked down there didn't think it was part of the refinery. They thought it was completely different and separate from it. The same rules at the refinery didn't always apply at the drum plant. Julius [Fifer] used to get on me all the time for having my own rules. … But how we made those drums!"

A few months before the war ended, Zallaq terminal received its last cargo on the 3rd of March 1945, a total of 1,831,053 net barrels of crude oil having been handled by the facility before it was abandoned and dismantled in favour of the now completed Arabia-Bahrain (AB) pipeline. Although it had taken longer to complete than planned, the aviation gasoline project was commissioned finally in mid-1945. By the autumn, 322,997 barrels of the product had been manufactured, but after the war ended, demand dropped so rapidly that all aviation gasoline tankage was full by the end of October. The plant was shut down and remained idle for the rest of 1945. Now Bapco braced itself for the worst. The rumours were that the refinery's future hung in the balance.

The drum plant (*Bapco archive*).

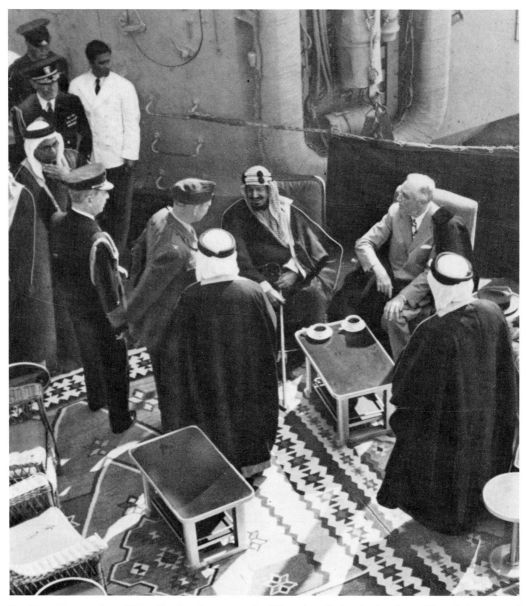

President Roosevelt and King Ibn Saud meet on the US cruiser *Quincy* in the Great Bitter Lake, part of the Suez Canal, in 1945. Admiral Leahy and Colonel William A. Eddy are standing to the left of the King (*Popperfoto*).

CHAPTER EIGHT

Modern Bapco Blossoms

The World Realigns

It was not until the end of the Yalta Conference in early 1945 [1] that President Roosevelt informed the British Prime Minister, Winston Churchill, that he had made arrangements to visit King Ibn Saud on his way home. According to Colonel William A. Eddy (the first resident US Minister in Jeddah, who had made the meeting's arrangements), Churchill was "thoroughly nettled" at being up-staged by the Americans. As a consequence, he "burned up the wires of all his diplomats" with instructions to arrange a similar audience, but, to his chagrin, he was too late to get there first [Mosley, 1973, p. 123]. On the 14th of February 1945, King Ibn Saud and President Roosevelt met aboard the US cruiser *Quincy*, anchored in the Great Bitter Lake in the Suez Canal area. Just three days later, on the 17th of February 1945, Churchill met King Ibn Saud at the Sayoum oasis.

It was during Roosevelt's audience that US-Arab relations were discussed, including the Palestine question. The American President had made the promise that there would be no change in policy without prior consultation with the parties concerned. Two months later, on the 12th of April 1945, Roosevelt died, and his pledge died with him. Soon afterwards, President Harry S. Truman announced a change in American foreign policy which was to have a profound influence upon post-war oil affairs in the Middle East.

Just a few weeks later, on the 7th of May 1945, [2] the German Army Chief of Staff and emissary, General Alfred Jodl, signed the instrument of Germany's unconditional surrender. The war in Europe, North Africa and the Middle East was over. In late June 1945 delegates from fifty countries assembled in San Francisco to sign the *World Security Charter*, an historic document which provided for the establishment of an international peace-keeping body and forum to be known as the United Nations Organization. Within a month, American President Truman, British Prime Minister Churchill and Russian Marshal Stalin had met at Potsdam [3] to chart a course for post-war Europe. They disagreed over where Germany's frontiers should be drawn; the western powers objected to Poland, with Russian encouragement, seizing huge areas of Germany; Stalin cast aside calls for free elections in Eastern European countries. Within a year, when in Fulton, Missouri, Churchill declared that an "Iron Curtain" had descended over Europe, a political barrier which was to divide the continent for the next four and a half decades. Meanwhile, on the 14th of August 1945, [4] President Truman and the newly-elected British Prime Minister, Clement Attlee, announced Japan's unconditional surrender to the Allies. On the 24th of October the United Nations Organization came into formal existence. Egypt, Iran, Saudi Arabia, the United Kingdom and the United States of America were among the countries admitted as members on that day. [5] Six years of armed combat and bitter struggles had changed the world map, particularly in Europe. Barely a country around the globe had been unaffected. The time had arrived for major reconstruction and a new economic beginning. However, the way in which this would take shape was by no means clear when World War II hostilities finally ceased.

The Bahrain Refinery Threatened with Closure

In late 1944 and during 1945, when numerous observers believed that the end of the war was in sight, many estimates of probable post-war oil demand and supply were made by various authorities. Almost without exception, these forecasts indicated a readjustment period during which the supply of petroleum and its products would greatly exceed the consuming ability of the post-war civilian industry. As a reflection of all these assumptions, the United States Government cancelled thousands of supply contracts during the twilight months of World War II. It also established a Senate Committee, chaired by Senator Joseph C. O'Mahoney of Wyoming, to investigate petroleum resources and future refining needs. As a knock-on factor of the contract cancellations, widespread unemployment in North America was predicted, accompanied with the belief that this would increase world-wide, after the war ended.

When C. Stribling Snodgrass, Director of the US Foreign Refining Division, Petroleum Administration for War, made his final summary before the O'Mahoney Oil Committee, his remarks included the statement: "There now exists a world-wide surplus of refining capacity in relation to the prewar consumption of finished products" [International Section, *The Oil and Gas Journal*, 29 December 1945].

However, this announcement was old news to the Bahrein Petroleum Company Limited. For, within two weeks of Victory over Japan Day being celebrated, Bapco's General Manager, Ward Anderson, had had just cause for concern. The future of the Bahrain refinery hung in the balance. On the 24th of August 1945, he had received a cipher cable from the company's Head Office in New York. Four days later he sent a translation to the Under Secretary of State for India.

"I understand that you wished to see this. I am sending it to you in longhand in order to keep it *strictly* confidential. Although there is nothing in the cable that is not pretty well recognised by serious minded persons, it would be apt to start a lot of wild rumours" [Anderson, letter to Bird, 28 August 1945].

The Bahrain refinery at night (*Bapco archive*).

The Head Office ciphered message referred to a prior exchange of cables regarding possible wage increases for Bapco employees in Bahrain. As a result, New York had cabled Bapco management in Bahrain:

"we will be very fortunate if all [the] new refinery in Bahrain is not closed down. ... Conditions are such that [the] present is no time to consider raising wage levels or making any adjustments other than possibly a few isolated cases of obvious inequity. ... For your personal and confidential information, [there is a] strong likelihood [that] present bonus may be discontinued December 31 with no compensating adjustments of any kind. ... Suggest you do not put this cable on [the] bulletin, but no objection to your informing individual employees [of the] situation prevailing at this time" [Herron, cable to Anderson, 22 August 1945].

Despite the uncertain future of the Bahrain refinery during the latter months of 1945, two years later its improved fortune prompted the trade press to remark:

Table 1

BAHRAIN REFINERY EXPORTS ACCORDING TO *PRODUCTS*

	1945		1946	
	Barrels	**%**	**Barrels**	**%**
Aviation gasoline	322,997 *	1.8	417,842	1.3
Motor gasoline	5,236,866	29.3	7,941,682	25.4
Kerosene	2,331,684	13.0	4,089,264	13.1
Diesel oil	4,356,653	24.4	6,478,022	20.8
Residual fuel oil	5,637,255	31.5	12,309,814	39.4
TOTAL	**17,885,455 ***	**100.0**	**31,236,624 ** **	**100.0**

* During November 1945, 360,650 barrels of aviation gasoline were sold to the US Navy. Of this amount, 130,414 barrels were shipped during 1945 and are included in the tabulation above. The balance of 230,236 barrels remained in Bapco's storage tanks at the end of 1945.

** During 1946, Bapco also shipped 118,275 tons of petroleum coke which had been produced during 1945's refining operations.

[Source: *The Bahrein Petroleum Company Limited Annual Reports,* 1945 and 1946]

"The significance of Bahrein, the small island in the Persian [Arabian] Gulf, as an oil refining centre has recently been greatly enhanced in connection with the production drive in the Middle East, notwithstanding the fact that Bahrein's own crude output has only grown at a comparatively moderate rate" [*Petroleum Press Service*, November 1947].

A comparison of the Bahrain refinery's exported products during 1945 and 1946, Tables 1 and 2, support the media statement. Bapco's *Annual Reports* recorded total runs-to-stills in the Bahrain refinery having risen to 22,875,136 barrels in 1945, of which approximately 32-per-cent was Bahrain crude. The remaining 68-per-cent comprised imports of Saudi Arabian crude, a total of 15,662,624 net barrels. By the end of 1946, Bapco's reported annual crude run was 34,066,194 barrels, 49-per-cent more than that recorded for the previous year. Imported crude from Saudi Arabia had increased to 25,879,686 barrels, a 65-per-cent jump. Of this, 25,017,463 barrels had been delivered through the new Saudi Arabia-Bahrain pipeline (commissioned in 1945), and 862,223

Table 2 BAHRAIN REFINERY EXPORTS ACCORDING TO *DESTINATIONS*		
	Millions of barrels	
	1945	**1946**
Australia	2.44	4.95
Malay States	0.00	2.55
China	0.07	2.30
South and East Africa	2.19	2.20
Egypt	0.00	2.12
India	2.08	1.61
Other countries	2.21	4.72
Military uses	8.02	8.56
Bunkers	1.14	1.99
TOTAL	**18.13 ***	**31.00 ****

* and **
These figures do not include the aviation gasoline and petroleum coke variations referred to in Table 1, hence the slight discrepancy in total figures between the two tables.

[Source: *Petroleum Press Service*, November 1947]

barrels had arrived by tanker at the Sitra marine terminal's extended and improved facilities (completed in 1946).

The *Petroleum Press Service* noted that during 1945, more than half of Bapco's recorded exports were oil supplies to "military authorities or to vessels calling at the local harbour". Although both of these export "markets" increased in 1946, their relative share had dropped to little more than a third, largely the result of shifting naval demands. However, the proportion of residual fuel oil exports rose in the same year, mainly at the expense of diesel oil and aviation gasoline.

Meanwhile the quantities of refined oil products shipped to foreign destinations rose from some 9 million barrels in 1945 to over 20 million barrels in 1946, a 128-per-cent increase. Following the pre-war pattern, most of the Bahrain refinery's markets for refined products were East of Suez, particularly those countries bordering the Indian Ocean which could be served well by the Caltex marketing organization. Although 1,389,000 barrels of gasoline and other products were shipped from Bahrain to the United Kingdom in 1946, Europe was only a minor market for the island's oil.

Comment on the immediate post-war fortunes of the Bahrain refinery would be incomplete without further mention of the remarkable growth of Saudi Arabia's burgeoning annual oil production, which by 1945 had risen to over 21 million barrels. A year later it had trebled. Remarkably, the Bahrain refinery was the beneficiary of 85-percent of Saudi Arabia's oil exports for 1946. During the following year, Saudi Arabian annual crude oil production rose yet again, to almost 90 million barrels. Also in 1947, a second pipeline, undertaken jointly by Aramco and Bapco, was added to the existing AB pipeline which ran under the sea from the east coast of Saudi Arabia to the west coast of Bahrain island, and then overland to the Bahrain refinery on the island's east coast. By 1948, further substantial oil reserves had been discovered in Saudi Arabia, drawing attention to the unprecedented development which had taken place in that country since the end of the war. For the foreseeable future, a spotlight was to remain on the Arabian Gulf region's relatively young oil industries.

Breaking the Red Line

It was against this background of immediate post-war market growth and the Arabian American Oil Company's rapid expansion of activities in Saudi Arabia that the *Red Line Agreement* came into focus once again. Within months of World War II ending, boardroom battles replaced military conflicts. Oil companies with a vested interest in the growth of Middle East oil exploration, but not party to the *Red Line Agreement*, challenged its legal enforceability. For almost three years, resolute positions were maintained in both conference rooms and lawcourts until the dispute was settled.

The problem began when Compagnie Française des Pétroles (CFP) and Calouste Gulbenkian, parties to the *Red Line Agreement*, sought a reaffirmation of its restrictive practices which prevented member companies of the Iraq Petroleum Company from operating elsewhere in the Middle East outside the IPC cartel. At the same time,

Standard Oil Company (New Jersey) [6] and Socony-Vacuum [7] were restless under the agreement's restrictions which prevented them from purchasing an interest in the Arabian American Oil Company (Aramco), [8] then owned 50:50 by the Standard Oil Company of California (Socal) [9] and the Texas Company. [10] Standard Oil Company (New Jersey) and Socony-Vacuum recognized that reaffirmation of the agreement would have guaranteed both the French and Gulbenkian increased oil production in Iraq, as well as entitlement to a percentage share in any acquisition which they *might* be allowed to acquire in Aramco. As a result, Eugene Holman, President of Standard Oil Company (New Jersey), confirmed their refusal to endorse the agreement's restrictions when, on the 29th of January 1946, he advised the Compagnie Française des Pétroles that under the present conditions reaffirmation of the agreement seemed inadvisable.

In October 1946, citing a need for increased crude oil from the Middle East to meet the greater demand of its expanded markets, the American group declared the *Red Line Agreement* to be dissolved. CFP reacted angrily and, in February 1947, filed a suit in the British courts against the American companies and other partners in the Iraq Petroleum Company. Four months later, Standard Oil Company (New Jersey) and Socony-Vacuum filed a statement of claim with the British High Court of Justice denying CFP's allegations concerning the validity of the restrictive provisions of the *Red Line Agreement*. Although they had abided by the terms of the *Red Line Agreement*, the American group declared that the document restrained trade and was unenforceable in law.

Negotiations towards a permanent solution continued until the 3rd of November 1948, when a new Heads of Agreement was signed and the suit and countersuit were withdrawn. Standard Oil Company (New Jersey) and Socony-Vacuum could now acquire an interest in Aramco. This became possible when, also in 1948, the Standard Oil Company of California and the Texas Company sold 40% of their shares in Aramco so that they could finance the construction of a pipeline [11] from the company's oilfields in the eastern province of Saudi Arabia to the Mediterranean port of Sidon. The result of this sale enabled Standard Oil Company (New Jersey) and Socony-Vacuum to buy shares in Aramco, 30 and 10-per-cent respectively, leaving Socal and the Texas Company with 30-per-cent each. [12] This also meant that the "Red Line" restrictions no longer applied to any oil company which might, in the future, wish to operate in Bahrain.

Post-War Refinery Expansion – A New Focus

While the post-war fortunes of both Aramco and Bapco had been improving during 1946, and the *Red Line Agreement* was being debated, the media had been monitoring economic recovery world-wide. At the end of the year, *The Oil and Gas Journal* reported optimistically:

"It is now apparent that petroleum has assumed a new and far more important place in the economy of practically every nation in the world. ... World-wide shortage of coal resulted in substitution of fuel oil on a huge scale. The wreckage of railroad and inland waterway transportation systems demanded new methods of transport, and outside the United States, the peoples of the world turned to truck and bus transport on a scale never previously approached in most of these countries. The seemingly enormous amount of military transport vehicles and vessels was quickly absorbed in the task of getting the wheels of civilian industry and distribution turning again" [Deegan and Burns, 1946, p. 154].

As part of his summary of findings to the O'Mahoney Oil Committee at the end of 1945, Snodgrass assessed foreign refining operations, construction and probable post-war trends. He noted that refinery expansion during World War II had been concentrated in the Middle East where the region's refineries had manufactured about 53-per-cent more products in 1943 than in 1938. In 1944, the corresponding increase had been 89-per-cent. Snodgrass had noted too that despite the many obstacles facing foreign refinery construction and expansion during the war, facilities built during that period contributed materially to wartime supply. In addition, international oil market economics made it particularly attractive to establish refining facilities at the source of crude-oil supply. In both respects, Bahrain and her immediate neighbour Saudi Arabia were well placed to satisfy the demand.

A view of a devastated refinery in the west of France, near to the River Gironde, photographed from the wrecked tower of a distillation plant (*Imperial War Museum*).

As world-wide post-war economic recovery was being planned, the pre-war focus on refinery construction in the United States of America was expected to swing away to other parts of the world. Four main reasons were identified for this shift. The first was that the reconstruction of bombed plants in Europe and the Far East was seen as a priority, at an estimated cost of half a billion dollars. Much of this funding was to be provided through the Marshall Plan, a scheme whereby the United States and Canada gave aid to European countries to assist their post-war economic recovery. [13] Marshal Stalin, however, eschewed this programme of financial assistance and forbade the eastern bloc countries to take advantage of the Plan. As Peter Tripp [14] remarked in a lecture which he presented in 1966: "there was plenty to occupy the minds and energies of war weary statesmen".

Secondly, it was believed that most, if not all, non-American refineries which had not been damaged by direct military action required complete rehabilitation to replace worn-out and obsolete plant. Although the Bahrain refinery was less than a decade old, its technical efficiency during the war had been determined to a great extent by the ability of engineers to improvise spare parts. The Director of the US Foreign Refining Division remarked that six years, the duration of the war, is a long time in the life of a refining unit, particularly when high rates of throughput between 1939 and 1945 had placed great stress on the physical condition of equipment. Additionally, as rapid technological development had rendered many pre-war installations out of date, modernization schemes to meet the growing gasoline demand had to include the construction of a Fluid Catalytic Cracking Unit (FCCU) in those refineries without one. Fortunately, Bapco had responded to the American State Department wartime policy and constructed a "cat cracker" in the Bahrain refinery during the second half of the war.

A third reason for a shift away from post-war refinery construction in the USA was a recognition that the expected increase in the world's consumption of petroleum products had much greater potential in countries where the per capita use of oil was a small percentage of that prevailing in the United States at the time.

Finally, as part of their national industrial development programmes, several host oil-producing countries had initiated a policy which required foreign oil companies, particularly those which were American and European-owned, to construct refineries within the host countries concerned.

Despite a world-wide surplus of refining capacity at the end of 1945 in relation to the pre-war consumption of finished products, the emphasis on post-war refinery construction was thus expected to move away from the USA particularly to Asia.

The Bahrain Refinery – Its Destiny Redefined

In November 1951, the Chairman of Standard Oil Company of California, R. Gwin Follis, addressing the American Petroleum Institute in Chicago, discounted any possibility of cheap Middle East oil flooding American markets.

R. Gwin Follis (*Chevron archive*).

"Because of the tremendous increase in US domestic consumption, the Western Hemisphere is no longer able to supply the bulk of the needs of the Eastern Hemisphere; it is fortunate that the Middle East can now meet the growing requirements of Europe, Africa and Asia and supplement US production as required" [*Petroleum Press Service*, December 1951, p. 401].

With the output of the Abadan refinery in Iran temporarily halted, [15] Follis concluded that there were only two major refineries in the Middle East: Bahrain and Ras Tannurah. As far as the rest of the world was concerned, Bahrain was now working above rated capacity and processing nearly 200,000 barrels of crude daily while Ras Tannurah, which lacked "cracking" facilities, was producing 170,000 barrels a day. Although the economic success of the Bahrain refinery made impressive reading, few people were aware of the human side to the story.

"Times were very hard for the operating and maintenance crews, most of whom spent their working lives on shift. During the summer, the refinery at that time was no place for the faint hearted. The only refuges with air conditioning were the processing area offices which were closed and locked at 4.30 p.m. each day when the Day Supervisors departed. So in the summer the next shift was left to sweat it out. For some, this started while walking down the path from the bunkhouse at 3.30 p.m. towards the waiting afternoon shift bus, and continued remorselessly until they returned at half past midnight. Salt tablets were swallowed by the handful. Prickly heat was an ever present source of discomfort and usually, by the fifth hour of the 8-hour shift, water had become tasteless and pretty useless" [Crombie, 1990].

Although some of the control rooms were fitted with air blowers, their use often made the situation worse than better. If, in desperation, the blowers were switched on, "the only effect was to produce a blast of hot air, lots of dust, dried insects and on one notable occasion, a dead rat" [Crombie, 1990]. At that time, no refrigerators, changing rooms or appropriate washing facilities were available in the refinery, although barrier cream and a rather abrasive soap were provided. The latter, being akin to sandpaper, had the advantage of being an excellent device for rubbing on prickly heat patches when the itching became unbearable.

A water chiller was located near the control room. This ingenious contraption consisted of a small round tank filled with well water. Butane was routed from the plant through one part of the system to cool the water. Drinking water for tea and coffee ran through a separate coil of copper tubing tied into the utility steam system. After the coil was immersed in a tea- or coffee-pot, the steam valve was opened to heat the water in just a few seconds. Some men preferred to add sugar, tea or coffee to the water before it was boiled, a practice which neither enhanced the condition of the "chai coil" nor the humour of all the "expert" tea makers!

"Apart from the heat, the industrial environment left a lot to be desired. Gloves were provided but apart from the odd asbestos suit, protective clothing was non-existent. Safety glasses or goggles were unknown except when firewatching and hard hats, for a few years, only appeared during plant shutdowns. Ear protection had yet to be invented. The lighting on the plants was generally poor with the dismal atmosphere projected into the control rooms where some grotesque abberation of design had specified that the control panels be painted black. ... When it was very cold, many men wore a pair of pyjama trousers under their work clothes and added a jacket or windbreaker. ... If it was bad at times for the operators, it was worse in certain ways for the shift maintenance crews. They may not have been tied to a particular plant or control room for a full shift ... and were dependent mainly on friendly operators for a cup of 'chai' and a place to sit down. At that time, what the shift crews may have lacked in experience, they made up for in numbers, so it was never lonely" [Crombie, 1990].

Laying a pipeline to the Sitra marine terminal, 1957 (*Bapco archive*).

Aerial view of the Bahrain refinery, circa 1960s (*Bapco archive*).

Bapco operated a bus service on the main routes to Awali, Manama and Muharraq towns, as well as to many of the villages. Other destinations were served by the "jungle bus", an imported American cab and chassis on which a wooden box-like structure had been built. Passengers entered through a wooden door at the rear to find a place to sit on one of the wooden benches which ran lengthways along the sides of the bus. But for many villagers, particularly those who lived on Sitra island, travel was either on foot or by donkey. The donkeys did not enjoy much in the way of creature comforts either since, for the duration of their masters' eight-hour shifts, they were left tethered to the east fence of the refinery, otherwise known as the "donkey park".

For many, it was a strange life, but one which most employees adapted to and accepted. In January 1954, the Bapco workforce comprised 8978 employees, of which 6720 were Bahrainis and 1190 were foreign contract employees (Indians, Goans and Pakistanis). The balance comprised a blend of "western" expatriates, a strange definition since these employees were recruited from far-flung countries such as Australia, Canada, Eire, South Africa, the United Kingdom and the USA. Later in 1954, Bapco's employees earned the praise of the international media which heralded the Bahrain refinery as one of the most significant in the world:

"Bahrain's oil production is modest by Middle East standards, but her refinery is one of the largest at present operating outside the Americas, and she is the most important trading centre in the Persian [Arabian] Gulf. Her orderly prosperity, now based directly and indirectly on oil, and fostered by a benevolent and able administration, is a factor for stability in an area of rapid change" [*Petroleum Press Service*, August 1954, p. 298].

Table 3	
REFINED PRODUCTS MANUFACTURED BY THE BAHRAIN REFINERY 1954	
Products	**US barrels**
Asphalt	692,620
Aviation turbine fuel	1,721,299
Diesel oil	13,839,870
Fuel oil	29,730,456
Gasoline	19,627,614
Kerosene*	4,587,175
Power kerosene**	2,062,873
TOTAL	**72,261,907**

* Domestic fuel such as for lamps and stoves.
** Mechanical equipment fuel such as for tractors.

[Source: *The Bahrain Petroleum Company Limited Annual Report for 1954*]

It was also at this time that Bapco made plans to expand its facilities in order to meet world-wide demand for higher-grade products. Although the programme was later modified substantially, in December 1955 a new 25,000 barrels per day vacuum unit went on stream. The reduction in the expansion programme reflected the growing competition Bapco faced from other refineries operating in the areas where Caltex was marketing Bapco's export products. On the other hand, Bahrain's local consumption of petroleum products, distributed through five service stations and eighteen village kerosene dealers, had risen by 30-per-cent that year. Demand for diesel oil had risen owing to the increased capacity of the Bahrain government's power station, and asphalt was in great demand for the construction of a new runway at Muharraq airport. Meanwhile, the Sitra terminal continued to be a scene of constant activity, for during 1955, 1522 tankers and freighters utilized the wharf's facilities.

T2 Tankers – A New Generation of Oil Transport

"Prosperity consequent on refining and shipping activities, and trade, directly connected with the oil industry, is now of comparatively long-standing in Bahrain – which since the decline of pearling has no other resource except its traditional substantial entrepôt trading and the financial activities of its merchants" [*Petroleum Press Service*, September 1956, p. 347].

For Bahrain, a relatively small archipelago of thirty-three islands in the Arabian Gulf, the availability of loading facilities and tankers has been essential to the development of the islands' oil industry since commercial production began. It will be recalled that, on the 7th June 1934, the *El Segundo* set sail from Bahrain, bound eventually for San Francisco and laden with Bahrain's first cargo of indigenous crude oil due for discharge in Yokohama, Japan. From 1938, following the discovery of oil in Saudi Arabia, Arabian light crude shipments were barged across the narrow strait from the Kingdom to Bahrain island's west coast jetty at Zallaq. From there the crude oil was transferred by pipeline to the refinery. In April 1945, barge traffic became obsolete when the AB pipeline went into service between the eastern province of Saudi Arabia and Bahrain.

Some two years before the outbreak of World War II in Europe, Bapco and Caltex had started international tanker operations. On the 24th of February 1937, Caltex incorporated the Balboa Transport Corporation in Panama. Then, on the 31st of July 1940, Caltex transferred Balboa's ownership to the Bahrein Petroleum Company Limited. Whilst Balboa's business affairs were administered by Caltex in New York and its tankers flew the Panamanian flag, they were manned by Scandinavian officers recruited by a Norwegian captain stationed in the Caltex Copenhagen office. In Asia, a Caltex Marine Superintendent in the company's Bombay office recruited Indian crews and administered the fleet's dry dock repairs, which were also effected in Bombay. During this formative period, 1937 to 1941, the tanker routes were to India and South Africa, with occasional trips to East Africa and Singapore transporting oil cargoes from Bahrain.

On the 7th of December 1941, at 2.22 p.m. New York time, the Associated Press teleprinters tapped out the fateful news which ended the USA's isolation from the Second World War. Just a few hours earlier, Japanese warplanes had bombed the US Pacific Fleet based at its home port, Pearl Harbor in Hawaii. By the end of the day Japan was at war with the United States and the United Kingdom. Japan's subsequent lightning advance into South East Asia and Germany's crippling U-boat attacks on North Atlantic convoys caused enormous Allied merchant shipping losses between 1941 and the spring of 1943. Throughout 1942, the U-boats maintained a punishing toll, consistently claiming over 600,000 tons of Allied merchant shipping per month. It was not until April 1943 that Allied convoy losses were curtailed, the result of escort ships being made available and the introduction of radar capable of detecting submarines. However, in spite of this turn-around, the Allies faced colossal tanker shortages. Thus, between 1943 and 1945, the American Government commissioned nearly 600 identical tankers designated the T2-SE-A1 type. They were large for their time (16,500 deadweight tons) and had a good speed (15-16 knots). They were also the world's first all-welded ships. At the end of the war, the T2 tankers became US Government surplus stock and were sold world-wide in response to the huge demand for industrial expansion.

One purchaser was the Overseas Tankship Corporation (OTC). Incorporated on the 12th of June 1946 to handle the tanker requirements of the expanded Caltex Group, OTC bought forty of the surplus T2-SE-A1 tankers. Bill Stine, who headed the Caltex marine group based in Bahrain, remembers that "this purchase was a Godsend in many ways. Forty nearly new, large, modern tankers at rock-bottom prices, giving rise to great standardization in spare parts, maintenance, etc."

Other than repair and refurbishment, one of OTC's first tasks was to rename the tankers, which had an odd assortment of obscure place names from American history. As Stine reflected, *Wagon Mound*, *Rum River*, *Yellow Tavern* and so on did little to enhance the Caltex image. Hence, OTC tanker names were changed to *Caltex* followed by a city name in the Caltex trading area. Thus a new generation of names emerged such as *Caltex Bahrain*, *Caltex Bombay*, *Brisbane*, *Capetown*, *Genoa*, *Lisbon*, *London*, *Singapore* etc.

Caltex Glasgow moored at Sitra Marine Terminal during the 1940s in the company of an Arabian dhow rigged with a lateen sail (*Bapco archive*).

Shanghai city and the River Whangpoo, circa 1948 (*Associated Press*).

"This proved to be an advertising bonanza with forty ships continuously sailing in and out of port with the huge Caltex star logo painted on either side of the smokestack, floodlit at night while in port."

The tankers were allowed to depart from the United States with Scandinavian officers but they had to have American crews. After sailing to Bahrain to uplift cargo, the tankers continued to Shanghai, where they discharged both oil and crews. Meanwhile, as Chinese crews were taken on board, the American crews were flown home to the US. Like Balboa Transport Corporation, OTC was overseen by Caltex in New York, and the same Norwegian captain recruited Scandinavian officers from his Caltex office in Copenhagen. On the other side of the world, an agent in Shanghai recruited the Chinese crews, while in Bahrain Bapco established offices for the Caltex marine staff – a Marine Superintendent, Port Captain, Port Engineer and Port Steward.

Since OTC had purchased the forty T2 tankers in American dollars and paid all their operating expenses in the same currency, the company demanded, and succeeded in obtaining, dollar freights for the voyages performed, even though most countries were hard pressed for foreign exchange in the aftermath of the war. On the 26th of May 1949 the situation changed abruptly when Shanghai fell to the Communists. Thereafter,

Table 4
CALTEX TANKER COMPANIES ESTABLISHED OUTSIDE THE USA

Nederlandsche Pacific Tankvaart Mij (NPTM)	*Tokyo Tanker Company (TTKK)*
• Based in The Hague • Flying Dutch flag • Dutch officers and crews • Trading to Holland only • 5 T2s	• Based in Tokyo • Flying Japanese flag • Japanese officers and crews • Trading to Japan only • 3 T2s
Outremer Navigaçion de Pétrole (ONP)	*Overseas Tankship (UK) Limited (OTUK)*
• Based in Paris • Flying French flag • French officers and crews • Trading to France only • 3 T2s	• Based in London • Flying British flag • British officers and Indian crews • Trading world-wide, including Bahrain • 22 T2s

One dollar (USA), *(Coincraft, London)*.

Chinese crews were replaced with Indian crews recruited by Caltex (India) Limited in Bombay, and Shanghai ceased to be a Caltex market.

By 1950 OTC could no longer command dollar freights, such were the foreign-exchange problems of most countries. As a result, tanker companies were established outside the United States, and thirty-three of the forty T2 tankers were sold to them, as Table 4 demonstrates. The remaining seven tankers were retained by OTC for trading to those areas of the world where dollar freights could still be obtained, and OTC continued to act as consultant and adviser to the four new tanker companies.

From Aspiration to Achievement in Post-War Bahrain

In many respects the end of the Second World War had been a turning point in the history of the oil industry world-wide. A steady decline in the control exercised by the "majors" began, although this did not become evident for at least a decade. Whilst the normal activities of the oil industry had been retarded by the war, the political consequences of World War II immeasurably accelerated change of another kind. As one eminent scholar noted:

28th July 1956: Egyptians cheer President Gamal Abdel Nasser of Egypt as he drives through the streets of Cairo after his return from Alexandria where he announced the nationalization of the Suez Canal (*Associated Press*).

"In particular, the war profoundly affected the relations between the dominant Western powers and the territories of the Afro-Asian world and hastened the processes that were soon to lead to the formation of new independent countries, altering not only the political map of the world but also the attitudes of both Western and Afro-Asian peoples and governments as well as the political balance of power. Added impetus was given to nationalist and independence movements everywhere, while at the same time the moral and political authority of the West was further weakened. Independence movements gave confidence to popular anti-(western) Colonialism and the evident political and economic strength of the USSR, non-capitalist power, encouraged anti-(western) Imperialism and anti-capitalist forces, causing some governments to take more aggressive attitudes towards the oil Companies than they had previously taken. ... The international petroleum industry could hardly be immune to these far-reaching psychological, political, and economic changes ..." [Penrose, 1968, p. 62].

With a focus on the Arabian Gulf nations, this view was echoed by former British Political Agent to Bahrain, J. P. Tripp, who reflected that "it was not until the early 1950s that the (British) Foreign Office began to take a closer look at the responsibilities they had inherited [16] and to evolve a policy for dealing with the Gulf Sheikhdoms in the conditions of the 1950s when the importance of the Gulf as an oil-producing basin was more clearly realized than in the 1940s, when Arab nationalism had found a new champion in Nasser, and when Communism was a very real threat". [17]

The widespread effects of the Second World War, associated with the advent of technologically-advanced communications, exposed the Arabian Gulf region more and more to foreign influences. Chords of pan-Arab unity struck in the hearts and minds of the Arab nations. At the beginning of the 1950s, Bahrain was the most economically advanced of the Arabian Gulf nations, having had the advantage of being the first country along the Gulf's southern shore to prospect for and to discover oil. A desire for reform was soon to be felt by Bapco management.

In 1953, the Ruler, His Highness Shaikh Salman bin Hamad Al-Khalifa, made one of the first moves to achieve change when he requested that the citizens of Bahrain should be informed regarding Bapco's activities and the company's contribution to the country. Therefore, on the 1st of July 1953, Russell Brown (Bapco's Chief Local Representative) informed the Political Agent by letter that:

"You will note that the 1952 [Bapco Annual] report has been prepared in an entirely new and different form. This was done to comply with the expressed wishes of His Highness, the Ruler of Bahrain, with the intent of producing a report that could be given widespread distribution. The Company has had prepared a sufficient number of copies ... to distribute a copy to each of its employees, to leading Bahrain merchants, Reading Clubs, schools etc. in order that the people of Bahrain can be informed of the Company's operations on Bahrain and its contribution to the Bahrain economy" [Brown, letter to the Political Agent, 1 July 1953].

The accompanying twenty-year review of Bapco's operations since the discovery of oil, the front and back covers of which appear on pages 210 and 211, was striking both in design and the message communicated on the leading page:

"This, in photographs and brief text, is the story of Bahrain, historically and economically important archipelago in the Persian Gulf. It is the story of an ancient people who, under wise, benevolent leadership, have adapted themselves to the demands of progress without sacrificing their traditions, the basic principles of their way of life, or their cultural heritage. It is the story of yesterday blending with today; of the development of a rich natural resource and the effects of this development upon a proud but friendly Middle Eastern race. It is the story of Bahrain."

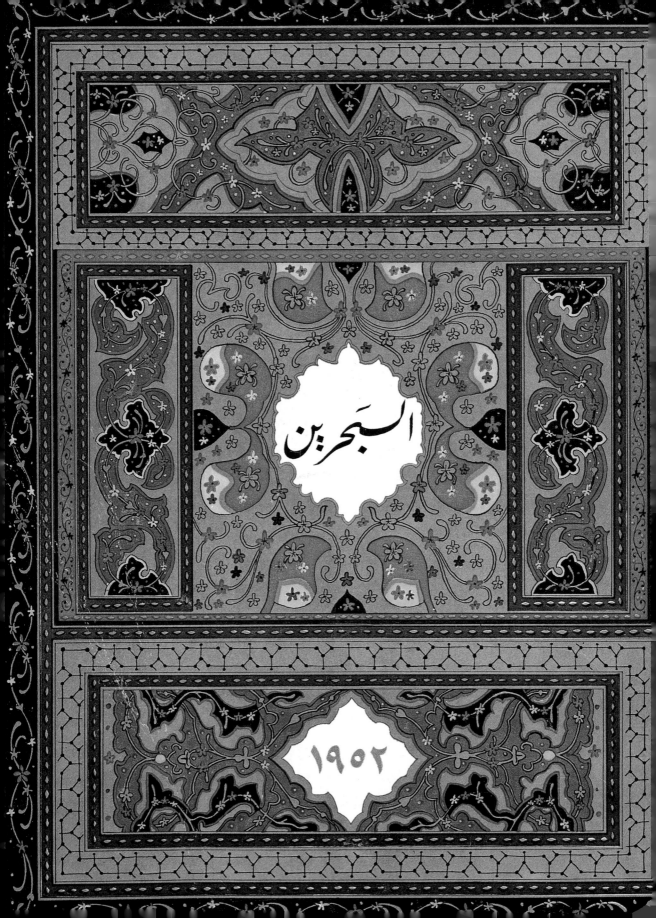

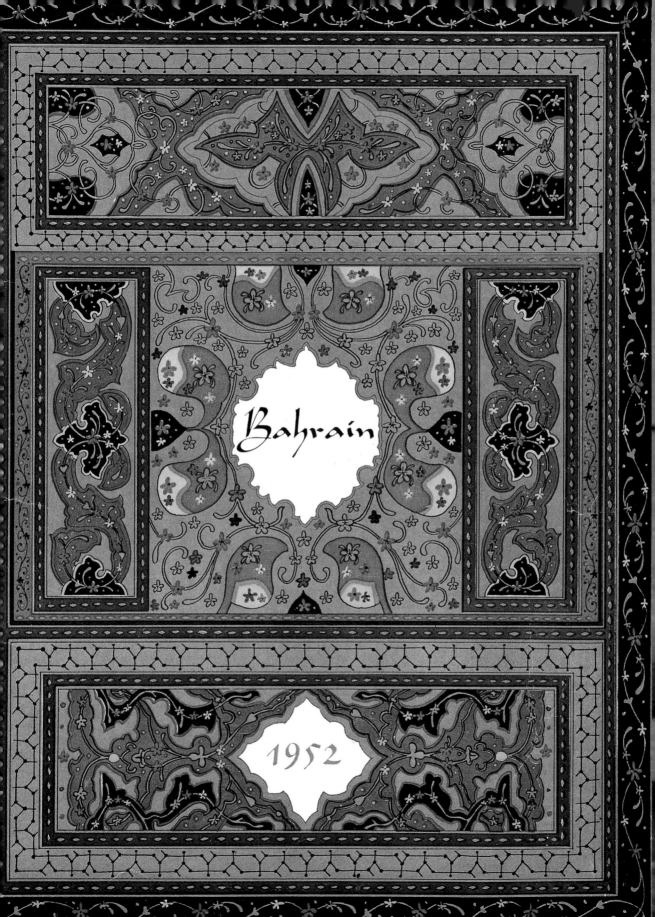

Bahrain

1952

Bahraini Advancement Programme Inspired

It was during the same period also that Bapco recognized that the aspirations of its Bahraini employees were inextricably linked with their wish to acquire advanced training qualifications as a foundation for achievement and promotion on merit, rather than as a privilege. The training of young Bahrainis to become skilled employees had begun after the Government of Bahrain's first Technical School opened in 1938. Part of the curriculum had been designed to meet the growing needs of the nation's young oil industry. The idea was that students should have the opportunity to combine classroom study, and practical work experience provided by Bapco, with a view to possible future employment with the company. Such was the success of this pioneering scheme that, within two years, three of the Technical School's first graduates joined Bapco as trainees, earning 60 rupees a month.

A group of young Bahrainis at work on an automobile engine as part of their training at the Bahrain Government's Technical School (*Bapco archive*).

"If they do well, their monthly wages in six months' time will be at least 100 rupees, and there is practically no limit to what a man may rise to, in the service of an oil company, if he proves himself a reliable and hard-working employee" [*The Bahrain Newspaper*, 13 February 1940].

Three years after World War II ended, Bapco introduced a new training programme to enhance employee education in English and arithmetic. Although it was not realized at the time, from this modest beginning Bapco developed an educational structure which was to have a profound impact on future generations of Bahraini employees. On the 1st of November 1948, a preparatory school comprising two classrooms opened at Zallaq, not far from the former barge terminal and close to several other villages where Bahraini employees lived. Here, thirty-six men attended the first four month course, in which they improved their job-related English vocabulary and mathematical skills. Within four years, the company's school on the west coast of Bahrain island proved to be too small for Bapco's needs, and too remote from the refinery located on the eastern shore of the island. It was decided, therefore, to construct a larger school in a more central location. In 1952, the new Bapco School building was opened in the Awali township, providing twice the classroom capacity previously available in Zallaq, and an auditorium which could seat up to 150 people for group lectures and meetings.

Bapco preparatory school in the early 1950s (*Bapco archive*).

The school continued to blossom. During 1953 alone, 284 Bahraini employees attended the courses. Another achievement that year was the expansion of Bapco's scholarship programme which funded Bahraini schoolboys to receive advanced education and training at the Bahrain Government Secondary and Technical Schools. It is interesting to note that 4500 boys and 2300 girls were recorded as having attended a total of twenty-eight Government schools during 1953, secondary education being available to both boys and girls. The Bapco scholarships provided both tuition and weekly-boarding at the Manama hostel for those students with no access to daily transport from their homes, especially those who lived in villages on Sitra and Nabiih Saleh islands. Also that year, Bapco extended its educational role within the community by building three new village schools and awarding scholarships to successful Bahraini students so that they could attend the American University of Beirut, particularly the summer school programmes.

However, as the Bapco school thrived, Bapco management began to note with great concern that many Bahraini employees left the company's service very soon after completing their craft training. It transpired that far from the Bahraini employees feeling ungrateful to Bapco, the reverse was the case. Recognizing the quality of the training they had received and the skills they had attained, many employees chose to set up in business as building contractors, electricians, plumbers and so on. Others, aware that skilled men of any discipline were scarce throughout the Arabian Gulf, realized that Bapco's intensively trained Bahraini employees were able to offer their services to other employers at higher wages than the company was offering at the time.

It was as a result of this that Bapco decided that the time had arrived to design an innovative programme to encourage and to assist young Bahraini men to make a career in their country's oil industry, rather than seek employment elsewhere, particularly abroad. The Bapco Apprenticeship Training Scheme, commissioned in 1955, became the cornerstone of the company's Bahraini advancement programme. The scheme offered craft training to Bahraini boys between the ages of fifteen and sixteen, the minimum academic requirement being that they had graduated from the Bahrain Government primary schools. Several Bahrainis who benefited from this training reflect that, at the time, it was regarded as a great privilege to be selected for the courses. For, unlike their contemporaries who attended Bahrain Government secondary schools, apprentices were paid throughout their training and were guaranteed employment with the company upon their graduation. As one senior Bapco manager remarked, "what better incentive to achieve could there have been, particularly since many teenage boys in those days were the breadwinners of their families". For many apprentices, Bapco's financial support throughout their training was the only way in which they could have attained their career successes of today.

Bapco – Pioneers of Professional Training in Bahrain

Trainee Mohammed Abdul Karim demonstrating equipment to the Ruler of Bahrain, Shaikh Isa bin Salman Al-Khalifa, during His Highness' visit to Bapco's Vocational Training Centre (VTC). The experiment shows a watt-hour meter being used to measure the electricity consumption of an air-conditioning unit (*Bapco archive*).

Before long, both the Bapco School and the Apprenticeship Scheme were renowned for their professionalism and the quality of their training. Hailed as a great success, the Apprenticeship Scheme was expanded in 1956 to enrol eighty-seven more apprentices. The scope of the training was broadened also to include two new trade groups, Commercial and Operator Apprenticeships, in addition to the Craft Apprenticeship. Such was the enthusiasm for the scheme that it was decided, in 1957, to provide the most outstanding second-year Craft Apprentices with the opportunity to enter a Technical Apprenticeship. This four-year course meant that the total apprenticeship training period was six years. The result was that the entire course provided a combination of classroom study and practical experience, enabling the apprentices to acquire both academic knowledge and the discipline required to practise their chosen crafts. At the same time, camaraderie was a byword, as each apprentice proudly wore the coloured shoulder-flash appropriate to his stage of course achievement on the epaulette of his uniform.

Also in 1957, the Apprenticeship Scheme was further enhanced by the implementation of Bapco's Development Programme. This was designed to provide company employees with new and improved skills and to equip them to gain accelerated promotion. It is now recognized that this impetus laid the foundations for the training of a new generation of Bahraini artisans, including boilermakers, carpenters, clerks, drivers, machinists, masons, pipefitters, tug-boat operators and welders, as well as those who wished to pursue studies which would give them entry into the commercial sector.

By the end of 1958, a total of 2803 Bahrainis had attended full-time training courses at the renamed Bapco school, the Vocational Training Centre [VTC] and the Industrial Studies Section [ISS]. During this year, for the first time, a group of successful VTC graduates was selected to attend short courses in the United Kingdom, whilst a number of Bahraini supervisors had the opportunity to visit British factories and industries as part of an Industrial Study Tour. This progressive move launched Bapco's overseas training programme.

The end of the 1950s was celebrated by the graduation from the first four-year apprentice training course of thirty-two trainees. At a special ceremony held in August 1959, the Heir Apparent, His Excellency Shaikh Isa bin Salman Al-Khalifa (now the Amir of Bahrain), honoured the students by personally presenting them with their qualifying certificates. This achievement was a major contribution towards the success of Bapco's intensive training programmes, typified by the growing number of Bahraini employees occupying positions of responsibility in almost every area of company activity.

His Highness Shaikh Isa bin Salman Al-Khalifa visiting Bapco students in London during 1964 (*Rashid Mubarak*).

1961 proved to be memorable for several reasons, particularly the accession of His Highness Shaikh Isa bin Salman Al-Khalifa as the Ruler of Bahrain. In the same year and for the first time, Bapco created a new course in its Industrial Studies Section [ISS] for those employees who wished to acquire professional qualifications appropriate to various company specialist operations. Bapco's 1961 Annual Report applauded the rapid progress of the Bahrainis who were attending the course and announced that the original three-year target for the programme had been reduced to two-and-a-half years. Upon completion of this level, successful candidates were eligible for overseas technical college and university education, from which many Bahrainis have benefited and returned to Bahrain after graduation to continue their career development. [18]

Secretarial Courses for Bahraini Girls

Much credit for the education received by Bahraini girls in the early decades of the century must be attributed to the foresight of Shaikh Abdulla bin Isa Al-Khalifa who founded the first school for girls in Bahrain. The pioneering work of Mrs L. P. Dame at the American Mission Girls' School, and the contribution of Mrs C. Belgrave towards the establishment of Bahrain Government schools for girls was recorded also in several *Annual Administration Reports of the Persian Gulf Political Residency* and *Bahrain Government Annual Reports*. Interestingly enough, the advancement of Bahraini girls' education received enthusiastic attention during Bapco's first year of operation, 1929-30.

"Only a few years ago the very idea of female education in Bahrain would have roused furious opposition. There are still many who disapprove of it. They say, seriously, as the main argument against it, that they do not wish their women to read or write because if they are able to do so they will correspond, privately, with persons who are not their legal guardians or masters. Apart from the ultra-conservative faction there is a large party of more enlightened Arabs, and foreigners, who of their own accord asked that there should be girls' schools. The first one was started in Muharrak, over a year ago, and owes its existence to Shaikh Abdulla bin Isa [son of the Ruler, Shaikh Isa bin Ali Al-Khalifa] and partly to the well known pearl merchant, Mohomed Ali Zainal whose chief interest is education. ... It is now a flourishing institution. The Manamah school was opened this year, in response to numerous requests. ... The lessons consist of religion, reading, writing and arithmetic and sewing and embroidery. ... The school is very popular and the attendance is as much as can be dealt with by the present staff." [19]

By the late 1960s the education of Bahraini girls had reached maturity. Already many had trained to become teachers. Nevertheless, until that time few Bahraini girls were able to serve in the commercial sector, simply because training in relevant skills had been unavailable in Bahrain. In response to this growing need, Bapco decided in 1965 that it would lead the field once again by offering secretarial training to Bahraini girls.

Since this was an innovative idea, the conditions imposed by the Ruler were that the girls had to travel in a female-only bus and be provided with an appropriate dining facility. The second requirement presented Bapco with a curious dilemma. Already, two separate dining rooms existed, a cafeteria for "blue collar" workers (Grades 1-6) and the other for "white collar" employees (Grade 7 and above). In order to avoid the creation of a third small dining facility, female secretarial trainees were allowed to dine in the "white collar" dining hall and were granted Bapco Club membership. Although the girls did not qualify for these privileges, Bapco considered it appropriate to make an exception in this case to ensure their privacy.

In mid-1967, ten trainees successfully completed a two-year secretarial training course, the first of its kind for Bahraini female employees. The course had included English, office methods, typing and Bapco procedures. Following their studies, the trainees were placed in line positions in various departments, including Marketing, Local Purchasing, Accounting, Personnel and Transport. L. D. Josephson, Bapco's Vice-President and General Manager, was one of the team's instigators. He remembers that about half of the girls were later offered jobs outside the company at much higher salaries than Bapco was able to afford. The fact that this occurred proved that posts in the commercial sector were available to Bahraini girls once they had been given the opportunity to acquire the necessary skills. A further two-year course, which included shorthand, was implemented in 1967 with an additional intake of seven trainees. Before long, several course graduates had been appointed to senior Bapco secretarial posts.

Good Neighbour

To engineers and economists, the hub of Bahrain's industry was the island's refinery. After all, oil was Bapco's business, and the Government of Bahrain relied upon the company to generate a large percentage of its income. But as Bahrain's post-war economic development gained momentum, it was recognized that Bapco expertise and good-will could serve the community well beyond the scope of employment and education within the company alone.

E. A. Skinner, Bapco's Vice President in charge of government relations, paid tribute in 1955 to the Ruler of Bahrain, His Highness Shaikh Salman bin Hamad Al-Khalifa, for his endeavours to bring progress to the people. It is also fair to acknowledge that Bapco played a significant role in this

J. R. Keith, His Highness Shaikh Salman bin Hamad Al-Khalifa and E. A. Skinner during the 1950s (*Caltex*).

C. R. Barkhurst, Bapco's General Manager, presents a mobile dispensary to the Ruler of Bahrain, His Highness Shaikh Salman bin Hamad Al-Khalifa (*Bapco archive*).

process throughout the 1950s and 1960s. A first step, as mentioned earlier, had been the donation of three schools in outlying villages to help the Government satisfy the growing demand for children's education. Health care had become a subject of attention also. Although the Bahrain Government maintained well-equipped hospitals, Bapco supplemented these facilities with five dispensaries and a 110-bed hospital for the needs of nearly 9000 employees. This health programme was a major factor in reducing down-time due to accidents or illness.

"A unique feature of the hospital staff is the small group of Bahraini women who work in the women's wards. This is a new departure in this Moslem country and has proved very successful. Women attendants are a matter of course in Western hospitals, but here the women are very seldom employed outside their own homes" [Roenigk, 1955, p. 105].

In addition to the clinics and the hospital, Bapco initiated several health benefit schemes for its Bahraini employees. The tuberculosis plan, reported to be the only one of its kind in the Arabian Gulf in the 1950s, provided leave of absence and financial assistance to employees during their recovery from the disease, fairly common to the region at that time. Malaria was prevalent too, since the islands' unprotected springs, swamps and irrigation channels were ideal breeding grounds for mosquitoes. As part of an anti-malaria campaign and in co-operation with the Bahrain Government, Bapco crews embarked on regular spraying sorties to eradicate the insects.

In 1954, Bapco instigated another special health service for villagers remote from the Government hospitals. A mobile dispensary, mounted on a six and a half ton truck base capable of being driven on unpaved desert tracks, was equipped with essential supplies and put into community service.

In addition, Bapco undertook a survey of the islands' underground water supply and provided the Government of Bahrain with technical assistance in the conservation of this vital and limited resource. Road building evolved as another contribution to the community, since Bapco depended upon a substantial communications network for the trucks which serviced its drilling, refining and wharf operations, as well as the buses

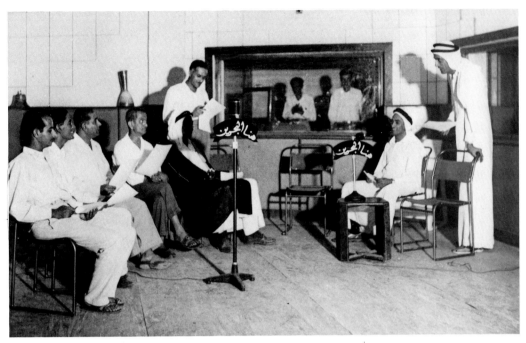

Opening of *Huna Al-Bahrain* (*Bapco archive*).

which ferried its employees to and from work. Communication by airwaves also became a reality when, on the 5th of May 1955, *Huna Al-Bahrain* (Bahrain Broadcasting Station) was opened, most of the equipment having been donated by Bapco Limited as a gift to the Government.

Until recent decades, many Bahrainis lived in *barasti* homes made from palm fronds, or in houses constructed with *faroush*, coral stone slabs cut from the sea. Ed Skinner remembered that when he first came to Bahrain in 1931 the company had tried to plan a special community for local employees which would incorporate modern housing and amenities. However, the scheme fell through when it was realized that Bahrainis did not like the idea of living in a strange new community, away from their relatives and familiar surroundings. By the mid-1950s the company felt that the time had come to help Bahraini employees acquire better homes in their own communities. As a result, it introduced a Housing Loan Plan in 1957 so that employees could remodel or enlarge their existing homes, or build new ones. Any employee with five years' service and aged between twenty-one and fifty-five years was allowed to borrow up to 100-per-cent of his equity in the Bapco thrift fund, which increased in value through employee savings and company contributions. Such was the popularity of the scheme that by the end of the year almost 4 million rupees had been loaned to Bahraini employees.

Eight years later, when the Government of Bahrain conducted the nation's fourth population census, it was revealed that although, overall, 71-per-cent of both stone and *barasti* houses were supplied with water and 81-per-cent with electricity, other essential

Aerial view of Isa Town, circa late 1960s (*Bapco archive*).

amenities such as mains drainage were relatively scarce. Even more concerning was the fact that only 315 (18.75-per-cent) of the islands' remaining *barasti* homes were connected to water, whilst a modest 3.3-per-cent were served with electricity.

In response to the growing housing needs of Bahraini citizens, His Highness Shaikh Isa bin Salman Al-Khalifa initiated his own personal brainchild, the creation of a modern town on land which he donated for the purpose. By the end of 1967, over twenty Bahraini Bapco employees had moved into new houses in Isa Town, having negotiated their purchase through the Bapco Home Ownership Plan which had been introduced in 1964.

A further major development in employee benefits occurred in January 1966, when a pension plan for Bahraini employees was introduced for the first time. All Bahraini national employees aged twenty and over with at least one year of continuous service became eligible to participate.

A year later, a major national event took place in Bahrain which required Bapco's assistance. To bring Bahrain traffic into line with that of most other Arab countries, a change from left to right-hand driving was introduced on the 17th of November 1967. To achieve a smooth and accident-free transition, Bapco set up a special "Go Right" Committee to co-operate with the Bahrain Government Traffic Department. Bapco's Driver Training School was adapted to right-hand driving instruction and made available to Government Departments, the Bahrain State Police (now known as Public Security), driving instructors and company employees. To help publicize the change-over, Bapco produced a tutorial colour film, news articles, photographs and strip cartoons in company publications, posters and bumper stickers. Meanwhile, the radio broadcast "Go Right" announcements and reminders on the art of safe right-hand driving.

Winds of Change

The early 1960s was a period when Bapco's parent company, Caltex, was forced to make radical cost-cutting decisions to ensure the company's economic survival. Not only did this action have a divisive impact upon Bapco's workforce, but also it precipitated a phase during which employees openly expressed dissatisfaction about the social imbalances within the company.

One of the factors which first promoted change was that Bahrain had been the training centre for the Caltex refining network world-wide. Its new refineries in Holland, Italy, France, Germany, Turkey, Lebanon, South Africa, India, the Philippines and Australia had been staffed by personnel trained in Bahrain. A product of Bahraini academic achievement was that the pool of highly qualified western expatriate personnel retained by Bapco to supply staff for Caltex's new refinery projects elsewhere could no longer be justified. Gradually, Bahrainis displaced expatriates, but the pace was slow, partly because it appeared to many that management was disinclined to accelerate the process of expatriate release, and partly because the expatriates themselves became over-protective of their own and their colleagues' positions. Meanwhile, aspiring Bahrainis, eager for promotion and the opportunity to prove themselves, began to interpret the slow pace of change as an imperialistic hold on their career development.

A second "wind of change" was the result of an examination of Bapco's fixed and variable costs. When compared with other Caltex refineries, as well as those belonging to the parent companies Socal and Texaco, Bapco was overmanned. This was the result of a conscious policy to ensure that enough expatriates would be available to keep the refinery operational during times of national disturbance. Nevertheless, an analysis of the fixed costs of refining oil, running Awali town, maintaining a transport system and other peripheral activities associated with a pioneering operation, together with the expatriate support system, comprising housing, schools, recreation facilities, medical care, a commissary, transport, annual home leave and so on, showed that Bapco was at a cost disadvantage when compared with competitors.

Additionally, Bapco was the "swing refinery"[20] for the Caltex marketing companies. It had the sprint refining capacity [21] to make up shortages in the overall supply network when less versatile refineries in the system were unable to do so. Maintaining this flexibility meant that refinery units were idle or under-utilized for some of the time. Although the Bahrain refinery was very versatile – able to supply a number of finished products to markets in Europe, Africa and the Far East, many of which were manufac-

Freighter *Elaine* being loaded with oil drums (*Bapco archive*).

tured in small quantities for just one country – the economics of this service were scrutinized.

It was in 1964 that Bapco's management in Bahrain received a directive from the company's head office in New York to the effect that its workforce in Bahrain should be reduced by 50-per-cent. As had happened almost twenty years earlier, New York's perspective was that without a reduction in staff, Bapco would not survive. Although in 1945 this threat was averted, nineteen years later Bapco was forced to implement a manpower reduction programme.

L. D. Josephson, Vice-President and General Manager of Bapco, recalled that everything went fairly smoothly until a study team arrived from New York. Even before its arrival at Bahrain's airport, news had circulated around the community that a "Chopper Team" was arriving. Soon afterwards, one or two team members aggravated an already tense and sensitive situation by making outspoken statements about the task force's objective to implement the 50-per-cent reduction of Bapco's workforce within six months. News of this proposal was not long in reaching the Ruler's palace, with the result that Mr Josephson was summoned to an audience with His Highness to apologize for the anxiety which had been caused generally and to provide some assurances as to the job security of Bapco's Bahraini employees.

By early spring, 1965, 200 Bahraini Bapco employees had been released. This action triggered numerous demonstrations amongst Bahrainis who took the opportunity to complain about their lot and to protest about their apparent lack of job security. Many Bapco Bahraini employees believed at the time that management had misjudged the situation. On the one hand, Bahrainis felt that, in spite of their overseas education and academic achievements, their jobs were in jeopardy and they would not be given a fair chance of promotion. Equally, they felt that western expatriates were anxious to protect the established positions they had held for many years. On the other hand, many expatriates had been recruited at the end of the war believing, and having been assured, that they had a lifetime career with Bapco. Now the promises which had been made to them by Bapco management were being broken.

Across the entire spectrum of Bapco employees, bitterness was directed towards those in New York, where it was felt that management had changed course with compassionless speed. Today, several senior Bahraini managers who witnessed these events as young Bapco trainees believe that the demonstrations were part of a complex situation which was both politically motivated and an eruption of many feelings against the company. This coincided with a period when newly-qualified Bahrainis were returning from overseas universities expressing other feelings of inequity. They sought vacations on the same terms as western expatriates, the opportunity to live in Awali, higher salaries and improved conditions of service. They did not wish to return to the same jobs they had before studying abroad.

By early March 1965, feelings were running high. Bapco employees became increasingly concerned at the number of redundancies affecting them. Rumours indicating that a demonstration was imminent added to the general atmosphere of unrest. As predicted, scuffles broke out in Manama on the 11th of March, causing the police to control the crowds with tear gas. In Muharraq, Bapco's buses which collected and returned shift employees to their homes during the day were stoned. An employee's diary which chronicled the events recorded that the slogans and attitudes appeared to be directed entirely towards Bapco.

These disturbances marked the beginning of what developed into an intense

Donkey carts delivering kerosene during the March 1965 strike (*Bapco archive*).

three-week strike by Bapco's Bahraini employees. Soon after the episode began, Bapco employees took their case to the Ruler for justice. At approximately 14.15 on the 16th of March, the Government broadcast a message to the effect that His Highness Shaikh Isa bin Salman Al-Khalifa had decreed that Bapco must offer to reinstate all those Bahraini employees who had been made redundant since the 1st of October 1964. This was to take effect on the following day, and the terms of employment should be reinstatement at the same rate of pay. A condition agreed between the Ruler and Bapco was that all applicants for reinstatement should be Bahraini nationals.

Nevertheless, demonstrations continued and reached a peak during the evening of the 8th of April. After darkness had fallen, a group of men entered Bapco's bus park in Manama, punctured several vehicle fuel tanks and then set the compound ablaze. Nine buses were destroyed and several more were severely damaged. By the end of the month calm had been restored, but for everyone the episode had been an enormous shock. Whilst some western expatriates immediately chose to think that the whole affair had been an unfortunate episode, many Bahraini employees were adamant that the recent events would have long-term implications for the enhancement of their stature in the Bapco organization. Once the dust had settled, it became very clear that "business as normal" was both naive in concept and unacceptable in practice. The 1965 strike had been a watershed in the life of Bapco Limited.

Relationships between management and Bapco's Bahraini employees were never to be the same again. Henceforth, management was required to take account of the aspirations and expectations of the Bahraini employees, for whom advancement became a matter of right rather than favour. Another response to the changing times was the formation of the Employee Discussion Committees Programme, a forum created as a two-way communication system so that Bahraini employees could air their concerns and participate in the mechanism of the company's management.

Pressure to accelerate the promotion of Bahrainis and reduce dependency upon western expatriate expertise was maintained. Thus, a dual-incentive redundancy package was offered to expatriates with the idea that they would train Bahrainis for promotion into the jobs they held, and then retire early. As 1965 ended, over 300 Bahrainis, representing some 5-per-cent of all Bapco employees, had been promoted to positions previously held by expatriates, and in that year alone twenty-five Bahrainis had assumed senior positions in seven de-

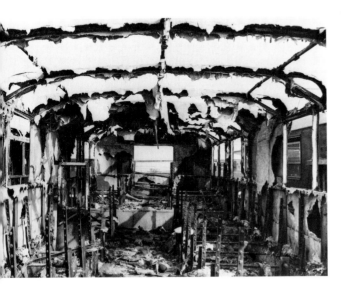

One of Bapco's buses destroyed by fire during the March 1965 strike (*Bapco archive*).

partments. In tandem with the programme of expatriate replacement and Bahraini promotion, natural attrition allowed Bapco to pursue a policy of "Lean Staffing" without causing further strife and turmoil within the company.

The 1965 Census and the Bahrain Development Bureau

As the "winds of change" concentrated the minds of Bapco management, the Finance Department of the Bahrain Government was preparing for the nation's fourth population census. During the preliminary stage, between the 26th October 1964 and the 12th February 1965, a Census Bureau was established to set up the necessary mechanisms. Supervisors and enumerators were selected from Government departments and the Bahrain Petroleum Company Limited was commissioned to design the questionnaires.

Abdul Hussain Faraj (now General Manager of Bapco's Administration Division) recalls his assignment to co-ordinate the design of the Census Schedules and to write the computer programs to analyse the data. The Census Schedules, comprising nineteen questions, were printed on large format paper, 19 inches x 13 1/2 inches (48 cm x 34 cm), coded according to locality and household serial number, and bound in books of fifty. The books also included twenty-five blank schedules to be used for houses containing more than one family. For a month, from the 13th of February, Government officials conducted the census fieldwork, completing their task on the 12th of March, the day after the Bapco strike had begun. Despite the disruption to company operations, the census was unaffected, for within two days of the fieldwork starting, Bapco had begun the arduous task of analysing the data. Abdul Hussain recalls that, in those days, it took four hours to load the computer with 182,203 punched cards – one for each person who had participated in the census. This was particularly significant since the IBM 1401 machine used for this task, installed in Bapco during the early 1960s, was also the first of its kind to be installed in the Middle East. Not only was it capable of reading 800 cards a minute, but also it is believed to have been the first of a new generation of computers to operate with transistor technology.

Answers to the census questions which focused on age, educational achievement and employment generated most concern. The results showed that 79-per-cent of the population – 143,814 people – were Bahrainis, of whom 16,667 were boys and girls between the ages of eleven and fifteen. It did not require a computer to work out that within five years, 12-per-cent of the Bahraini population could be seeking employment, unless they chose to continue their studies. Looking no further than one decade ahead, virtually a third of the current Bahraini population – 46,840 young boys and girls – would have reached school-leaving age. To many this was a startling revelation. To others, it was confirmation of what they had realized already. Due to Bapco's current "Lean Staffing" programme, Bahrain's oil industry was hardly in a position to contribute towards new job creation. Bapco's President, Roy L. Lay, introduced the company's 1966 *Annual Report* with the observation:

"Intensive competition has produced conditions of stringency in the oil industry throughout the world, yet I am able to report another year of achievement. This has been accomplished by planning and teamwork, by the co-operation of all employees and, in particular, by the help and encouragement received from Your Highness and members of the Government. ... It is my hope that this spirit of co-operation and understanding, whose fruits may be seen on every hand in far-reaching social and economic developments and well-planned works, will continue in the years to come."

It was in such a spirit of co-operation that His Excellency Shaikh Khalifa bin Salman Al-Khalifa, Head of the Finance Department of the Bahrain Government at that time and now Prime Minister, implemented informal discussions with senior executives of Bapco and Caltex in November 1966. The Bahrain Government's concerns about the immediate and future problems facing the country focused on the expected rapid growth in the population of young people completing their schooling and entering a labour market with few opportunities for employment. The Government's anxiety was expressed on another equally important matter, namely the levelling off of oil revenues at a time when the Bahrain Government needed funds to continue the welfare pro-grammes it had established.

IBM punch card machine, acquired by Bapco in 1947 (*Bapco archive*).

The talks resulted in the creation of the Bahrain Study Group, a Bapco-Caltex task force with broad powers of authority to investigate and recommend activities which might stimulate the economy of Bahrain. Between late December 1966 and early January 1967, two of the Study Group's members, Yousuf Shirawi (later appointed Minister of Development and Industry) and Hugh Story (Assistant General Manger, Operations, Bahrain), discussed their task. During this formative period, it became obvious to them that a full-time Bahrain Government team would be required to follow through and be responsible for the successful implementation of each industrialization idea. It was as a result of this recommendation that the concept of the Development Bureau was launched.

Table 5

ABSTRACTS FROM THE BAHRAIN POPULATION CENSUS 1965

AGE	BAHRAINIS			ALL NATIONALITIES		
	Boys	Girls	Total	Boys	Girls	Total
5	2,609	2,691	5,300	3,094	3,129	6,223
6	2,475	2,574	5,049	2,865	2,972	5,837
7	2,825	2,730	5,555	3,216	3,129	6,345
8	2,692	2,618	5,310	3,035	2,971	6,006
9	2,048	2,120	4,168	2,347	2,371	4,718
10	2,474	2,317	4,791	2,787	2,619	5,406
11-15	8,669	7,998	16,667	9,991	8,889	18,880
Ages 5-15	23,792	23,048	**46,840**	27,335	26,080	**53,415**
All Ages	72,368	71,446	**143,814**	99,384	82,819	**182,203**

Sir Anthony Parsons, former British Political Agent at the time, wrote in his memoirs:

"The principal questions on my mind through 1966 were the interlocking problems of the internal political situation and the economy. There had been no outbreaks since the rioting of 1965, but the atmosphere was uneasy. … Meanwhile the Development Bureau [formed in July 1967] was wrestling with the associated problem of rising unemployment. The programme submitted the previous year by the British government offered little comfort. Such measures as increased taxation and the establishment of handicrafts and small cottage industries would do little to absorb the thousands of urban school-leavers whose job prospects, or the lack of them, would do much to influence their political frame of mind. The government turned to the American-owned Bahrain Petroleum Company for an alternative Economic Survey. The BAPCO management, long established in Bahrain and conscious of the link between political stability and the continuing success of their operation, agreed. They tackled the job from a fresh perspective, basing their researches not on the limited size of the domestic market but on Bahrain's central position in communications between the Western world and the Far East and on the existence of large quantities of cheap and unexploited natural gas. By the end of the year, their report was beginning to generate optimism" [Parsons, 1986, p. 126].

Once the Bahrain Study Group completed its fieldwork, its report, to which Sir Anthony referred, was submitted in May 1967 with its findings and recommendations. On oil-related matters, the conclusion was that maintaining sprint refining capacity was only a short-term possibility. This would depend upon the refinery's future markets, the likely demand for "spot" [22] business and the economics of manufacturing certain petroleum products. Until 1970, it was assumed that Bapco would continue to supply products to markets east of Suez at existing levels, but that supplies to North Europe would decrease gradually. Previous experience had shown that short notice requirements tended to be winter demand in North Europe for diesel and in Japan for fuel oil. Non-seasonal *ad hoc* demand focused again on Japan, whose demand for naphtha varied, whilst the United States Military demanded a full product range from time to time. Whilst it was hard to predict how often Bahrain would be called upon as a swing refinery, it was assumed that this capacity would be required up to three months a year. However, after 1970, it seemed likely that sprint refining demand would decline. If this occurred it would be hard to justify the economics of retaining such capacity.

As far as new non oil-related projects were concerned, after studying the possibilities of over 200 industries, thirty-six projects were short-listed, of which twelve were eventually implemented. Ranging from heavy and light industry to servicing facilities, these innovative ideas became the core of Bahrain's economic development during the following decade. Two months after the Bahrain Study Group had presented its report, a new era for the country was announced in Bapco's company newspaper:

"Bahrain's economy stands poised to move rapidly ahead on a number of major projects aimed at increasing the gross national product and opening up a wide range of employment opportunities for Bahrainis. That is the significance of last week's official announcement that the Bahrain Government has created a new Development Bureau under the auspices of the Finance Department, headed by His Excellency Shaikh Khalifah bin Salman Al Khalifah, President of Finance" [*The Bahrain Islander*, Vol. 27, No. 49, 12 July 1967].

The Development Bureau, attached to the Petroleum Affairs Bureau, was led by Mr Yousuf Shirawi. The officers assigned to the Bureau were Mr Habib Qassim, Economic Adviser (later appointed Minister of Commerce and Agriculture), and Mr Denis Jones, a process engineer who was seconded from Bapco as Technical Adviser.

Celebrating Change

1971 was a year for celebration in which several significant events took place. The first occurred on the 14th of August 1971, when His Highness Shaikh Isa bin Salman Al-Khalifa declared the creation of the State of Bahrain in a proclamation broadcast to his subjects. On the following day the United Kingdom recognized the new State, agreed to rescind all previous treaties and signed a new Treaty of Friendship between Bahrain and Britain which provided for consultation and co-operation on matters of mutual interest such as trade relations.

Within a month the new State was recognized by Arab and other nations and diplomatic relations were established with other countries. Bahrain joined the League of Arab States on the 11th of September 1971 [23] and just ten days later, 500 guests attended a reception held in New York to celebrate its admission as the 128th member of the United Nations Organization (UNO). In November 1971, Bahrain joined the World Health Organization (WHO) and the Food and Agriculture Organization (FAO) – both sponsored by the UNO.

As an appropriate finale to this memorable year, Bahrain celebrated her first National Day on the 16th of December, the tenth anniversary of the accession of His Highness Shaikh Isa bin Salman Al-Khalifa.

Without doubt, 1971 had been a year of jubilation in Bahrain, and there were great expectations for the future. Not only was the country's oil industry about to experience immense change throughout the decade of the 1970s, but also every sector of Bahraini society was to experience dramatic economic change in some way or another. These ten years were to be accompanied by rapid social, cultural and technological changes. However, before considering aspects of these changes, it is appropriate to pause and review how the wealth was generated over a period of forty years to fuel the transformation of the State of Bahrain's economy in a quarter of that time. Oil revenues, their evolution and enhancement, are the subject of the following chapter.

CHAPTER NINE

Oil Revenues:
Evolution and Enhancement

When Bapco celebrated its sixtieth anniversary in 1989, it had good reason to look back over six decades of achievement and identify many ways in which the company has made a major contribution to the development of modern Bahrain. However, the purpose of this chapter is not to assess the nature and scope of that achievement, but rather to analyse and understand the evolution of the Government revenues which have made it all possible.

Charges and Royalties Defined

As a beginning, particularly for those unfamiliar with the oil industry's financial arrangements, it will be helpful to understand the basic concepts of charges, royalties and taxation, which are often confused. Since corporate income taxation was not introduced to Bahrain until 1952, it is appropriate to give first consideration to the concepts and application of charges and royalties.

The basic premise of a royalty, in oil industry parlance, is fairly simple. Broadly defined, the owner of a natural resource, or the holder of a right to exploit a natural resource, may grant the right to exploit that resource, usually for a specified period of time, in return for a royalty. In the case of Bahrain, the Government is the owner of the country's oil and gas resources. In most cases, the royalty is expressed as either an amount per unit of production or a percentage of the revenue which the exploiter derives from the exploitation. Both situations applied in Bahrain at one time or another, as will be discussed shortly.

In addition to the royalty, separate charges may be made. Frequently they relate to the use of land for the exploration of, or the production of, natural resources. In the oil industry, this charge is called rent, not royalty. Sometimes a charge may be made in consideration of certain services rendered by the Ruler, as explained in the following

His Highness, Shaikh Hamad bin Isa Al-Khalifa, circa 1930s (*Bapco archive*).

paragraph. In this instance, the charge was neither royalty nor rent.

On the 2nd of December 1925, under the terms of the *Bahrein Island Concession* [1], the Deputy Ruler of Bahrain, His Excellency Shaikh Hamad bin Isa Al-Khalifa, granted the Eastern and General Syndicate Limited an exclusive exploration licence for a period not exceeding two years, [2] with a right of renewal for two more years. [3] In consideration of the "assistance and protection to be afforded" to the Syndicate's employees, it was mutually agreed that the company would pay to the Ruler an annual sum of 10,000 rupees. [4] This assistance took the form of various privileges to be enjoyed by the company, such as free access to all parts of the territory under the Ruler's control, "whether private or public property, saving only sacred buildings, shrines and grave-yards", [5] free use of water and fuel lying on the Ruler's property, the right to purchase all necessary food and fuel, as well as imported materials, apparatus and machinery being exempt of duty. [6] Protection from theft, assault and so on was to be provided by guards supplied by the Ruler, paid for by the company. [7]

Should exploration prove to be promising, which in this case means investigation of the *surface* characteristics of the terrain, the concessionaire could then apply for a prospecting licence with a reasonable expectation that the Ruler of Bahrain would grant the request. [8] In that event, the concessionaire would enjoy the same privileges under the terms of a prospecting licence as those granted during the currency of the exploration licence, [9] with the addition of the right to prospect which, in practice, meant the drilling of test wells. [10]

If granted a prospecting licence, the concessionaire would not be bound to make any payment for work carried out on *uncultivated* land, the proviso being that the land would be restored, as far as possible, to its previous state unless a *Mining Lease* had been granted subsequently. However, if prospecting were to be carried out on *cultivated* land, the concessionaire was liable to pay a rent for the land occupied. [11] A further annual charge of 10,000 rupees for that privilege would be payable by the company. [12] The terms of the prospecting licence also allowed the company "the right to win up to 100 tons of oil free of payment and further quantities of oil on payment of the royalty per ton provided in the *Mining Lease*", [13] the Second Schedule of the *Bahrein Island Concession.*

The *Mining Lease* made provision for the time when the company might succeed in finding oil in commercially exploitable quantities. In this event, the concessionaire agreed to "pay half yearly to the Sheikh' a royalty of 3 rupees and 8 annas per ton of net crude oil extracted from the ground. [14] This would replace the annual charge of 10,000 rupees required under the conditions of Article III of the Concession's Third Schedule. The royalty *rate* would be subject to review after ten years of payment of royalty.

For reasons already explained in earlier chapters, the Bahrain oil concessionaire, namely the Eastern and General Syndicate Limited, assigned its option on the *Bahrein Island Concession* to the Eastern Gulf Oil Company on the 30th of November 1927. Under the conditions of the assignation, Eastern Gulf Oil undertook to pay the annual charge of 10,000 rupees, due under the terms of the concession, to the Syndicate, which in turn would pay the Government of Bahrain.

An additional financial arrangement, namely the concept of an overriding royalty, was introduced into this agreement. If Eastern Gulf Oil were to exercise its option on the *Bahrein Island Concession*, then the Syndicate would assign and transfer the concession to Gulf or its nominee, together with all the privileges. It was also agreed that if Gulf's nominee should discover oil in paying commercial quantities, then the Syndicate would receive an overriding royalty of one shilling (1/-) sterling per ton on crude oil in excess of 750 tons per day produced from the territories covered by the concession. [15] Thus, by orchestrating agreement on this provision, the Eastern and General Syndicate Limited (and its corporate descendants) received substantial royalties from the mid-1930s (when oil *was* found in commercial quantities in Bahrain) until 1973 (when Bahrain Government equity, management and operational participation began and the overriding royalty was terminated).

To continue the sequence of corporate changes which affected the royalty arrangements, it is relevant to note that on the 27th of December 1928, the Eastern Gulf Oil Company assigned its option on the *Bahrein Island Concession* to Standard Oil Company of California, in return for actual expenses incurred by Eastern Gulf Oil amounting to $157,149. On the 11th of January 1929, the *Charter* to form the Bahrein Petroleum Company Limited as a wholly owned subsidiary of the Standard Oil Company of California was signed in Ottawa, Canada. On the 1st of August 1930, the 1925 *Bahrein Island Concession* was assigned from the Eastern and General Syndicate Limited to the Bahrein Petroleum Company Limited and the financial arrangements under the concession remained unchanged.

Thus, from this time, the Bahrein Petroleum Company Limited was liable to meet payments required by the Bahrain Government under the terms of the concession, until such time as oil might be discovered in Bahrain in commercial quantities and revenues could be generated from the export of crude oil. Less than two years later, oil *was* discovered in Bahrain, on the 1st of June 1932. By the middle of 1934, six wells had been drilled, the field had demonstrated that it was capable of sustained commercial production over a large area and the first shipment of Bahraini crude oil left the country on the 7th of June, bound for Yokohama, Japan.

Evolution of Royalties

The next major change occurred on the 29th of December 1934, when the *Bahrein Island Concession* was converted to a *Mining Lease*, provision for which had been made in the 1925 document. [16] The *Mining Lease*, crucial to an understanding of the Bahrain oil royalty arrangements, formed the basic document to which all subsequent amendments and agreements referred, that is until 1973, when the process of Government participation began as just mentioned.

According to the terms of the 1934 *Mining Lease*, the Ruler of Bahrain granted Bapco Limited an exclusive right for a period of fifty-five years from the 1st of January 1935 "to prospect and drill for, extract, treat, refine, manufacture, transport and deal with petroleum products, naphtha, natural greases, tar, asphalt, ozokerite and other bituminous materials"[17] within the area described in the lease.

This clearly defined area,[18] in which Bapco Limited was granted exclusive rights, including the rights to drill for and to produce petroleum, comprised 100,000 acres on Bahrain and Muharraq islands. This area was indicated by a blue line on a map attached to the Second Schedule of the 1934 *Mining Lease*. The municipal boundaries of Manama (the modern capital, situated on Bahrain island) and Muharraq town (located on the island of the same name to the north), indicated by the red lines on the map,[19] were excluded. The land within a radius of 300 yards (274.32 metres) from the north east corner of the Ruler's palace at Sakhir (defined by a green circle on the map) was also excluded. At the same time, Bapco Limited was granted the right to perform all ancillary functions necessary to make commercial use of oil production within the broader territory of Bahrain, outside the delineated 100,000 acres.

The Ruler also granted Bapco Limited[20] the right, but not the *exclusive* right, to prospect for, collect and use free of charge (but not to export or sell) stone, gypsum, salt, sulphur, clay and wood anywhere within the leased area, as well as water anywhere within the territory under the Ruler's control.

In consideration of the privileges outlined above, the company agreed to pay a royalty at the rate of 3 rupees and 8 annas per ton of net crude oil, as well as a similar rate per ton of casing-head petroleum spirit extracted by Bapco from natural gas (although the company was under no obligation to extract the latter). These royalties were calculated

Indian 10 rupee note (*Coincraft, London*).

and payable to the Ruler of Bahrain half-yearly. The royalty rates were subject to revision after fifteen years, namely in 1950. However, no royalties were payable for oil and casing-head petroleum spirit required for the company's customary operations in Bahrain. In addition, if Bapco Limited chose to recover and sell natural gas, the company was obliged to pay a royalty equal to one sixth of the field price received.

The 1934 *Mining Lease* also stipulated that the Ruler of Bahrain should receive a guaranteed minimum royalty of 75,000 rupees per calendar year during the period of the lease, [21] with the exception of circumstances beyond the control of the company which delayed its work. [22] Any period of such delay would extend the term of the lease by an equal period. At this point, it is important to emphasize the distinction between royalty and other charges. In particular, the annual charge of 10,000 rupees was not included in the principle of minimum and maximum royalties.

To explain this principle more clearly, the following example illustrates the guaranteed minimum royalty concept. To simplify the calculations, the royalty rate of 3 rupees and 8 annas per ton has been expressed as Rps.3.5 (there being 16 annas to a rupee):

If:

28,571 tons of crude oil were to be extracted, each ton would attract a royalty of Rps.3.5 = Rps.100,000

In this case, the Ruler of Bahrain would receive Rps.100,000, that is, Rps.25,000 *more* than the guaranteed annual minimum of Rps.75,000.

But if only:

17,143 tons of crude oil were to be extracted, each ton still subject to a royalty of Rps.3.5, then the royalty would amount to only Rps.60,000.

Since that figure is Rps.15,000 *less* than the guaranteed annual minimum royalty, Bapco Limited would be required to "top up" the figure to the required Rps.75,000.

Effective from the 14th of November 1935, [23] *less than one year* after the date of the 1934 *Mining Lease*, the guaranteed minimum annual royalty was doubled to Rps.150,000, [24] whilst the rate of 3 rupees and 8 annas per ton of net crude oil was maintained. This new arrangement was later documented in an agreement between the Ruler of Bahrain and Bapco Limited, known as the first *Supplemental Deed* modifying the *Mining Lease*, dated the 3rd of June 1936. This change recognized the increase in Bahraini crude oil output for the first three years of the oilfield's commercial production. The daily average output for the years 1934, 1935 and 1936 was 781, 3465 and 11,545 barrels respectively.

The two most important provisions in the first *Supplemental Deed* (1936) were the obligation imposed upon Bapco Limited to build a refinery as soon as practical [25] (thereby

A new shipping storage tank under construction during 1969 (*Bapco archive*).

adding to the industry and employment in Bahrain) and, in return, the grant to the company of exemption from all taxes and other governmental impositions except royalties and certain specified minor impositions. [26] These included duties on the personal requirements of employees and taxes on imported oil sold for consumption in Bahrain.

It is relevant at this point to mention that on the 30th of June 1936 California Texas Oil Company Limited (Caltex) was incorporated as a subsidiary of the Bahrein Petroleum Company Limited. The purpose of this event, also described in an earlier chapter, was to provide Bapco Limited with marketing capability now that oil had been discovered in commercial quantities.

On the 19th of June 1940, within a year of the outbreak of World War II, the *Deed of Further Modification* to the 1934 *Mining Lease* was signed. This document was important since it extended the area of the company's *exclusive* rights - primarily to drill for and to produce petroleum, and to construct and operate refineries and storage tanks - from the original 100,000 acres to all of the Ruler of Bahrain's then existing and future domains. This extended area was known as the "Additional Area". The exceptions were sites selected by the Ruler of Bahrain for military airplane, seaplane, wireless or telegraph installations, or harbour developments. All rights granted by the first *Supplemental Deed* (1936) were correspondingly extended, and the term of the *Mining Lease* (1934), as amended, was extended to the 19th of June 1995. For this additional area, a one-time payment of Rps.400,000 was made to the Ruler of Bahrain upon the execution of the *Deed of Further Modification*. [27]

However, Bapco Limited was not allowed to drill on Umm An-Nassan island, nor within the boundaries of Manama and Muharraq (as stipulated in the 1934 *Mining Lease*), nor within 300 yards (274.32 metres) of the Ruler's hunting lodges at Romaitha, Amar and Al Mamtala without first obtaining the permission of the Ruler of Bahrain. [28] As these areas were covered by the 1940 *Deed*, no further payment would be required from Bapco Limited, if permission for access were to be granted.

As to the oil royalty *rate*, this was to be maintained until the 1st of January 1950 at 3 rupees and 8 annas per ton, after which time it would be subject to review.

In any event, the *guaranteed minimum* royalty was increased to not less than twelve and a half lakhs of rupees (Rps.1,250,000) per annum. [29] However, should oil be found in commercial quantities in the Additional Area, then, from the date of that finding, the minimum royalty per annum would be increased to seventeen lakhs (Rps.1,700,000 rupees), provided that "good oilfield practice" created sufficient production to generate that amount of royalty. [30] These guaranteed minimum royalties were to apply for fifteen years, that is until the 19th of June 1955.

After this date, and in the event of oil *not* having been found in commercial quantities in the Additional Area, the guaranteed minimum royalty would revert to Rps.150,000 per year. However, if oil *were* to be found in commercial quantities in the Additional Area, during or after the expiry of the fifteen-year period in 1955, then Bapco Limited would be required to pay a minimum annual royalty of Rps.300,000. [31] Since these obligations depended upon the ability of Bapco Limited to continue normal operations, and in recognition of the uncertainties of the Second World War, provision was made for circumstances when the minimum royalty would not be applicable. The *force majeure* provision [32] freed the company from responsibility for discharge of all duties and obligations – not just minimum royalties – for the interval it might be prevented from discharging its obligations. Also, in these circumstances, the term of the 1934 *Mining Lease* would be extended for a period equal to the duration of the moratorium.

Meanwhile, since 1938 and following the discovery of oil in Saudi Arabia two years earlier, Saudi Arabian light crude oil had been shipped in barges from the Eastern Province to the Zallaq jetty on the west coast of Bahrain. From there it was pumped through a pipeline to the new Bahrain refinery, completed in 1936. Within a decade, this development had provoked discussion in both Bahraini and British Government circles and led to an

The first barge to arrive at Zallaq terminal, 1938 (*Bapco archive*).

intense debate between 1948 and 1950 which resulted in the next major step in the evolution of oil revenues for the Government of Bahrain.

Prelude to Increasing Government Revenue

As will be clear at this point, the only oil-related revenues which had been received by the Government of Bahrain by 1948 were early payments derived according to the terms of the 1925 *Bahrein Island Concession*, the 1934 *Mining Lease* and the two supplemental deeds dated 1936 and 1940. As discussed, these took the form of both charges and royalties. In July 1948, it became apparent to R. J. Ward, acting Bahrain Government

Inspector of Oil, [33] that in view of the development of the refinery and the export sale of petroleum products, this situation from the Bahrain Government's point of view was no longer equitable.

Ward, having travelled from London to review Bapco's operations in Bahrain, reported to J. E. Chadwick at the British Foreign Office that he had found certain deviations from what he believed to be good oilfield practice. [34] As a result of this and other observations, he considered that an overall structural review of the Bahrain Government revenues from its oil industry was required. This assessment had led Ward to conclude that the Bahrain Government was receiving perhaps half, or even less than half, of the revenue to which it should reasonably be entitled. Ward communicated these views to a senior Bahrain Government official who had shown considerable interest in the matter. However, just as some progress was being made in the discussions, the regular oilfield inspector, whom Ward had relieved during his leave of absence, returned unexpectedly to Bahrain a month early. As a consequence, the said Bahrain Government official dropped the matter on the grounds that it would be impolitic, at that time, to interfere with the existing financial arrangements.

Nonetheless, Ward expressed his concerns to Chadwick and they were duly minuted in the Foreign Office records [Chadwick, *Minute*, 25 November 1948]. The first point on which the visiting inspector had taken exception concerned the large quantity of Saudi Arabian light crude Bapco was importing from Dhahran, Saudi Arabia, which together with Bahraini indigenous crude was utilized in the manufacture of finished petroleum products. The quantity of imported crude oil amounted to about four times the production from the Bahrain oilfield at that time. Chadwick minuted that "no rent was paid to the Bahrein Government for this refinery and in Mr Ward's opinion the Bahrein Government were entitled to some revenue from the refining of imported oil", presumably meaning more correctly that Bapco Limited was paying no rent for the land occupied by the refinery.

The next point which had concerned Ward was that the crude oil produced in Bahrain was being fed straight into the refinery's storage tanks, which were treated as though they were royalty tanks and sealed by the Government inspector. In Ward's view, this method of control and issue of royalty oil was not entirely satisfactory since the crude oil storage tanks were under the control of Bapco's refinery and not the Bahrain Government.

The Government inspector was also required by Bapco Limited to sign certificates which confirmed the quantities of imported Saudi Arabian light crude oil. Strictly speaking, this was a mechanism of control over which the Government inspector had no jurisdiction, from which the Bahrain Government derived no revenue and from which the British Treasury received no payment for carrying out the service.

Although Ward accepted that it was normal oilfield practice for an exploration company to deduct from the royalty oil such oil as was required to drive machinery to extract oil from the ground, in practice, very little was used for this purpose in Bahrain. According to Ward, Bapco was improperly deducting from the royalty oil, the oil used to refine the crude oil produced in Bahrain, a practice which was quite contrary to normal procedures elsewhere, such as in Egypt. Ward claimed that these deductions caused a loss to the Government of Bahrain of royalties which should have been paid on 5000 tons of crude oil per week. He also claimed that Bapco had sunk a gas-well to produce natural gas for refinery use, a facility for which no royalty was being paid to the Bahrain Government. He believed too that the process was reducing the underground pressure of the oilfield, thus lowering the productivity and capability of the wells.

Finally, Ward suggested that while Bapco was importing large quantities of Saudi Arabian light crude, it was not exploiting the Bahrain oilfield to its full potential. In his view, it seemed to be in the Government of Bahrain's interest to press Bapco to *increase* the production of Bahrain's royalty-generating oil.

W. L. F. Nuttall at the Petroleum Division of the Ministry of Fuel and Power in London did not share Ward's views. In a "restricted" letter to J. E. Chadwick, he suggested that the visiting inspector's observations had been made without full acquaintance with the Bahrain concession and that Ward appeared to have stated that the company was quite improperly deducting from the royalty oil, the oil used to refine the crude oil produced in Bahrain.

"In this connection, Article 7 (a) of the concession [35] states that royalty is payable on 'net crude oil got and saved (i.e. after deducting water and foreign substances and oil required for the customary operations of the company's installations in the Ruler's territories other than the oil used in connection with the refining of imported oils)'. The last part of this sentence, from the word 'other' to the end, was added by a deed signed by the Ruler and witnessed by Political Agent on the 3rd of June 1936. From this, it can be inferred that the Company is not under an obligation to pay royalty on oil used in connection with the refining of crude oil produced in Bahrein" [Nuttall, letter to Chadwick, 6 December 1948].

To support his case, Nuttall remarked that the Anglo-Iranian Oil Company [AIOC] adopted a similar practice and did not pay royalty on oil used for internal consumption in refinery operations. He also rejected the suggestion that royalty should be paid on natural gas used in Bapco's installations, adding that there was no evidence to suggest that the underground pressure in Bahrain had been reduced by the extraction of gas for refinery use, nor that Bahrain's oil production was being deliberately restricted by Bapco. On the contrary, Nuttall concluded that the Bahrain oilfield was being produced at a reasonable rate relative to its proven oil reserves and that the loss of royalties to the Bahrain Government was far less than the 5000 tons of oil per week claimed by Ward.

Within three weeks of receiving Nuttall's letter, Chadwick wrote to Sir Rupert Hay, the Political Resident Persian Gulf. The extent to which Ward and Nuttall were correct or incorrect in their analyses was a matter which the Foreign Office was pleased to pass over to Sir Rupert "for whatever discreet use you may care to make of them". Chadwick looked forward to Hay's thoughts on the matter and added that Ward had left Bahrain "with an uneasy feeling that the present Inspector is in the oil company's pocket. He [Ward] did not give the impression of being a man with a bee in his bonnet, but seemed a serious and competent engineer" [Chadwick, letter to Hay, 31 December 1948].

Now that Ward and Nuttall had made their concerns and observations official, an active debate ensued which involved all parties and subsequently led to the creation of a more sophisticated mechanism to generate additional oil revenues for the Bahrain Government.

Royalty Rate Revision

In early 1949, following Chadwick's letter to Sir Rupert Hay, the Ruler of Bahrain drew attention to the fact that according to Article VII (b) of the 1934 *Mining Lease*, oil royalties were subject to revision at the end of fifteen years' payment of royalties by the company, the first payment having been received by the Bahrain Government in October 1935. It appeared that the revision should be effective from January 1950. The letter conveying the Ruler's wishes to the Political Agent explained that as negotiations about the matter would take time it was desirable that Bapco should be informed now of the Ruler's desire to revise the Agreement [Belgrave, letter to the Political Agent, 17 January 1949]. Furthermore, the Ruler was "of the opinion that the present revenue from oil is far below what it should be, taking into consideration the amounts which are

Caltex Bahrain at Sitra wharf (*Bapco archive*).

Table 6

**STATEMENT OF ROYALTIES PAYABLE BY OIL
COMPANIES OPERATING IN THE ARABIAN GULF REGION IN 1949**

Bahrain Petroleum Company	3 rupees and 8 annas per English ton
Petroleum Concessions Limited in respect of Qatar, the Trucial Shaikhdoms and Muscat	3 rupees per English ton
Kuwait Oil Company	3 rupees per English ton
American Independent Oil Company (Aminoil) in respect of the Kuwait share of the Saudi Arabia Neutral Zone	$2.57 per English ton based on the gold equivalent when the concession had been signed. At the free-market rate of 4 rupees and 12 annas to the US$ this was equivalent to about 12 rupees per ton
Arabian American Oil Company (Aramco)	4 gold shillings per English ton paid equivalent rate of $2.20 (about 10 rupees and 8 annas at the free-market rate in January 1949*)
Anglo-Iranian Oil Company (AIOC) and **Iraq Petroleum Company (IPC)**	4 gold shillings per English ton, believed to have been the equivalent of about 8 rupees*

* The conversion discrepancy may be explained by the omission of several words in the source document.

[Source: *Comparative table* attached to the Bahrain Government's Letter 691-210, Belgrave to the Political Agent, dated 17 January 1949]

received by other oil producing countries. Important social and economic programmes for the benefit of the people of this country cannot be carried out through lack of funds. His Highness wishes to know what new terms the Bahrain Oil Company intends to propose which will bring Bahrain's oil revenue in line with conditions as they exist today."

The letter conveying these thoughts to the Political Agent enclosed a comparative statement, Table 6, showing the royalties payable by Bapco Limited and other oil companies operating in the Arabian Gulf and adjoining countries in 1949. This communication precipitated sixteen months of complex and often very contentious negotiations. Proposals and counter-proposals were presented by all parties, namely the Bahrain Government, the British Foreign Office, the British Treasury, the Political Agent in Bahrain and Bapco Limited. On the 13th and 17th of March 1949, a delegation from Bapco, led by the President, held preliminary discussions with a senior representative of the Bahrain Government and the British Political Agent in Bahrain. Just under three months later, the Political Resident Persian Gulf, Sir Rupert Hay, wrote to his colleague Bernard A. B. Burrows at the Foreign Office in London:

"we have now received a sumptuous document in English and Arabic entitled *Statement Regarding Oil Royalties Paid to His Highness Shaikh Sir Salman bin Hamad Al Khalifah, KCIE, Ruler of Bahrain* and signed by Pinckhard, the Bahrain [*sic*] Petroleum Company's president" [Hay, confidential letter to Burrows, 3 June 1949].

After much argument the conclusions reached were as follows:

1. Regarding the Additional Area only, the existing royalty rates were to be continued for a period of four years to the 31st of December 1953.

2. In respect of the Original Area, the existing royalty rates were to be increased on the 1st January 1950 "by a moderate amount" to be effective for the same four year period to the 31st of December 1953.

   "This would give all concerned a period of time in which to learn what the present uncertainties will bring forth, and to determine what may be the results of the present seismological survey. The Company feels that important and great changes may take place during the next few years which cannot be fairly assessed today."

3. The royalty rates relating to *both* the Original Area and the Additional Area should be subject to further review at the end of 1953.

After five more months of deliberation, the Treasury agreed on the 25th of November 1949 that Bahrain's royalty rates *were* low when compared with those obtaining elsewhere in the Middle East, with the exception of Kuwait. Mary Ashe, writing from the Treasury, suggested to T. E. Rogers at the Foreign Office that "you will no doubt consider it our duty as protecting power to help them [the royalty rates] to be raised". However, she added that, in view of likely "repercussions on rates payable by other British companies" operating in the Arabian Gulf at that time, the Treasury did not wish the matter to be pushed with undue haste [Ashe, letter from the Treasury, 25 November 1949].

On the 26th of December 1949, a senior Bahrain Government official wrote to Bapco's Chief Local Representative:

"I have the honour to state that I am instructed by His Highness Shaikh Sir Sulman bin Hamad Al-Khalifah, Ruler of Bahrain [*sic*], to make known to the Bahrain Petroleum Company that the royalty revision for which provision is made in the Lease dated 29th December 1934, granting an Oil Concession to the Bahrain Company should be effective as from 1st January, 1950, as provided, and that the revised royalty rate should be Rs 10/- (ten rupees) per ton instead of Rs 3/8 (three rupees and eight annas) per ton as under the present agreement" [Belgrave, letter to Bapco, 26 December 1949].

The letter went to some length to explain the reason for the substantial, rather than modest, increase which had been expected by Bapco. It was argued that the revised rate was comparable to payments being made by other oil companies to the governments of neighbouring oil-producing countries. Bapco's attention was drawn to the fact that it was difficult to make precise comparisons between the foregoing figures and the royalty payments made by other oil companies for corresponding rights, because of differing circumstances and terms of payment. However, it was well known that the host governments of the two longest established concessionaires in the region, the Anglo-Iranian Oil Company and the Iraq Petroleum Company, were currently negotiating improvements to their royalty terms.

The Bahrain Government's 26th of December letter proposed that the revised royalty rate should remain effective for ten years, with the proviso that it would be subject to a further revision at any time during the first five years of that decade if the total production of royalty oil (including that from the Additional Area if oil were to be found there) dropped below a daily average of 25,000 barrels during any royalty period.

The British Treasury saw the situation somewhat differently. Quite apart from her perception of the unexplained haste with which the revisions were being negotiated, Mary Ashe wrote to the Foreign Office on the 16th January 1950, with other comments:

"I am sorry we are making such heavy weather about this question of the revision of the Bahrain Petroleum Company's royalty rates.

"I am afraid I have never quite understood just why it has been felt that the royalty rates must be increased, soon and substantially, when the concession does not fall due for revision until 1951.

"I take it you and Sir Rupert Hay [PRPG] feel that it is our duty as the protecting power to help the rates to be raised since they are admittedly low at present. But from our point of view, bearing in mind the interests of the UK oil companies in Iran and Iraq, this is really quite a dangerous precedent" [Ashe, letter to A. Leavett, 16 January 1950].

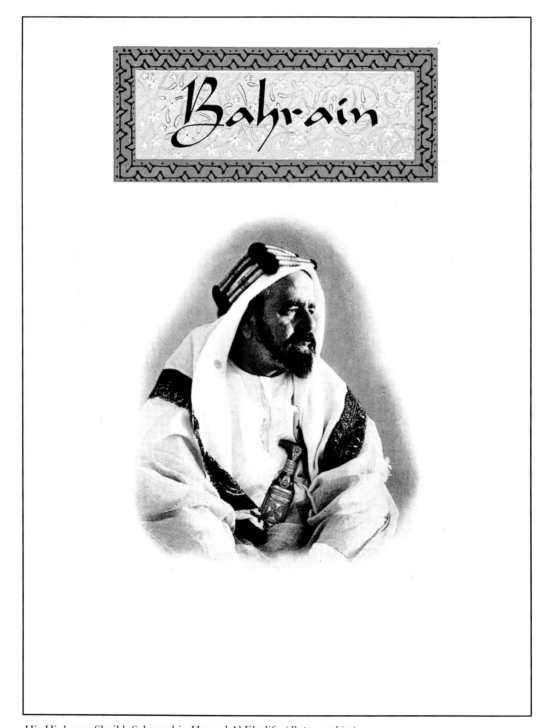

His Highness Shaikh Salman bin Hamad Al-Khalifa (*Bapco archive*).

On the 30th of March 1950 Sir Rupert Hay sent a cable to the Foreign Office in London in which he said that, two days earlier, Edward Skinner (Bapco's Vice-President) had had a "stormy interview" with the Bahrain Government. The Ruler of Bahrain had maintained that the royalty rate should be increased to 10 rupees per ton. Hay's cable reiterated that the royalty rate was less than that being paid in neighbouring Arabian Gulf countries and that the Ruler was neither asking for payment in gold or dollars nor requesting revenue from the import of Saudi Arabian light crude. He was asking simply for an increased royalty rate to bring Bahrain's oil revenues more in line with those pertaining in neighbouring countries. In consideration of these factors, the Ruler of Bahrain had made it clear that he was not prepared to compromise and he insisted furthermore that, should the matter remain unresolved, it should be presented for arbitration.

Within two weeks, after further vehement debate which involved all parties, Bapco acceded to the Government of Bahrain's request for an increase in the royalty rate to 10 rupees per ton. On the 12th of April 1950, Russell Brown, Bapco's Chief Local Representative, wrote to the Ruler of Bahrain:

"Reference is made to the proposal of Your Highness for revision of the terms of oil royalty provided in the Lease Agreement ... dated 29th December 1934. ...

"Your Highness has stated that the proposal ... includes the minimum terms Your Highness will voluntarily accept. The Company has for many years enjoyed Your Highness' friendship, and it desires and will ask for nothing that would impair the excellent relationship and mutual understanding which have always existed between Your Highness and the Company.

"With this in mind, the Company hereby accepts Your Highness' proposals as set forth in paragraphs 1 and 4 of letter No. 626-20... dated 26th December 1949" [Brown, letter to the Ruler of Bahrain, 12 April 1950].

The agreement was retroactive to the 1st January 1950.

Six days later, on the 18th of April, the Ruler of Bahrain expressed "our very great pleasure and satisfaction at the Company's action in acceding to our request and we look forward to a continuation of the friendly relations which have existed between us and the Company for so many years" [Al-Khalifah, letter to Bapco's Chief Local Representative, 18 April 1950].

As a postscript to this round of strongly-argued negotiations, the Political Agent declared in his Economic Report for the period June-July 1950 that Bapco's royalty payments had gone up by a "remarkable jump", far in excess of the modest increase that originally had been envisaged.

Table 7
SUMMARY OF ROYALTY AND BAHRAIN OILFIELD OBLIGATIONS
1925-50
prior to the introduction of corporate taxation

Documents

Bahrein Island Concession, 2 December 1925
Option on the Bahrein Island Concession, 30 November 1927
Option on the Bahrein Island Concession, 27 December 1928
Mining Lease, 29 December 1934
(First) Supplemental Deed, 3 June 1936
Deed of Further Modification, 19 June 1940
Political Agreement, 29 June 1940
Government of Bahrain Letter, 626-20, 26 December 1949

Areas Covered

Original Area	100,000 acres on Bahrain island (*Mining Lease*, Article I and Schedules I & II).
Additional Area	All other lands, water and submerged lands under the Ruler of Bahrain's dominion (*Deed of Further Modification*).
Exceptions	Sites selected by the Ruler of Bahrain or the British Government for military airplane, seaplane, wireless or telegraph installations or harbour developments.

Resource Rights Granted

The exclusive right to prospect, drill for, extract, treat, refine and market oil products, naphtha, natural gases, tar, asphalt, ozokerite and other bituminous materials (*Mining Lease*, Article I, and the *Deed of Further Modification*, Articles 3b, 4a, 4c).

Ancillary Rights

- The exclusive right to construct and operate a refinery and storage tanks
- The right to construct and operate pipelines, wharves, jetties, roads, buildings, machinery, radio and telegraph apparatus (*Deed of Further Modification*).
- The free use for operational needs, but not for export or sale, of water, stone, gypsum, salt, sulphur, clay and wood
  (*Mining Lease*, Article Ia, and the *Deed of Further Modification*, Article 3b).

Surface Use of Land

- The right to the free use, for operations, of uncultivated land, except land belonging to the Ruler in a private capacity, land within the municipalities of Muharraq and Manama, and a 300 yard (274.32 metre) radius of Sakhir Palace (*Mining Lease*, Article II).
- The right to the use of private lands belonging to the Ruler of Bahrain, subject to a reasonable rental (*Mining Lease*, Article II).

Exploration Obligation

The pursuit of oil in the Additional Area "with diligence and without unnecessary delay" (*Deed of Further Modification*).

Drilling Obligation

Original Area   Continuous operation with two drilling rigs until the requirements of good oilfield practice for the productive area were met (*Mining Lease*).

Additional Area   Operations for a test well had to commence by the 19th December 1946 if a favourable area were found (*Deed of Further Modification*). If oil were to be discovered in this area, then drilling must continue until oil was produced in commercial quantities (500 tons per day for thirty consecutive days), or until the area was deemed no longer favourable (*Deed of Further Modification*, Article 5c). Upon oil being found in commercial quantities, drilling had to continue in the productive area until the requirements of good oilfield practice had been met.

Forbidden Area   Drilling could not take place on Umm An-Nassan island, within the municipal boundaries of Manama or Muharraq, or within 300 yards (274.32 metres) of the Ruler's hunting lodges at Romaitha, Amar and Al Mamtala (*Deed of Further Modification*).

Overriding Royalty

One shilling (sterling) per ton on crude oil extracted (produced) exceeding 750 tons per day, payable to the Eastern and General Syndicate or its corporate descendants. This amount of oil was allowed on a royalty-free basis for operational needs, as opposed to that utilized for conversion to finished petroleum products (1927 *Option*, Article X, & 1928 *Option*, paragraph 6).

Oil Royalties and Charges

1925-34   10,000 rupees per year (*Bahrein Island Concession*)

1934-49   3 rupees and 8 annas per ton, subject to revision on 1 January 1950 (*Mining Lease*, Article VII, 1936 *Supplemental Deed*, paragraph B3, and *Deed of Further Modification*, Article 6)

1949   10 rupees per ton (Government of Bahrain letter, 626-20, dated 26 December 1949, effective on the 1st of January 1950).

Guaranteed Minimum Annual Royalty

1934-36   75,000 rupees (*Mining Lease*)

1936-40   150,000 rupees (1936 *Supplemental Deed*)

1940-55   1,250,000 rupees per annum, subject to increase to 1,700,000 per annum should oil be found in commercial quantities in the Additional Area (*Deed of Further Modification*, Article 7a).

One-Time Payment

1940   Upon the execution of the *Deed of Further Modification*, a one-time payment of 400,000 rupees was due (Article 3a).

Taxes

The only taxes payable in Bahrain until 1950 were those imposed on motor vehicles, boat and fish trap registration, driving licences, stamped paper and leased forms, porter fees on imports, customs shed rentals, and crane user fees, fees for alterating customs documents, house tax, shop and market stall taxes, and donkey cart taxes.

Government Revenue based on the Importation and Refining of Foreign Crude

From 1929 to 1952, Bapco Limited evolved into a fully-integrated oil company, organized into divisions according to three operational functions: the *producing* division, which was responsible for oilfield exploration, prospecting, the extraction of crude oil from the ground and its transfer to the refinery; the *refining* division which converted the Bahraini indigenous crude oil, co-mingled with imported Saudi Arabian light crude oil, into finished petroleum products; the *marketing* division which undertook the local and international sale of refined petroleum products.

Since 1935, the Bahrain Government had received revenue in the form of royalties on the extraction of indigenous crude oil. From a strict legal standpoint, this was a mutually *agreed obligation* to make payments and was not a government *imposed tax*.

Regarding the refining and marketing activities carried out by Bapco in Bahrain, the British Government oil inspector, R. J. Ward, had pointed out that Bapco Limited (an American-owned oil company registered in Canada) was largely in business to refine and market crude oil imported from Saudi Arabia. It had become evident that Bapco was making a significant profit on these operations, but was making no payment to the Bahrain Government with respect to those activities.

The Bahrain Government finally decided that this situation was unacceptable and initiated a new round of negotiations to reach an agreement giving it appropriate revenue from the importation and refining in Bahrain of foreign crude oil. On the 8th of December 1952 an important agreement was signed by Bapco Limited, sealed by the Ruler of Bahrain and witnessed by the British Political Agent, retroactive to the 1st of January that year. This profoundly significant document amended the 3rd of June 1936 *Supplemental Deed* to the 1934 *Mining Lease* and contained four significant provisions.

1. An import fee was introduced at the rate of two and a quarter (2 1/4) pence sterling per barrel in respect of all foreign crude oil imported into and refined in Bahrain. [36] An important point was that, irrespective of what Bapco paid for each barrel of imported crude, the company was obliged to pay the flat-rate import fee of two and a quarter (2 1/4) pence per barrel.

2. In return for this agreement to pay a fee, the Ruler of Bahrain agreed to an extension of the term of the *Mining Lease* to terminate on the 31st of December 2024. Bapco Limited also agreed that for the next five years it would produce oil from the then known Bahrain field to the limit permitted by good oilfield practice, [37] subject to *force majeure*, and if *force majeure* should at any time compel suspension of operations, the Ruler of Bahrain would receive a minimum of Rps750,000 per year (recoverable from royalty and fee obligations after *force majeure* ceases). [38]

3. The currency of the monetary obligations changed from rupees to sterling. This was not intended for general use in Bahrain, but to determine the currency in which

Bapco's obligations to the Bahrain Government would be computed and paid. From the 8th of December 1952, the royalty was to be paid in sterling at the exchange rate of one shilling and sixpence (1/6) per rupee. [39]

4. The Government of Bahrain would be supplied with 150,000 imperial gallons of petrol per year for its own use, free of charge. [40]

Texaco tanker at Sitra wharf (*Falcon Cinefoto*).

Introduction of Corporate Income Tax on Bahrain Crude Oil

Simultaneously, a vigorous debate had developed in other countries regarding the taxation on the profits derived by oil companies from the export of indigenous crude oil. Aramco, of which Socal and Texaco (the ultimate owners of Bapco Limited) were major shareholders, submitted to a "fifty-fifty" corporate income tax in December 1950, effective from January 1951. Once this had occurred, Bahrain could not be denied the same benefit. It is generally agreed that the introduction of corporate income tax in Bahrain was a direct result of the Saudi Arabian experience. However, unlike Saudi Arabia, Bahrain's indigenous crude oil was not being marketed as crude but was being sold in the form of finished petroleum products from the Bapco refinery.

When the concept of corporate income taxation was mooted in Bahrain, a problem had arisen since no price – sometimes known as the tax reference price – had been assigned to Bahrain's indigenous crude oil. Therefore, in a separate document dated the 8th of December 1952, the Ruler of Bahrain and Bapco Limited agreed that the prices of *Bahrain* crude oil, the principal factor in computing the company's income, would be the weighted average of the prices of *foreign* crude oil brought into and refined in Bahrain. This was the key element necessary for the imposition of a corporate income tax on profits from the sale of Bahrain crude oil. At the same time, Bapco was very careful to ensure that the terms of any corporate income tax law imposed upon the company would not prevent tax payments from qualifying for US foreign tax credits (as distinguished from agreed payments such as royalties). This was important to the Bahrain Government, since Bapco's ability to operate successfully in Bahrain provided a valuable source of revenue and of economic well-being to the country.

Thus, on the 15th of December 1952, the *Income Tax Regulation*,[41] retroactive to midnight on the 31st December 1951, was issued. On the following day, R. M. Brown, Bapco's Chief Local Representative, advised the Ruler of Bahrain by letter that Bapco Limited recognized and submitted to the decree which imposed a tax upon income "from the sales of crude petroleum and other natural hydrocarbons produced and extracted from the ground and from under the seabed belonging to Bahrain" [Brown, letter to the Ruler of Bahrain, 16 December 1952]. This tax was determined by starting with 50-per-cent of income and then deducting from the 50-per-cent portion a sum equal to the aggregate of all taxes other than this corporate income tax, and all royalties and "other exactions" mentioned in Article 1 of the decree.

It will be apparent by now that a relationship between royalties and taxation had begun to evolve. The notional example in Table 8, illustrating this relationship as it existed in 1952, has been applied to 100 barrels of Bahraini indigenous crude oil.

Extending Taxable Income to Include Refining Profits

Within three years of the introduction of corporate income tax on the production of Bahrain indigenous crude oil, the next major tax change occurred. On the 28th

November 1955, the Bahrain Government issued the *Bahrain Income Tax Regulation, 1955,*
[42] effective on the 1st of January of that year. This repealed the *Income Tax Regulation*
published on the 15th of December 1952.

At the time of the 1952 corporate income tax decree, it was realized that since Bapco
Limited was selling its oil to affiliated companies there *could* be concern that it *might* sell
at prices below fair market prices and thereby shift some of the taxable profits from the
jurisdiction of the Bahrain tax laws. It was for this reason that, as part of the 8th of
December 1952 agreement, the weighted average of the prices which resulted in
Bapco's gross income from crude produced in Bahrain would be equal to the prices it
had to pay for foreign crude oil imported into and refined in Bahrain.

Following the issue of the 1955 tax decree, [43] the Bahrain Government and the company recognized that it might cause concern similar to that expressed in 1952 about

His Highness Shaikh Salman bin Hamad Al-Khalifa visits *Caltex Bahrain* (*Bapco archive*).

Table 8

RELATIONSHIP OF ROYALTY AND INCOME TAX TO COMPANY MARGIN

	US Dollars
INCOME TAX CALCULATION	

Step One

"Market" value of 100 barrels of Bahrain crude oil @ US$2.00 per barrel	200
Less:	
Production/extraction costs (exclusive of royalties and other exactions paid to the Ruler)	(20)
BAPCO'S NOTIONAL INCOME	180

Step Two

50-per-cent of BAPCO'S NOTIONAL INCOME	90
Less:	
Royalties and other exactions paid to the Ruler	(27)
TOTAL INCOME TAX	63

Therefore:

BAHRAIN GOVERNMENT TAKE:

Income Tax	63
Royalties and other exactions paid to the Ruler	27
TOTAL	90

BAPCO TAKE:

Bapco's notional income		180
Less:		
Income Tax	(63)	
Royalties and other exactions paid to the Ruler	(27)	(90)
TOTAL		90

This even division of "take" is reflected by the term "fifty-fifty".

		Table 9
		ROYALTY AND TAX PAYMENTS TO THE GOVERNMENT OF BAHRAIN – SELECTED YEARS FROM 1933 TO 1952
Year	**Dollars**	**Circumstances**
1933	8,800	
1938	305,000	1934 *Mining Lease* + 1936 first *Supplemental Deed* increased the annual guaranteed minimum royalty.
1940	1,000,000	1940 *Deed of Further Modification* increased the annual guaranteed minimum royalty
1945	780,000	Reduced throughput because of World War II
1946	1,200,000	Post-war expansion
1948	1,508,876	
1949	1,500,000	
1950	3,300,000	Royalty rate revision, effective 1 January
1951	3,800,000	Increased Bahraini crude production
1952	6,300,000	Introduction of import fee (8 December) + corporate income tax (15 December), both retroactive to 1 January 1952

appropriate crude oil prices, regarding the prices of refined products and services that would affect Bapco's taxable income from refining operations in Bahrain. This applied to profit both from the sales of refined products and from any refining operations Bapco Limited might undertake on behalf of other parties. Therefore, on the 29th November 1955, the Ruler of Bahrain and Bapco Limited agreed by letter [44] that the weighted average of Bapco's prices of refined products, and of fees charged by the company for refining crude for other parties, regardless of the sources of crude involved, would always include a refining profit element of seven (7) pence sterling per barrel of crude refined. Thus, the law and the agreement together assured that Bapco Limited's profits subject to Bahrain income tax would always be appropriate.

The Evolution Continues

Prior to the next major pricing agreements implemented in the early 1960s, two important amendments were made to existing financial arrangements, namely an increase in the allocation of free petrol to the Bahrain Government and a royalty rate review.

Bapco Limited's former office in Manama (*Bapco archive*).

The first of these was effected on the 7th of December 1958 when M. H. Lipp (Bapco's Chief Local Representative) wrote to G. W. R. Smith (Secretary to the Bahrain Government) stating that the company "desires to afford all reasonable assistance to the Government in its operations". This letter responded to a previous advice from the Ruler of Bahrain that the amount of free petrol provided to the Government of Bahrain under the provisions of the 8th of December 1952 *Supplemental Agreement* was insufficient to meet the Government's requirements. Therefore, commencing from January 1959, Bapco would increase the amount of petrol which it supplied free of charge to the Government of Bahrain for its own internal use and consumption from 150,000 imperial gallons to 200,000 gallons annually. [45] Almost exactly a year later, the second amendment was made. On the 1st of December 1959, Lipp wrote to the British Political Agent proposing that the existing royalty rate applicable to the company under the terms of the 1934 *Mining Lease*, as amended, be extended for an additional period of ten years. The present rate which had been established by an exchange of letters and made effective for a period of ten years from the 1st of January 1950, was set at 10 rupees per ton. [46] In his review of the existing arrangement, Lipp added that:

"At the time the present rate of royalty was established it was a prime factor in determining the amount of income the Ruler would receive as a result of petroleum production in Bahrain. Subsequently, however, an income tax was imposed, in the computation of which the royalty is taken fully into account. As a result, the rate of royalty is no longer a major factor in determining the income which the Ruler derives from this enterprise. The existing rate of royalty is, however, consistent with that in effect in other oil producing areas" [Lipp, letter to the Political Agent (E. P. Wiltshire), 1 December 1959, C/PA-359].

Three weeks later, His Highness Shaikh Salman bin Hamad Al-Khalifa, Ruler of Bahrain, informed Bapco's Chief Local Representative that:

> "We agree to the rate of royalty to continue in effect for an additional ten years period and to the royalty being paid in pounds Sterling (United Kingdom) provided that if States other than Bahrain bordering the Arabian Gulf in which oil is now being produced should receive substantially better terms, the Company will be willing to review the situation" [Al-Khalifah, letter to Bapco, 22 December 1959].

The Ruler had, in this letter, reiterated the regional "equal treatment" philosophy which became the cornerstone of Bapco/Bahrain Government financial relationships as they evolved over the following years. In the light of developments in the international oil industry during the late 1950s, this was an important and timely statement. It assured the Bahrain Government that it would enjoy equal treatment from Bapco in respect of revenue rate enhancements achieved by the other regional host governments with their potentially greater leverage over the international oil companies.

OPEC – Instrument of Pricing Practice

By the late 1950s, oil production and prices were controlled to a great extent by the major international oil companies [47] – often referred to as the "seven sisters" – which owned most of the free world's internationally traded oil. The host governments of the oil producing countries had virtually no say in production and pricing decisions. During 1958-9, under host government pressure, crude oil posted prices set by the major producing companies remained constant despite the fact that *actual* market prices received by sales companies were declining because of a substantial glut of crude oil available for sale. Since taxes and royalties levied by the host governments on crude oil production were based on sales at the posted prices and not on the lower market prices, the profit margins of the oil companies were being severely eroded.

In 1959, in the face of this profit squeeze, the "majors" unilaterally announced a reduction in posted prices. Host governments of the producing countries expressed dismay at the resultant drop in their tax and royalty revenues, but took no action against the oil companies concerned. The following year, in August 1960, the "majors" acted again. Without consulting the host governments, they cut the price of Middle East crude oils by 10 cents a barrel. This time, the oil-producing countries reacted. Perez Alfonso of Venezuela and Abdulla Tariki of Saudi Arabia as leaders, together with representatives from the other three major oil-producing countries (Iran, Iraq and Kuwait), assembled in Baghdad on the 14th of September 1960 to found the Organization of the Petroleum Exporting Countries. OPEC's five founder-member countries had one thing in common: their national economies were almost entirely dependent upon the revenue derived from the production and export of crude oil from their countries by the "majors". As a result, the new organization successfully challenged the right of the multinationals to reduce posted prices arbitrarily and thus affect the host governments' national revenues.

To achieve this, OPEC implemented a philosophy comprising three main objectives. First was the stabilization of the price of oil, insisting that no future oil price changes would be made without prior consultation with the host governments of the producing countries. The second goal sought to examine ways of achieving assured national income and stable prices through mutual action, such as voluntary regulation of production. Third, the OPEC members agreed to support each other against tactics by the major international oil companies deemed divisive among the producing bloc.

In its formative years, OPEC's attention focused on the accounting complexities of royalties, taxation rates and associated allowances. Before long, the organization's effective economic data-gathering system provided an impetus to create an information office, primarily for the benefit of the host governments of oil-producing countries.

Following the formation of OPEC, the official posted price of the benchmark Arabian light crude oil remained unchanged throughout the 1960s (at that time in the Middle East oil producing countries income tax and royalty rates on oil production were calculated on the basis of posted prices), though the realized market price declined steadily to US$1.35 by 1970. This continuity of post/tax-reference prices enabled OPEC member countries for the first time to sustain their national incomes at a more or less stable and somewhat predictable level. Whilst the Organization itself became the defender of the official posted price *status quo*, it did not attempt to change the basic power structure of the industry. This permitted the OPEC members to achieve their oil-related revenue objectives without OPEC itself becoming involved in the complex business of owning, managing and operating the oil industries in the areas of the member nations. Additionally, the OPEC members were able to avoid the daunting prospect of penetrating the complex jungle of international oil trading in competition with long-established major oil companies which controlled the downstream consumer markets. It was against this background that the next major steps in the evolution of Bahrain's oil revenues occurred.

Power Shift Towards Bahrain Government Control

Before moving forward to discuss the next tax development, it is worth highlighting several concurrent events which, in effect, comprised the next major shift towards the Bahrain Government taking over control of its administrative and economic affairs, a continuation of the progressive elimination of the British Government's role in Bahrain's affairs.

On the 20th of May 1961, in response to a recent request by Bapco Limited and agreed by the Bahrain Government, the Political Agent in Bahrain informed Bapco that the British Government had no objection to the company corresponding directly with the Bahrain Government's legal advisers, providing that Bapco supplied the British Government with copies of all correspondence which passed between the two parties involved.

The next change occurred on the 5th of August 1964, following a meeting held in London during the previous July. His Excellency Shaikh Khalifa bin Salman Al-Khalifa wrote to Bapco's Chief Local Representative, L. D. Josephson:

"It was, I understand, agreed that the channel for all routine communications between the Government of Bahrain and your Company should, until further notice, be from myself as Head of Finance to yourself as Vice President in charge of the Bahrain Petroleum Company. I accept this arrangement gladly, and take the opportunity in this letter of initiating the procedure" [Al-Khalifah, letter to Bapco, 5 August 1964].

A year later, on the 9th of August 1965, a significant new document was signed. This was an *Amendatory Agreement* to three previous documents: the 1934 *Mining Lease*, the 1936 *Supplemental Deed* and the 1940 *Deed of Further Modification*. The 1965 agreement formalized several requests made by the Government of Bahrain, amongst which was the requirement that, in future, Bapco was to refer to "the Sheikh" as "the Ruler" [48] and that the company should delete the phrase "acting on the advice of the Political Resident in the Persian Gulf" from all future documents. [49] In addition, Article XV of the *Mining Lease* was to be replaced with the words: "At all times one of the Directors of the Company shall be a person who has been selected in consultation with the Ruler and whose appointment is acceptable to Him". [50] The *Amendatory Agreement* also modified other aspects of existing contracts to avoid terminology which the Bahrain Government considered to be inconsistent with current circumstances.

Fair Pricing Agreements

On the 8th of February 1961, Bapco Limited confirmed an agreement into which the Ruler of Bahrain and the company had entered on that day, in effect amending the 29th of November 1955 document, whereby the refining element used in the company's pricing of refined products would be increased from seven (7) pence sterling per barrel of Bahrain indigenous and imported foreign crude refined, to eight and a half (8 1/2) pence sterling. [51]

On the 27th of October 1962, Bapco Limited informed the Ruler by letter that the company had made a careful study to establish the means whereby the Bahrain Government might obtain additional net income from refined petroleum products, sold by the company for consumption in Bahrain, without imposing any additional financial burden on the citizens of Bahrain. [52] As a result, after an exchange of clarifying communications, the Ruler and Bapco's Chief Local Representative, W. A. Schmidt, signed a further pricing agreement, dated the 10th of February 1963, [53] but retroactive to the 1st of January 1962. This new arrangement removed products sold for local consumption in Bahrain from the scope of the 29th of November 1955 agreement concerning the refining profit margin to be included in the sales prices of products refined in Bahrain.

One pound (sterling), in circulation until 1960 (*Coincraft, London*).

Products sold for consumption in Bahrain were primarily gasoline, kerosene and light products and more valuable than the average of products sold for export (a much larger proportion of which were fuel oil and heavy products of lower value). Thus the removal of products sold for consumption in Bahrain from the overall 8 1/2 pence refinery profit margin (RPM) requirement placed the burden of satisfying that RPM requirement entirely on exported products. This meant that average taxable profits derived from exported products had to be increased, thereby providing additional tax revenue for the Ruler. (In this regard, it is relevant to note that the earlier 8th of February 1961 agreement had increased the refinery profit margin requirement from seven (7) pence sterling to eight and one-half (8 1/2) pence, effective as of the 1st of January 1961).

Also retroactive to the 1st of January 1962, but not agreed until the 11th of August 1964, a third pricing agreement was confirmed which related solely to export sales of refined products derived from Bahrain crude.[54] Simplified, this 1964 agreement provided that sales of refined products made to buyers for export would be at the buyers' posted prices, less a marketing allowance of three-sevenths (3/7) pence sterling per barrel. However, where the buyers (or their affiliates) then resold the products to non-affiliates at less than posted prices, then Bapco's sales prices would equal the prices actually realized by the original buyers (or their affiliates), less a marketing allowance of three-sevenths (3/7) pence. In other words, a distinction was made between the sale of refined products to Bapco/Caltex affiliates[55] for their own use, and refined product sales to affiliates for subsequent resale to non-affiliates.[56] It was agreed also that the refinery profit margin would no longer apply to Bahrain crude oil after the 1st of January 1962, but would be retained on imported crude oil.[57]

1966 – Royalty and Corporate Tax Payment Schedules

The next change took place on the 11th of June 1966, when Bapco confirmed its agreement [58] to the payment of royalties on a monthly basis, fifteen days in arrears, by the 15th of February 1967. Equally, tax was to be paid monthly during the taxable year on the basis of estimated income tax declarations, in twelve equal instalments, with estimates revised quarterly. Although these tax payment provisions could not amend the law, they did constitute agreement to pay on an accelerated schedule.

Two days later, on the 13th of June 1966, the *Bahrain Income Tax (Amendment) Decree, 1966* was published. In summary, this lengthy document comprised redefinitions of the term "exploratory well", and the accounting treatment of the costs related thereto for tax calculation purposes. It also redefined the accounting treatment, for tax purposes, of tangible and intangible oilfield assets.

The third significant aspect of the decree amended the treatment of royalty payments, which, up until that time, had been deducted from 50-per-cent of income to determine the amount of corporate income tax. The amendment required that for taxable years beginning after December 1963, royalties with respect to Bahrain crude oil refined and sold for export, not exceeding 12.5-per-cent of the value of the crude oil, be treated as expense and deducted from income in the calculation of taxable income.

New Royalty Agreement for Bahrain Crude Oil

The Middle East War in June 1967 was a critical episode in world affairs, particularly those concerning the international oil industry. The following month was also a watershed in Bahrain's oil affairs, for in July 1967 a new concept for the basis of calculating royalties on Bahrain indigenous crude oil was introduced.

Five Gulf rupees (*Coincraft, London*).

Until 1967, the value of Bahrain crude oil had been based upon the market price of similar regional crude oils brought into Bahrain for refining. On the 24th of April 1967, Bapco Limited confirmed an agreement [59] into which the company and the Ruler of Bahrain had entered on that day whereby royalties would no longer be calculated solely on a *volume* basis (at the rate of 10 rupees per long ton), but as a percentage of the *value* per barrel of Bahrain crude oil. Therefore, effective from the 1st of July 1967, the amount of royalty per barrel was considered equal to twelve and a half (12.5) per-cent of the value of Bahrain crude, including casing-head petroleum spirit. [60]

Currency Changes

It was also in 1967 that corporate income tax and royalties began to be *computed* in United States of America dollars, although both continued to be *paid* in pounds sterling. [61] For the purposes of such payments, the sterling equivalent of the dollar amount of each instalment of tax and each royalty payment was determined by using the bank's buying rate for United States dollars, as quoted by the company's London bank on the bank's last working day before the date of payment. [62] This change was effected partly for convenience, since the new Bahrain Dinar [63] was linked to the United States of America dollar, and as a response to the declining value of sterling. [64]

At the beginning of 1968 two other currency changes became effective, although the agreements were not confirmed in writing until August of that year. On the 1st of January 1968, the marketing allowance of three sevenths ($3/7$) pence – agreed on the 11th of August 1964 – was substituted by the words "one half ($1/2$) US cent". [65] Effective on the 1st of February 1968, the eight and a half ($8\ 1/2$) pence sterling refining profit margin was substituted by ten (10) United States of America cents. [66]

Revenue Enhancement to Participation

This situation continued until November 1970, when the Bahrain Government increased the rate of taxation on Bapco's producing, refining and marketing income from 50 to 55-per-cent, a move which marked the beginning of a rapid acceleration in the enhancement of Bahrain Government oil revenues. It was not until 1974 that the tax and royalty rates had reached a peak of 85-per-cent and 20-per-cent, respectively, by the end of that year. However, the most significant percentage increase in the Bahrain government's revenue during 1974, 39-per-cent, was the result of the revenue from the Abu Saa'fa oilfield being shown in the national budget.

The background to this began in early 1966, when the offshore oilfield between Saudi Arabia and Bahrain was brought into production by Aramco at an initial rate of 30,000 barrels per day. Its six wells were connected to a production platform in the field, from which a 28-mile long (45 km) 18-inch diameter (0.46 m) pipeline ran to the Ras Tannurah terminal, where the gas-oil separator was located. As early as 1958, the Governments of Bahrain and Saudi Arabia had agreed on the demarcation of their offshore boundaries, with provision for the two nations to share equally the revenues from what was hoped would become a productive field in the Abu Saa'fa area. In the event, this oilfield generated BD40.046 million for the Bahrain Government in 1974, 39-per-cent of the year's BD193.456 million total. Abu Saa'fa thereafter continued to make a major contribution to the State of Bahrain's annual budget.

Many international and national events contributed to the changes which occurred in Bahrain's oil industry during the decade of the 1970s. It may be concluded from Table 10, that the unprecedented rise in crude oil prices was an important factor which influenced this accelerated rate of change. It will be noted too that the evolution of Bahrain Government oil revenues during this period was associated with the incremental participation of the Bahrain Government in the ownership of the Bahrain oilfield and its assets. A discussion of the *process* of participation, rather than its financial impact, together with a brief analysis of OPEC's activities in response to world affairs during the same period, will appear in the next and final chapter.

Before studying Table 10, it will be helpful to highlight a few points which are not self-explanatory:

- *Crude Oil Prices:* from 1960 to 1970, the deemed posted price of Bahrain crude oil remained at US$1.60 per barrel. From 1970 to 1980, it rose from US$1.60 to US$31.40 per barrel, a dramatic 1,862.5-per-cent increase.

- *Marketing Allowance:* this was discontinued in November 1970.

- *Oil and Gas Wells:* effective on the 1st of January 1973, the Bahrain Government acquired 25-per-cent of the exploration and producing rights, operations and facilities of Bapco. The second phase, when the Government acquired a further 35-per-cent of the oilfield and its assets, took effect on the 1st January 1974. The combined effect of these two phases was documented in an *Agreement* dated 23rd November 1974.

  Following the second phase, it was agreed that all existing and new wells drilled for *oil production* would be owned by the Bahrain Government and Bapco Limited on a 60:40 basis. However, all new wells drilled solely for the *production of natural gas* would be paid for and owned 100-per-cent by the Bahrain Government.

- *Bapco Limited's payments to the Bahrain Government,* derived from the production of Bahrain crude oil, were based on the company's equity share of the Bahrain oilfield, which until the end of 1972, was 100-per-cent. On the 1st of January 1973, its equity share was reduced by a quarter to 75-per-cent and again, exactly a year later, still further to 40-per-cent. On the 15th of December 1979, an agreement was signed whereby the Bahrain Government completed its progressive ownership of the Bahrain oilfield and its assets by acquiring Bapco Limited's remaining 40-per-cent equity share. Although the financial impact of the 15th December 1979 agreement was made retroactive to the 1st of January 1979, the other terms of the agreement were applicable as from the 1st of January 1980.

It should be noted that six months later the next major step in Government participation occurred. On the 1st of July 1980 the Bahrain Government acquired a 60-per-cent equity share in the Bahrain refinery and became its majority shareholder.

- *Royalties*: rates were increased several times during 1974. In July, the royalty percentage of the deemed posted price of Bahrain crude oil was increased to 14.5-per-cent, in October to 16.6-per-cent and in November to 20-per-cent. Royalty payments ceased on the 1st of January 1980, having applied continuously for fifty-five years.

- *Refinery Profit Margin*: discontinued on Bahrain indigenous crude oil between the 1st January of 1962 and the 7th August 1977.

As a development of Table 10, the comparison in Tables 11 and 12 illustrate the percentage impact of the peak royalty and taxation rates at the end of 1974. *Example One* reflects the financial impact *without* 60-per-cent participation, whilst *Example Two* demonstrates the financial impact *with* 60-per-cent Government ownership. Both examples are based on 100 barrels of Bahrain indigenous crude oil.

To appreciate the precise dimension of the *Government's cumulative share of income* from the producing function of Bahrain's oil industry, it should be remembered when looking at *Example Two* that Bapco Limited was paying royalty and tax on its 40-per-cent share of Bahrain indigenous crude oil, as well as buying the Bahrain Government's 60-per-cent share of the oilfield's production for the manufacture of refined products.

One Bahrain Dinar (*Bahrain Monetary Agency*).

Finally, Table 13 illustrates the percentage ratios between Bahrain Government and Bapco revenues in 1974, derived from the production of Bahrain indigenous crude oil, expressed *without* and *with* participation. These are compared with 1952, the year in which taxation of corporate income was introduced, and 1980, the year in which the Bahrain Government became the 100-percent owner of the country's oilfield and its assets.

A 100 fil coin (*Bahrain Monetary Agency*).

A beam pump in the Bahrain oilfield, decorated as a giraffe (*Joe C. Torres*).

Table 10

BASIS FOR THE CALCULATION OF BAHRAIN GOVERNMENT OIL INDUSTRY

YEAR END	Year-end deemed price of Bahrain indigenous crude oil $ per barrel	Tax rate on producing/ refining/ marketing %	BAHRAIN INDIGENOUS CRUDE OIL
			Royalty as a % of the deemed price of Bahrain crude %
1970	1.6	55	12.5
1971	2.2	55	12.5
1972	2.4	55	12.5
1973	4.9	55	12.5
1974	11.0	85	20.0
1975	11.0	85	12.5 local/ 20.0 export
1976	11.4	79	ditto
1977	12.4	78	ditto
1978	12.4	78	ditto
1979	23.6	46	ditto
1980	31.4	46	Discontinued

*See preceding note: *Bapco Limited's payments to the Bahrain Government*

REVENUES RECEIVED FROM BAPCO LIMITED UNTIL 1 JULY 1980

BAHRAIN INDIGENOUS CRUDE OIL	REFINERY OPERATIONS		YEAR END
Participation (Government ownership) of the Bahrain field and its assets %	Refinery profit margin on Arabian light crude oil cents per barrel	Participation (Government ownership) of the Bahrain refinery and related facilities %	
0	9.0909	0	1970
0	9.0909	0	1971
0	9.0909	0	1972
25	11	0	1973
60	11	0	1974
	11	0	1975
60	11	0	
60	11	0	1976
60	11	0	1977
60	11	0	1978
60	11	0	1979
100*	11	60	1980

Table 11

COMPARISON OF BAHRAIN GOVERNMENT REVENUE IN 1974 WITHOUT AND WITH THE IMPACT OF 60

EXAMPLE ONE: *Without* 60% Bahrain Government ownership

BAHRAIN INDIGENOUS
CRUDE
US$ per 100 barrels

BAPCO REVENUE

"Market" value of 100 barrels of Bahrain crude oil @ a deemed posted price of US$12.00 per barrel

1,200.00

BAPCO EXPENSES
Production costs @ 30 cents per barrel x 100

(30)

Royalty (20% of deemed posted price) payable by Bapco to the Bahrain Government and deducted as an expense for tax calculation purposes

(240)

TOTAL EXPENSES

(270.00)

TAXABLE INCOME

930.00

INCOME TAX PAYABLE (85%)

(791.00)

BAPCO'S TOTAL NET REVENUE

139.00

BAHRAIN GOVERNMENT REVENUE

US$

Income Tax — 791

Royalty — 240 — 1,031.00

BAHRAIN GOVERNMENT'S TOTAL NET REVENUE

1,031.00

BAHRAIN GOVERNMENT OWNERSHIP OF THE BAHRAIN OILFIELD AND ASSETS

EXAMPLE TWO: *With* 60% Bahrain Government ownership

BAHRAIN INDIGENOUS
CRIDE
US$ per 100 barrels

BAPCO REVENUE

40 barrels x US$12.00 deemed posted price		480.00
BAPCO EXPENSES		
Production costs @ 30 cents per barrel x 40	12.00	
Royalty (20% of US$480, i.e. Bapco's 40%)	<u>96.00</u>	
TOTAL EXPENSES		<u>(108.00)</u>
TAXABLE INCOME		372.00
INCOME TAX PAYABLE (85%)		<u>(316.20)</u>
BAPCO'S NET REVENUE		55.80

Plus:
Income from discount allowed on Bahrain crude purchased
by Bapco from the Bahrain Government (60% of Bahrain's
net production). The difference between the deemed posted
price of US$12.00 and the price paid by Bapco is a discount
equal to 84 cents per barrel.

<u>50.40</u>

BAPCO'S TOTAL NET REVENUE   <u>**106.20**</u>

BAHRAIN GOVERNMENT REVENUE

Income Tax paid by Bapco on Bapco's 40%	316.20	
Royalty on Bapco's 40%	<u>96.00</u>	412.20

Plus:
60 barrels @ US$11.16 per barrel (price received by the
Bahrain Government from Bapco for the sale of Government
crude to Bapco at 93% of the deemed posted price). Bapco
purchased the Bahrain Government's 60% share of Bahrain's
oil production at a price equal to 93% of the deemed posted
price of Bahrain crude, which in this example is US$11.16
per barrel.   669.60

Less:
Production costs @ 30 cents per barrel x 60   <u>(18.00)</u>

BAHRAIN GOVERNMENT'S NET REVENUE FROM
BUY-BACK CRUDE   651.60

BAHRAIN GOVERNMENT'S TOTAL NET REVENUE   <u>**1,063.80**</u>

Table 12 DIVISION OF NET REVENUE FROM 100 BARRELS OF PRODUCTION

Without 60% Government ownership

BAPCO TAKE	139.00	(12%)
BAHRAIN GOVERNMENT TAKE	1,031.00	(88%)
TOTAL TAKE	**1,170.00**	(100%)

(BAPCO AND GOVERNMENT NET REVENUE COMBINED)

Table 13 PROGRESSIVE PERCENTAGE DIVISION IN REVENUES DERIVED

1952		1974	
		without 60%* Government ownership	
Government %	Bapco %	Government %	Bapco %
50	50	88	12

* Only for demonstrating the notional comparative figures,

Reflection

The enhancement of Bahrain's oil revenues evolved during several phases. The first, 1925-52, was a period during which the Government of Bahrain's oil-related revenues were derived from charges and royalty payments as a result of mutual agreements between the company and the Government. The second phase, 1952-5, saw the imposition of corporate income tax. After the formation of OPEC in 1960, the next decade witnessed a succession of fair pricing agreements which further enhanced the Government of Bahrain's oil revenues. It was also during this decade that currency changes were implemented to reflect the Bahrain dinar/United States of America dollar exchange rate established in 1965 and the devaluation of sterling in 1967.

WITHOUT OR WITH 60% BAHRAIN GOVERNMENT OWNERSHIP

With 60% Government ownership

BAPCO TAKE	106.20	(9%)
BAHRAIN GOVERNMENT TAKE	1,063.80	(91%)
TOTAL TAKE	**1,170.00**	(100%)

(BAPCO AND GOVERNMENT NET REVENUE COMBINED)

FROM THE PRODUCTION OF BAHRAIN INDIGENOUS CRUDE OIL

1974		1980	
with 60%* Government ownership			
Government %	Bapco %	Government %	Bapco %
91	9	100	0

as illustrated in Table 12

Following the declaration of Bahrain as an independent State in 1971, the Government increased its participation in the ownership and management of Bahrain's oil industry, and the royalty agreements were terminated. It is appropriate, therefore, that the final chapter should highlight Bahrain's economic achievements of the last two decades, and consider how oil revenues were progressively channelled into the consolidation of the nation's infrastructure, the diversification and expansion of the country's established industries, and the creation of new commercial enterprises, particularly in association with regional partners throughout the Gulf Co-operation Council member countries. [67]

Beam pump in the Bahrain oilfield decorated as a Hoopoe bird (*Joe C. Torres*).

CHAPTER TEN

The Years of Achievement

A Dynamic Decade of Change

His Highness Shaikh Isa bin Salman Al-Khalifa's proclamation, broadcast on the 14th of August 1971, declaring the creation of the State of Bahrain formally ushered in a decade of enormous political, social and economic change. To many Bahrainis, this statement signalled the dawn of a new age when both national and personal aspirations were transformed into tangible achievements. The future held both promise and prospects for everyone.

Bahrain was thus poised to emerge as a new, full member of the community of nations and a significant player on the world stage of international affairs. For the first time, Bahrain was entirely free to seek its own national destiny, to shape its own economic, legal and political institutions and to establish its own international relationships. The 1970s were to be a period of challenge, a time for dramatic change. Every segment of society and every aspect of life entered a period of transition which, in this single decade, manifested itself more intensely than at any time since the discovery of oil, forty years earlier. In the sphere of economic and industrial development, a total redirection and transformation took place, centring on the Bahrain Government's acquisition of the nation's oilfield and related assets, the creation of a national oil industry, the evolution of industrial diversification and the creation of an international offshore banking sector which placed Bahrain in the forefront of the world's leading financial institutions.

Already, many young Bahrainis had gained higher education qualifications and had returned home with ambitious expectations for their career development. By the beginning of the 1970s, many Bahraini Bapco employees in their early twenties were being groomed as future managers, gaining experience and further expertise in their chosen specialist disciplines. However, as many successful Bahrainis reflect today, one aspect of this rapid change generated real concern, namely the realities of a promotion pyramid which narrowed sharply towards the top, combined with a hungry enthusiasm for rapid career advancement. Fortuituously, Bahrain's burgeoning new industries of the 1970s – such as aluminium smelting, banking, insurance, commerce and hotel management – generated new employment opportunities and alternative career paths for eager young men and women. One Bahraini, representative of that generation, offered his tribute to the first four decades of oil industry operations in Bahrain from 1931 to 1971:

> "Bahraini society was divided into different trades. Before the oil industry came, we had farmers, divers, carpenters, welders, but they were all segregated into different communities. The oil industry brought them all together under one umbrella, Bapco, which also found and developed new practical and administrative skills for the long term, for which many industries in Bahrain are very grateful now" [Extract from interview with Yousif Khalil, 3 April 1988].

Through a process of almost continuous negotiation and modification of the royalty and corporate taxation structure, the Bahrain Government had, by 1970, ensured that its percentage share of the profits derived by Bapco from the production, refining and marketing of Bahrain's indigenous crude oil was equal to, or better than, that enjoyed by its Gulf neighbours – a "most favoured nation" relationship. [1] Its economic position was thus assured for the remaining productive life of Bahrain's oil and gas fields. With these financial arrangements achieved and the Bahrain Government's position secure, the nation's oil industry entered a new phase. As the 1970s began, a far-reaching new dimension emerged in the relationship which the Bahrain Government enjoyed with Bapco and Caltex, a reflection of the changing world in which they had hitherto comfortably co-existed.

Already, during the 1960s, the seeds of independence from both political and economic foreign influence had been sown, and the nations of the Arabian Gulf began to foresee the day when they would assume full control of their national destinies. It was also during this time that the Bahrain Government had become increasingly sophisticated and confident in articulating and achieving its economic objectives in negotiations with Bapco, Caltex and the British Government.

As the nation's oil and associated natural gas reserves began to decline, non-associated natural gas from the Khuff zone assumed an increasingly important role in the islands' economic and industrial development. Some twenty years earlier, in March 1947, Khuff gas (a non-associated gas) had been discovered at a depth of approximately 10,500 feet (3200 m). It was not until 1969 that commercial development from the Khuff structure began when two gas wells were drilled to supply gas-turbine fuel to the island's new aluminium smelter being constructed as a result of the Bahrain Study Group's recommendation in 1967 to form Aluminium Bahrain.

In 1970, almost as if to celebrate the new decade, Bahrain's oil output peaked at 77,000 barrels per day. From 1971 onwards the output began to decline steadily, dropping to 50,000 barrels per day by 1980, finally stabilizing at around 43,000 barrels per day throughout the following decade. In Arabian Gulf terms, this represented a very modest output. Nevertheless, it was on this relatively small production base of indigenous crude oil and associated gas, together with the production of non-associated Khuff natural gas, that Bahrain successfully built a thriving modern society enjoying regional and international respect in the world community.

OPEC Power Play

In Bahrain, the major events about to unfold during the 1970s had their origin in a number of dramatic political and economic developments which had occurred in the Arab world in the final years of the previous decade. For the international oil industry, the relative stability of the 1960s was shattered by the Arab-Israeli Six-Day War in June 1967. One of the far-reaching consequences of this conflict was the closure of the Suez Canal and the disruption of short-haul oil traffic to Europe. Although a new generation of very large crude carriers (VLCCs) was available to transport crude to Europe around Southern Africa, the incremental costs incurred often made the economics of this operation relatively unattractive. Within a year or so of the 1967 Arab-Israeli conflict, Arab oil-producing countries began to develop a new perspective of their relationship with the major international oil companies. Their interest began to focus on equity participation, on the ownership and control of their major natural resource – crude oil.

Some oil industry analysts suggest that a fundamental shift took place in 1969 when two significant events occurred in the Middle East. Tapline, an oil pipeline passing through the Syrian territory of the Golan Heights, by then under Israeli occupation, was sabotaged. This pipeline, which transported half a million barrels per day of Saudi Arabian crude oil across Jordan and Syria to the Mediterranean port of Sidon in southern Lebanon, was shut down for 112 days from the 30th of May 1969 to the 18th of September 1969 whilst it underwent repair. This additional disruption to the transfer of crude oil from the Arabian Gulf region's oil producers, in conjunction with the closed Suez Canal, meant that Libya and Algeria became the foci for supplying short-haul crude to the Mediterranean markets and the rest of Europe.

The second incident occurred on the 1st of September 1969 when King Idris of Libya was deposed by a group of military officers led by Colonel Muammar Gaddafi. The new Government that resulted insisted that oil companies in Libya should raise their posted prices of crude oil by 40 cents per barrel, apparently to reflect the quality of Libyan crude and to make the point that the country was well-placed as a short-haul location for the supply of crude oil into the nearby European markets.

Reasonably, one might ask how this affected Bahrain and its oil-producing neighbours. At the time, economic growth world wide had created an increased demand for energy which could best be satisfied by the increased supply of oil, thus increasing the oil sector's share of the total energy market at the expense of other primary energy sources such as coal. By June 1970, the new Libyan Government exerted pressure on the oil companies operating in the country, particularly Occidental. Although, for several months, the oil companies had strongly resisted the Government's pressure to increase their posted prices of crude oil by 40 cents per barrel, first one oil company and then, finally, all of them conceded to a 30 cent increase.

More significantly for the rest of the industry, the Libyan Government decreed an increase in the rate of income taxation from 50-per-cent to 58-per-cent. Occidental created a precedent by submitting to that increase, even though its contract with the Government under the 1955 petroleum law had given it the right to limit its tax liability to 50-per-cent. With that precedent, Colonel Gaddafi gave the other independent oil companies producing in Libya no choice but to submit to the increase in the tax rate. These companies included subsidiaries of Standard Oil Company of California and Texaco. Socal and Texaco were, in turn, linked to Bahrain through their equities in Caltex and the Bahrain Petroleum Company Limited.

OPEC swiftly capitalized on the Libyan breakthrough. At the organization's Caracas conference in December 1970, a common minimum 55-per-cent tax rate was agreed by the member governments. This was implemented in Bahrain on the producing, refining and marketing functions, effective from November 1970. In the following year, two further important OPEC conferences were held, one in Teheran

King Idris of Libya (*Popperfoto*).

and the other in Tripoli, both convened to exert greater government control over posted prices and also to agree between the member governments on their further increase. It was in this context of growing OPEC strength and its members' demand for control over the industry that the independent oil companies found themselves under mounting pressure to enter into participation negotiations which, at the time, they anticipated would span a decade or more. Thus, along with the rest of the international oil industry, the concept of national ownership of Bahrain's oilfield and its assets entered a serious stage of negotiation between the Bahrain Government, Bapco Limited and Caltex regarding the assets which belonged to Bapco. Although both Bapco and Caltex were separate companies, many of the persons involved in the negotiations were directors of both companies. Therefore, these persons made decisions for, and issued directives to, operating divisions of both Bapco and Caltex.[2]

The atmosphere during the initial phase of these negotiations was strongly influenced by the supply and demand situation which had developed in the late 1960s and continued throughout the early 1970s. On the 15th of February 1971, the Teheran Agreement provided for certain oil price increases to be phased over the next few years. The posted price of crude oil rose from US$1.80 (prior to the Agreement) to US$3.011 by the 1st of October 1973, just before the outbreak of the Middle East October War.

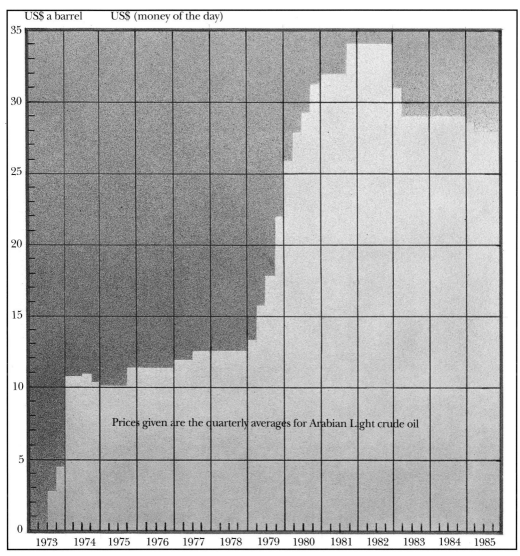

The oil price explosion between 1973 and 1985 (*artwork: Jane Stark; source: BP Statistical Review of World Energy*).

Meanwhile, early in 1973, realized market prices had begun to exceed the official posted prices for the most widely traded crudes. This was the result of a growing tightness in crude oil supply as oil companies found it difficult to expand their producing capacities at a rate sufficient to cope with rising world wide demand. This situation was clearly reflected by OPEC oil exports which had increased steadily over eight years from 15 million barrels per day in 1966 to 29.5 million barrels per day in 1973.

Bill Tucker, a retired Chairman of Caltex Petroleum Corporation, recalled that situation. Late in 1972 he visited some of Caltex's associates in Japan and explained to them that, for the first time in the history of their relationship, which went back to the end of the Second World War, Caltex would be unable to supply all their forecasted crude oil requirements for 1973.

> "We simply did not have enough crude oil to supply the whole system. That was a revolutionary idea. Caltex had never been crude-limited. There had always been plenty of it. Whatever we thought we might do, whatever market we saw that we might exploit, whatever refinery expansion we might build, crude oil supply was the one thing we never worried much about. ... We'd now reached a time when there was a shortage, a foreseen shortage" [Extract from interview with W. E. Tucker, 24 October 1987].

OPEC oil ministers meet in Vienna on the 24th of September 1975 to set a new oil price (*Popperfoto*).

This situation was further exacerbated when, on the 6th of October 1973, Egypt and Syria launched a joint military campaign to repossess part of the Arab territory which Israel had captured and occupied in 1967. This new round of Arab-Israeli warfare had a profound impact on the future of the international oil industry. Within two weeks, two events occurred on successive days which accelerated crude price increases and contributed significantly to the speed with which the governments of oil-producing countries assumed control of the major sources of OPEC crude supply.

The first event occurred at a meeting held at the Sheraton Hotel in Kuwait on the 16th of October. Based on the findings of a previous OPEC working group and in response to the Arabian Gulf states having demanded a substantial rise in the posted prices of crude oil from some members, OPEC made a dramatic announcement. The posted price of Saudi Arabian light crude, used by OPEC as a pricing benchmark for crude, was to be increased on a "take it or leave it" basis by 70-per-cent to US$5.119 per barrel. Faced with a tight supply position, the oil companies had no choice but to accept the OPEC price increase. Within less than three months, under the impact of the Arab oil embargo measures, the direct result of the October War, the posted price of Arabian light crude had again more than doubled to US$11.65 per barrel by the 1st of January 1974.

The second equally dramatic announcement was made on the 17th of October 1973, when the oil ministers of the OAPEC members, including Bahrain, met to decide how oil could be used as an economic and political weapon to support the Arab war effort. Whilst recognizing that it would be unwise to disrupt the economies of the western powers to the extent that the oil-producers themselves would suffer, measures had to be sufficiently severe to have the impact which the ministers desired. Thus, with the exception of Iraq, the OAPEC members [3] each agreed to cut their crude oil production by a minimum of 5-per-cent with immediate effect, using the September 1973 production levels as the base. At subsequent meetings held in November and December, further significant production cuts were announced. Thus OAPEC/OPEC ability to regulate a significant share of internationally traded oil emerged as a major factor to be reckoned with both in the international oil industry and in international diplomacy.

"Not since the launching of the first Soviet sputnik in 1957 did Western powers feel as vulnerable as they did in the face of the Arab oil embargo of the autumn and winter of 1973-4. The sense of crisis was deepened and prolonged by the steep price increases decreed unilaterally by members of the Organization of Petroleum Exporting Countries" [Rustow and Mugno, 1976, p.v].

For a group of nations which had so recently emerged from powerful external political and economic influence, these were heady days indeed. With this newly defined, albeit unexpected, power base, OPEC members were able to articulate more forcefully their growing national aspirations for ultimate political control and full ownership and management of their respective oil reserves.

The Emergence of Bahrain's National Oil Industry

Once the State of Bahrain came into formal being in August 1971, an aura of excitement and confidence emanated throughout the country. Almost immediately Bahrain initiated discussions for the Government's participation in the ownership of the nation's oilfield and its assets. As complex and ardent discussions between Bapco, Caltex and the Bahrain Government gained momentum, the new State increased its stature and its power at the negotiating table by expanding its membership and active participation in various international and regional organizations.

In January 1972, the State of Bahrain joined the United Nations Educational, Scientific and Cultural Organization (UNESCO). During the summer on the 11th of July, Bapco Limited became a founder member of the Gulf Area Oil Companies Mutual Aid Organization (GAOCMAO), an association formed to provide a co-ordinated response mechanism to a major oil spill in the Arabian Gulf. Two months later, in September, Bahrain joined the International Monetary Fund (IMF) and the International Bank for Reconstruction and Development (IBRD), more usually referred to as the World Bank.

Meanwhile, discussions to effect the Bahrain Government's participation in the ownership of the nation's oilfield and its assets were well under way. Negotiations finally culminated in an *Agreement* signed on the 23rd of November 1974 [4] between the Government of the State of Bahrain and the Bahrain Petroleum Company Limited whereby, in one document, the first two phases of the the Government's acquisition of the nation's oilfield and assets were formalized.

In the first phase, which took effect as of the 1st of January 1973, the Bahrain Government acquired from Bapco 25-per-cent of the exploration and producing rights, and related oilfield assets. The second phase, whereby the Bahrain Government acquired a further 35-per-cent of the oilfield and its assets, took effect on the 1st of January 1974. The combined effect of these two phases, expressed in the 23rd November 1974 *Agreement*, was announced by Bapco's President, Walter Stolz, in his introduction to the company's 1974 *Annual Report*:

> "We would like to begin our report by alluding to the arrangements developed in 1974 for the Government to acquire 60-per-cent ownership of The Bahrain Petroleum Company Limited's gas and oil facilities. It is our hope that the mutual understanding and co-operation existing for so long between Bapco and the Government will thereby be enhanced and that we may work together even more closely in the future."

Also effective on the 1st of January 1974, and formalized in the 23rd of November *Agreement* of that year, the Bahrain Government acquired the sole right to install *new* gas wells and related facilities at its sole expense, and to own the entire resultant gas production. [5] This move recognized that the Government considered gas to be a vital component of the economy of Bahrain.

Storage tanks under construction, early 1970s (*Bapco archive*).

As a result of the Bahrain Government's participation in the production and owner-ship of Bahrain indigenous crude oil, the arrangement under which Bapco was obliged to pay Berry Wiggins and Company Limited an overriding royalty was reviewed. On the 11th of March 1974, Berry Wiggins (a successor in interest to the Eastern and General Syndicate Limited), was notified that the overriding royalty payable to the company would end, a move which was resisted. However, on the 25th March 1975, the Board of Directors of Bapco Limited ratified an agreement to cease royalty payments.

By the beginning of 1976, an earnest view was expressed in Bahrain that a new Government entity in the form of a national oil company should assume full ownership and control of the nation's oil industry, thus replacing the Bahrain Petroleum Company Limited, an American-owned oil company registered in Canada. [6] The concept was that a fully integrated national oil company should be formed, totally responsible for all aspects of the development and management of Bahrain's oil industry. Differences of opinion arose as to the appropriateness of such a corporate entity which would have such broad-ranging responsibility and authority over the entire oil industry, the nation's primary source of Government revenue. Although discussions as to the ultimate role of the Bahrain National Oil Company BSC (Banoco) continued for several years thereafter, on the 23rd of February 1976, Banoco, wholly owned by the Bahrain Government, was incorporated in accordance with an Amiri Decree issued on that date. Under the terms of the Articles of Association published in the *Official Gazette* on the 4th of March 1976, [7] Article II states quite clearly that the company's objectives shall include its involvement *in any stage* of the oil industry, an implication for the existing industry which many people may not have fully appreciated at the time. Exploration for oil, natural gas and other hydrocarbon substances; the production, refining, purification, process, transport and storage of these substances as well as any refined products were stated as being within the new national oil company's objectives, together with the trading, marketing, distribution, sale and export of the refined products.

Following the formation of Banoco in 1976, the next step in the development of Bahrain's national oil industry took place on the 28th November 1978, when the Bahrain National Gas Company *Founders Agreement* was signed. Hitherto, no facilities had existed for the utilization of gas associated with the production of oil. As a result, the associated gas had been vented to the atmosphere, a necessary but nevertheless wasteful disposal of potentially useful hydrocarbons. The creation of Banagas would provide a processing plant in which associated gas would be converted into economically useful products, create new jobs and generate additional revenues for the Bahrain Government. To facilitate the development of this project, Bapco waived its remaining right to 40-per-cent of the associated gas produced from the Bahrain oilfield.

Upon the incorporation of the Bahrain National Gas Company BSC (Banagas) on the 22nd of March 1979, Banoco assumed 75-per-cent of the equity on behalf of the Government of Bahrain. The remaining 25-per-cent was shared equally between Caltex

Gas from 10,000 feet (3048 metres) below ground roars out under pressure during the testing of the first development well drilled in the Khuff zone formation (*Bapco archive*).

and the Arabian Petroleum Investment Corporation (APICORP) – an OAPEC entity – each holding 12.5-per-cent of the equity. The construction of the gas-processing plant on a site at the foot of Jebel Ad-Dukhan, not far from Bahrain's first oil well, was completed at a cost of US$95 million, just thirteen months after the *Founders Agreement* had been signed.

The plant, commissioned in late December 1979, was designed to process 110 million standard cubic feet per day of associated gas, and to produce daily 2900 barrels of propane, 2400 barrels of butane and 3610 barrels of naphtha, with 85 million standard cubic feet per day of residual dry gas being piped to Alba for use as gas turbine fuel. This presented a major economic opportunity, for it will be recalled that, hitherto, Bapco had been supplying Khuff gas to Alba. As Table 14 shows, this residue gas, previously vented into the atmosphere, replaced a major portion of Alba's Khuff gas consumption and so conserved Khuff gas for other uses.

The Government of Bahrain had made enormous progress towards the creation and consolidation of a national oil industry, an improbable vision perhaps in 1970, but one which had become a reality by the close of the decade. Before considering the key events of the 1980s, discussion of the successful oil and gas downstream projects, and of oil-related enterprises established in the 1970s, will help to show how these ventures formed the foundation for Bahrain's future development.

The Aluminium Bahrain smelter under construction (*Bapco archive*).

Seismic equipment (*Bapco archive*).

Expanding the Economy

In the mid 1960s gloomy predictions could be heard in the community regarding the estimated remaining productive life of the Bahrain oilfield. In an economy almost totally dependent upon a single depleting natural resource, this was a matter of very real public and private concern.

When oil was discovered in Bahrain on the 1st of June 1932, Well Number One was the first to have been drilled on the archipelago as well as the neighbouring Arabian peninsula. This remarkable success, cautiously predicted by Fred Davies during his 1930 survey, led to the discovery of vast oil reserves, first in Saudi Arabia and later in other Arabian Gulf countries. Bahrain's onshore oilfield proved to be small by comparison with the enormous reservoirs of her neighbours. Even more disappointing was the failure of extensive seismic work and exploratory drilling to discover new oil reservoirs off shore within Bahrain's territorial waters. Many people were thus convinced that Bahrain's oil reserves and related revenues would be exhausted within fifteen years. As a result pessimism concerning Bahrain's near-term economic outlook prevailed in many quarters.

An added concern, as discussed already, came from analyses of the 1965 census which showed an urgent need to expand the nation's commercial and industrial activity to create thousands of jobs for young men and women who were about to seek employment. These economic and social imperatives required that the Bahrain Government should develop alternatives to reduce the country's dependence upon the oil industry as the nation's primary source of revenue and employment.

The first major achievement in industrial development occurred on the 9th of August 1968 when Aluminium Bahrain (Alba) was incorporated, with the Bahrain Government as a substantial shareholder. One of the last official acts of Mr Roy L. Lay before he retired as Chairman of American Overseas Petroleum Limited and a Director of Bapco at the end of 1968,[8] was to sign a contract[9] under which Bapco would supply 60 million cubic feet of gas per day[10] from Bahrain's prolific Khuff gasfield. Since Alba's commercial feasibility was a function of the availability of low-cost electrical power generation, the company planned to install its own gas-turbine driven generators to utilize competitively priced fuel gas supplied by Bapco. In 1969, Mr L. D. Josephson, Bapco's newly appointed President, advised the Ruler:

Table 14

GAS SUPPLY TO ALBA

Year	Khuff gas mm.s.c.f.d.*	Residue gas (from Banagas) mm.s.c.f.d.	Total mm.s.c.f.d.
1971	15		5 (Alba commissioned)
1972	74		74
1973	104		104
1974	112		112
1975	112		112
1976	114		114
1977	115		115
1978	115		115
1979	118		118 (Banagas commissioned)
1980	49	62	111
1981	31	92	123
1982	41	100	141

* mm.s.c.f.d. = million standard cubic feet per day

[Source: Producing and Technical Division, The Bahrain National Oil Company]

"In Company [Bapco] operations we are happy to report that the first development well drilled into the Khuff Zone, a sweet gas field underlying the oil reservoir, was completed in September. Gas from this field will be used to produce electrical energy for Aluminium Bahrain (Alba) for whom a large aluminium reduction plant is currently under construction. It is hoped that the availability of this large reserve of sweet gas will attract further industry to Bahrain and lead to rapid expansion and diversification of the economy" [*The Bahrain Petroleum Company Limited, Annual Report, 1969*].

Less than two years later, on the 11th of May 1971, the Ruler of Bahrain, His Highness Shaikh Isa bin Salman Al-Khalifa attended a celebration ceremony on the 11th of May 1971 and conducted Alba's first "pour". By the end of 1972 and the first eighteen months of operations, the smelter had produced 77,567 tonnes [11] of aluminium ingots, with an ultimate production target of 120,000 tonnes per year for sale to world markets.

By the end of its first year of operation, Alba had provided new jobs for 1282 employees, of whom 1136 (89-per-cent) were Bahrainis. This welcome addition to Bahrain's labour market and career opportunities in a modern high-technology industry, peaked in 1974 when the company's workforce comprised 2992 employees, of whom 2308 were Bahrainis.

Several spin-off industries were created as a result of Alba's success. These included Bahrain Atomisers in 1972, Bahrain Aluminium Extrusion Company (Balexco) in 1977 and Gulf Aluminium Rolling Mills Company in 1981. Whilst the Government of Bahrain wholly owns Balexco, it has a substantial shareholding in the other three companies just mentioned. Another spin-off, this time in the private sector, was the Midal Cable Company, formed in 1978 as a Bahrain-Saudi Arabia joint venture. The equipment and technology for the Midal plant was transferred to Bahrain from a factory in Australia. The unique feature of Midal's plant design was that molten aluminium was to be transported by truck in four ton capacity crucibles, direct from Alba's pot room, to a holding-furnace in the cable factory. The molten aluminium was converted into rod of $^3/_8"$ (10 mm) diameter. Approximately two thirds of Midal's rod production was marketed to manufacturers in other Arabian Gulf states and in the Indian sub-continent for the production of electric cables. The remaining one third was used by Midal for the manufacture of cables for high-voltage overhead power transmission and electrical distribution systems. With a consumption of 36,000 tonnes per year, Midal Cables became a very convenient major outlet for Alba's annual production.

Meanwhile, along with the rapid development of Bahrain's aluminium industry, the foundation stone of the Arab Shipbuilding and Repair Yard Company (ASRY) was laid on the 30th of November 1974. The shareholders in this major oil-related diversification project were seven of the nine OAPEC members, including Bahrain,[12] with Lisnave of Portugal being assigned a sole management contract. The ultimate dry-dock design was increased to accommodate a supertanker of 500,000 dead-weight tons.[13]

The idea for this enterprise emerged in 1966 when the Bahrain Economic Development Study team's research had shown that 25 million tons of new tanker construction were under way in the world's shipyards. Of this, 64-per-cent comprised vessels of 100,000 tons or more, 31-per-cent of vessels between 100,000 and 60,000 tons, and the balance in smaller tonnage. This massive surge in supertanker construction caused a severe shortage of dry-docks capable of handling them. Only Naples and Singapore, west and east of the Suez Canal respectively, were capable of handling these large vessels, which were operating in increasing numbers in the Arabian Gulf and the Gulf of Oman areas.

It was noted that Bahrain's geographical position was about halfway between Naples and Singapore, the distance from Bahrain to Naples being 4251 nautical miles, and from Bahrain to Singapore 3652 nautical miles. Also at the time, the majority of these supertankers transporting crude oil to Europe and the Orient traversed the Arabian Gulf, passing in close proximity to Bahrain. Thus a dry-dock, together with four supertanker wet-berths, was seen as a very attractive commercial proposition. It also had the very real added advantage of requiring an anticipated workforce of about 1000 employees, most of whom would be highly skilled craftsmen.

On the 23rd of October 1977 ASRY's first customer, an Italian supertanker, 1230 feet (375 m) long, 246 feet (75 m) wide, and 40 feet (12.24 m) deep, sailed into the new dry-dock. Less than two months later, on the eve of National Day, the dry-dock was officially inaugurated on the 15th of December 1977.

As the 1970s drew to an end, the last of the major industrial projects launched in that decade was the Gulf Petrochemical Industries Company (GPIC). Created as the country's first petrochemical venture, GPIC utilized Khuff gas as feedstock in the production of two basic petrochemical products, ammonia and methanol. Initially, GPIC was incorporated on the 5th of December 1979 with the Bahrain National Oil Company (Banoco) and Petrochemical Industries Company (PIC) as shareholders on behalf of the Governments of Bahrain and Kuwait respectively. Almost six months later, on the 29th of May 1980, GPIC was re-incorporated to include Saudi Basic Industries Corporation (SABIC), representing the Government of Saudi Arabia. Each participant acquired one third of the company's equity. The project plan required GPIC's output to reach 1000 tonnes of ammonia and 1000 tonnes of methanol per day. The US$450 million complex, sited on 600,000 square metres of reclaimed land east of Sitra island, went on stream on the 19th of May 1985 with the production of methanol, followed two months later by the production of ammonia on the 18th of July.

ASRY dry-dock, Bahrain, July 1977 (*Bapco archive*).

Bahrain – The Banking Centre of the Gulf

At the start of the most dynamic decade in the recent history of Bahrain, the entire world faced serious financial problems when the post-war fixed exchange rate structure collapsed at the beginning of the 1970s. Throughout this period of turbulence in world foreign exchange markets, the Bahrain Dinar showed remarkable stability and strength against the US dollar. In January 1974, the posted crude oil prices had increased to US$11.65 per barrel (from US$1.80 prior to the 15th February 1971 Teheran Agreement), with a corresponding increase in the prices of refined petroleum products.

Partly in response to the volatile exchange markets world wide and, more particularly, to meet a need for more sophisticated management of the financial sector in Bahrain, the Government decided to replace the Bahrain Currency Board with a Monetary Agency. Thus, on the 5th of December 1973, the Bahrain Monetary Agency Law was signed, a move which gave the BMA full central banking powers over the banking and monetary sectors of the country's economy, although technically the Agency remained subordinate to the Ministry of Finance.

Alan E. Moore, appointed at the end of 1974 as the Agency's first Director-General, reflects that it is doubtful if those who came to Bahrain in 1921 to open the Eastern Bank, now the Standard Chartered Bank, could have foreseen that banking would become one of Bahrain's major industries in terms of Bahrainis employed and local value added. This process began in January 1975 when the Bahrain Monetary Agency took over the function of purchasing dollar oil revenues from the Bahrain Government, previously handled by the National Bank of Bahrain. This move enabled the BMA to establish full control of the dinar exchange rate, which it set slightly below 395.80 fils per US$, at which level it was held steady for the next three years.

As it turned out, the creation of the Bahrain Monetary Agency at the end of 1973 was particularly timely. As demonstrated on page 298, a dramatic seven-fold increase in Bahrain Government revenue occurred during 1974, a 685-per-cent jump from BD13.173 million in 1973, to BD103.456 million the following year. Of this remarkable enhancement, 39-per-cent of the 1974 total – BD40.046 million – was revenue from the Abu Saa'fa oilfield being recorded in the national budget. The remaining 61-per-cent of the Govern-

The Bahrain Monetary Agency as depicted on a one dinar note (*Bahrain Monetary Agency*).

ment's total revenue for 1974 reflected the impact of various crucial factors: the Government acquired 60-per-cent equity in the producing field, effective on the 1st of January 1974; crude oil prices rose to unprecedented levels; the royalty rate increased from 12.5-per-cent to 20-per-cent; and, finally, the tax rate on Bapco's producing, refining and marketing activities increased from 55 to 85-per-cent.

Against this background and that of other events in the Middle East, many international banks indicated in 1975 their wish to establish a branch in the Arabian Gulf, possibly in Bahrain. Moore suggests that this was in addition to the fact that existing banking systems in Bahrain were equipped to handle the investment and recycling of the rapidly growing surpluses of the major oil-exporting countries of the Gulf. "Oil revenues not required for immediate conversion bypassed the local banking systems and were invested directly in the international money markets in Europe or America. That is why, for much of 1974, the local banking statistics had shown little evidence of the growing wealth of the region" [Moore, private communication, 1980].

The Bahrain Monetary Agency concluded that the creation of an offshore banking system would have three direct advantages. Firstly, the banking systems throughout the Arabian Gulf region would benefit from it. Secondly, it would attract international

Bank dealing room, Bahrain (*Gulf Colour Laboratories, Bahrain*).

banks which sought a Middle East link in their world wide network. Thirdly, it would encourage the diversification of Bahrain's economy away from its limited oil base by stimulating foreign investment in new projects.

Furthermore, the establishment of an active and progressive banking industry in Bahrain would have a multiplier effect in the other industries being developed simultaneously, some of which have been discussed. By this time, Bahrain had established a very good telecommunications system operated by Bahrain Telephones, now Bahrain Telecommunications Company (Batelco), and excellent airline links through Gulf Air's expanding network. Since bank dealing rooms depend upon complex telephone and telex networks, and bankers rely on a reliable reservations communication system for regional transportation and hotel accommodation, Bahrain offered an attractive "package" to foreign investors. This included freedom from exchange controls and the fact that Bahrain's time zone "bridged" those of Europe and Asia. Thus, from 7 a.m. local time, Bahrain could offer the Hong Kong, Singapore and Tokyo markets at least three hours' afternoon trading, before the European markets began their morning trading on the same day. [14]

Banking was seen too as a potentially large employer of young people starting their careers, as well as an employer of skilled labour. By the end of 1979, over fifty offshore banking licences had been granted to Arab-based, Arab-connected banks, as well as those with head offices in other countries world wide. Expressed in US dollars, the volume of assets of the offshore banking units (OBUs) grew from US$3.5 billion in June 1976 to US$28 billion at the end of December 1979, a volume which approached that of Singapore.

Bahrain Government Completes its Acquisition of the Producing Field

Meanwhile, following the 1974 agreement which had given the Bahrain Government 60-per-cent ownership of the oilfield and its assets, negotiations continued throughout the latter half of the 1970s regarding the Government's desire to complete its full acquisition of the Bahrain oilfield. William E. Tucker, representing Bapco Limited, recalled that during these negotiations "we were talking about reaching agreements which would put Bahrain on an equal footing with arrangements which had been completed, or were evolving, in Saudi Arabia" [Extract from interview with W. E. Tucker, 24 October 1987].

By June 1976, five key issues had emerged from the discussions: interest payable on amounts due to the Bahrain Government and Bapco as a result of retroactive changes in 1974 arrangements; the feasibility of continuing with the 60:40 ownership of the producing field and related assets; arrangements for processing the Bahrain Government's crude oil through the Bapco refinery; the basis for valuing Bahrain crude processed by Bapco; and, finally, services to be provided by Caltex to the oil-producing operations after the Government had acquired 100-per-cent ownership of the oilfield.

These, and a number of related issues, formed the basis for extensive negotiations over the next two years. By early 1979, the general format of a settlement had begun to emerge. Another critical factor which had to be addressed by the negotiators was re-examination of the basis for arriving at the notional refinery profit margin for the purposes of calculating the Bapco refinery's taxable income. Between 1973 and 1979, the notional refinery profit margin (RPM) had remained unchanged at the fixed amount of 11 cents-per-barrel. By early 1979, it had become apparent that the agreed fixed per-barrel *notional* RPM mechanism no longer bore any relationship to the much higher *actual* profit margins being achieved at that time. One reason for this situation was that refined product prices were forging ahead in a very volatile market, with crude oil prices dragging behind.

Tucker reflected that "we were out in no man's land. We could no longer seek the protection of the arrangements that were being made in the volumetrically much more important regions such as Saudi Arabia and the other countries involved. We had to find a mechanism of our own to suit the Bahrain operation." Whilst travelling to Europe during a lull in the debate, it occurred to Bill Tucker that one way to find an "anchor" might be to examine the basis for the RPM calculation which had been first implemented in 1955. In an effort to develop a new basis for arriving at Bapco's taxable income, he suggested to Yousuf Shirawi that the Bahrain Government and Bapco should revert to a time when the Government could agree that 11 cents-per-barrel was an appropriate notional refiner's profit margin.

The glass globe of a Bapco Limited petrol pump.

After much discussion, it was agreed that 1977 would be an acceptable reference year for this purpose. In developing a new formula for tax calculation purposes, the 11 cents-per-barrel was expressed as a percentage of the 1977 profit margin (product revenues minus product cost, including the cost of crude). This turned out to be 13.444% of the profit margin. Both parties accepted this as the new basis for arriving at the taxable income of the Bapco refinery and agreed to apply the new mechanism when the Government completed its 100-per-cent acquisition of the Bahrain oilfield.

This milestone was finally reached on the 15th of December 1979, when an *Agreement* was signed by His Excellency Yousuf Shirawi, on behalf of the Bahrain Government, and William E. Tucker, representing Bapco Limited. The terms of the agreement provided for the Bahrain Government to "acquire the Company's 40-per-cent remaining interest in the exploration and producing rights, operations and facilities and the related production"[15] of the Bahrain field as from the 1st of January 1980. However, the financial effects of the agreement to the Government were made retroactive to the 1st of January 1979.[16]

A number of other significant changes were made by, or concurrently with, this 15th of December 1979 document. In particular, all previously existing agreements between Bapco Limited and the Bahrain Government were cancelled, except for a short list of "Agreements to be Preserved", such as those relating to Banagas. Notable among the cancelled agreements were the 1934 *Mining Lease*, together with its various amendments and supplements. In addition, a new lease was executed to cover areas expected to be required for Bapco's ongoing operations, such as the refinery, the wharf, pipeline rights-of-way and so on. This new lease, now jointly-held by the Bahrain Government and Caltex Bahrain, is licenced to Bapco BSC. Also, a totally new Income Tax Law was enacted by Amiri Decree which set a 46-per-cent tax rate and conformed to modern

His Excellency Yousuf Shirawi, Minister of Development and Industry for Bahrain (right), and William E. Tucker, President of Caltex, sign the *Participation Agreement* on oil and gas production and local marketing facilities, 15th December 1979 (*Bapco archive*).

corporate income tax practice. Like the previous tax law, it applied only to a limited class of companies in, but not all, oil related activities.

As for the marketing of refined petroleum products manufactured by Bapco Limited and sold for consumption in Bahrain, the 15th December 1979 *Agreement* provided for the Government to assume responsibility for this activity by acquiring ownership of the company's assets in service stations and the related transport. [17] Although the terms of the document did not say so specifically, in practice this meant that Banoco assumed full responsibility for the local marketing of refined products in Bahrain.

An *Interim Operating Agreement*, in the form of a side letter to the main agreement, was signed simultaneously on the 15th of December 1979. This provided for the producing facilities and operations to be managed and conducted by Bapco Limited on behalf of the Bahrain National Oil Company (Banoco). In November 1981 the Banoco Phase-In Committee was formed with the Government's intention that Banoco should take full responsibility for the Bahrain producing field. The procedure for orderly transfer was promptly agreed to and implemented, effective from the 1st of January 1982. On the same day, the Bapco exploration and producing division employees were transferred to the Banoco payroll.

In the final letter of this compendium of 15th of December 1979 documents, the Government reassured Bapco that "we are happy with the way in which the refinery is run". On this subject, it seemed as though the Government's future focus was to be on consultation rather than negotiation:

> "The Bapco refinery is an important and an integral part of the life of Bahrain. It is, of course, in both our interests that it should operate with efficiency and profitability, but there are wider political and economic implications. ... In many countries of the Gulf, the Government has taken over the refinery. We have not done so. What we wish to do is to be brought into its affairs on a consultative basis" [Shirawi, unnumbered letter to the President of Bapco, 15 December 1979].

By anyone's standards, the 1970s had been remarkable years of achievement and change, culminating in the final month of the decade with the Bahrain Government's 100-per-cent acquisition of the nation's oilfield and related assets. As the ink dried on the 15th of December 1979 *Agreement* and side letters, apparently the negotiators could now relax. Little did the Bahrain Petroleum Company realize how events were to unfold.

The Birth of a New Bapco

Bill Tucker recalls that during the following morning, Bahrain's National Day, the Shirawis invited himself and his wife to their home for coffee. Once the children arrived, there were not enough chairs to go round, so:

"I remember Yousuf sat me down on a cushion on the floor. He told me that he was under extreme pressure and that we would have to talk about further negotiations promptly. We had to find a formula for the refinery and draw up new agreements. … My heart sank, because I'm saying to myself, we're going to start out completely from scratch" [Extract from interview with Bill Tucker, 24 October 1987].

Clearly, consultation was not the *only* "name of the game". The Bahrain Government had now declared its intention to acquire an equity interest in the Bapco refinery. In fact, as Bill Tucker soon discovered, they did not "start out completely from scratch". Unlike the somewhat protracted, just completed, oilfield acquisition discussions, the refinery negotiations moved forward swiftly. Clearly, the Government's negotiators had learnt a great deal during the previous three years. Industry sources in the Arabian Gulf had also informed them that refinery profit margins had reached a financially dazzling all-time high. Cabinet colleagues thus urged the negotiators to conclude the Government's participation in the Bahrain refinery promptly so that the Government could expedite its access to the then booming refinery cash flow.

In cabinet circles, the main area of debate focused on one question, how much equity in the refinery should the Government insist upon, the popular view being that it should acquire 100-per-cent participation, thus ending foreign ownership in the nation's oil refinery. Others recognized, however, that serious consideration had to be given to the high level of technology required for refinery operations; the ongoing need for ready access to reliable offshore technical assistance from technology leaders in the refining industry; and the complexities and uncertainties of trading refined products in world markets.

The weight of dialogue moved in favour of a partnership which would be sufficient to retain Caltex's active participation, but which would place the Government securely in control of the enterprise. Thus, a decision was reached that the Bahrain Government should acquire a majority shareholding of 60-per-cent, while Caltex would retain 40-per-cent of the refinery's equity ownership. Concurrent with the refinery participation discussions, consideration had to be given to the provision of technical services to the new joint venture. The result was a five-year Technical Services Agreement (TSA) in which Caltex undertook to provide technical personnel and services to the refining company to ensure plant performance, reliability and safety, in keeping with international oil industry standards.

Remarkably, within six months, these arrangements were formalized on the 26th of June 1980 when the Bahrain Government and the Bahrain Petroleum Company Limited executed a document entitled *Agreed Principles*. These set forth the basic terms of the Bahrain Government's acquisition of a 60-per-cent interest in the Bahrain refinery.

Just a few weeks later, on the 19th of July 1980, the *Bahrain Refinery Participation Agreement* was signed which provided for the Bahrain Government to acquire from Bapco Limited the "beneficial ownership of a 60-per-cent interest in the Bahrain refinery fixed assets, equipment, materials and supplies", and to assume responsibility for 60-per-cent of the refinery operations' obligations and liabilities. The Bahrain Petroleum Company Limited continued to own a 40-per-cent interest in the Bahrain refinery assets and a corresponding 40-per-cent responsibility for its operations.

Together, these two documents comprised the *Initial Agreements* and paved the way for the preparation of a further document, the *Participants' and Operating Agreement*, which was signed almost ten months later, on the 4th of May 1981. All three documents became effective on the 1st of July 1980.

Thus, between the signing of the first oilfield participation *Agreement* on the 23rd of November 1974 and the *Bahrain Refinery Participation Agreement* on the 19th of July 1980,

Company logos representing the oil industry in Bahrain today (*Jane Stark*).

the Bahrain Government had achieved its plan of acquiring control of Bahrain's oil industry, a remarkable *tour de force.*

It was not until the 21st of May 1981 that the new Bapco, a Bahrain Government-Caltex joint venture, was incorporated as a Bahraini registered company: the Bahrain Petroleum Company BSC (closed). The company was formed solely to manage and operate, on behalf of the Bahrain Government and Caltex, the Bahrain refinery and its related facilities. As such it did not own the refinery assets, did not purchase crude oil and did not market its refined products. This situation continues to apply today. Soon after the incorporation of Bapco BSC (closed), the old Bapco Limited was dissolved on the 8th of December 1981.[18]

Speculation among Bapco employees as to what the new company would be called had generated a certain amount of discussion. At Government level there had been but one question and one answer. When the Minister of Development and Industry sought the Prime Minister's advice, he replied unequivocally that the name of the Bahrain Petroleum Company should be retained. And so to all outward appearances, with the exception of the logo, Bapco was still the old Bapco. It was felt that to give the refining company any other name would eliminate an important part of the previous fifty years of Bahrain's industrial and social heritage. Many people had worked for or with Bapco for all their working lives, and many families in Bahrain had some form of connection with the company. As one Bahraini former Bapco employee observed: "Bapco's footprints can be found everywhere throughout Bahrain."

A senior Bahraini member of Bapco BSC's management expressed the same feeling another way. He observed that although the company's shareholders had changed, and the role of the company in the nation's oil industry had diminished over the recent years, to change the company name would have denied the next generation the opportunity to be identified with this aspect in the evolution of modern Bahrain. If the name Bapco were to disappear, the perception would be that the continuity of the nation's economic development had been broken, even though obviously this would not have been the case.

Meanwhile, on the 9th of December 1980, Caltex Bahrain Limited was incorporated in Bermuda as a new wholly-owned subsidiary of Caltex Petroleum Corporation (CPC). The new company was to be responsible for Caltex's continuing presence in Bahrain, namely as a 40-per-cent shareholder in Bapco, as a 12.5-per-cent shareholder in Banagas, and as an active supplier of aviation fuels at Bahrain International Airport. Caltex Bahrain was also to become an important petroleum products trading centre within the Caltex international logistics and trading network.

For operational convenience, Caltex Bahrain Limited was dissolved less than two years later, on the 26th May 1983. In its place Caltex created a division of Caltex Trading & Transport Corporation, a 100-per-cent owned subsidiary of Caltex Petroleum Corporation. This was undertaken with the consent of the Bahrain Government and registered in Bahrain as Caltex Bahrain (a Division of Caltex Trading and Transport Corporation).

National Oil Policy

The changes which were taking place in the oil industry world wide, particularly in the Arabian Gulf area, and the speed with which they had occurred, resulted in governments taking an increasingly active role in deciding the future of the industry. Bahrain was no exception. The establishment of Banagas and GPIC, the expansion of power plants requiring natural gas, questions relating to the use of natural gas as a source of energy or as a raw material for the petrochemical industry and other downstream ventures, determined a need for closer examination of industry policies and the laying down of longer term development priorities.

As a result, a sub-committee of the Council of Ministers was formed in late 1979, initially to provide guidance to the Council and to the Minister of Development and Industry. By early 1980, it became obvious that such a committee, on a permanent basis, would be extremely beneficial for the future management of this vital sector of the economy.

On the 3rd of November 1980, an Amiri Decree was promulgated to create the Supreme Oil Council under the Chairmanship of the Prime Minister, His Highness Shaikh Khalifa bin Salman Al-Khalifa, and attached to the office of the Council of Ministers. The Supreme Oil Council's function, as stated in the *Official Gazette*, [19] was to formulate general oil policy so as to ensure the conservation of petroleum resources and to develop alternatives to them. The Council was also charged with the responsibility of supervising any petroleum organizations which might be established and to develop petroleum-related industries, thereby ensuring optimum investment of oil resources and the greatest possible returns on them.

In pursuit of these objectives, the Decree made provision for the Supreme Oil Council to decide on the creation of appropriate new companies, to determine the participation of the State of Bahrain in the founding of those companies and organizations, and to decide the takeover of existing companies by the State of Bahrain, together with their reorganization and the Government's participation in them. The Supreme Oil Council had the additional brief to set the maximum limit on the companies' borrowing from the Government or its financial institutions, and to guarantee their loans. It also had the right to appoint the boards of directors of those companies.

At the same time, the Minister of Development and Industry, His Excellency Yousuf Shirawi, as a member of the Council, [20] was charged with the responsibility for implementing the Council's decisions, for the day-to-day oversight of the oil industry as a whole, and the management of oil and gas related companies in which the Government held an equity interest.

Within weeks of its formation and after much debate, the Supreme Oil Council arrived at a far-reaching policy decision regarding the future structure of the oil industry in Bahrain. Under this policy, it was the intention that each major segment of the industry would be a separate corporate entity, functioning under the overall policy direction of the Supreme Oil Council and the routine oversight of the Ministry of

Development and Industry. Each corporation would be established with such partners and corresponding equity participation as would be appropriate to its corporate objectives and business activities. Each company was to have its own board, its own Chief Executive and a management structure that would be held fully accountable for the economic and operational success of the enterprise. Thus the Council had chosen to create the concept of a group of efficient, profit-motivated business enterprises, each able to compete successfully in the international oil industry. Under this corporate structure, each company would be capable of being accurately assessed as to its corporate performance, and its contribution to the nation's revenue and/or economy. The alternative choice would have been to create a single holding company with operating divisions to administer the country's oil and gas industry.

Today, the Supreme Oil Council, under the Chairmanship of the Prime Minister, provides decisive corporate leadership which continues to have a profound impact on the nature and scope of the Bahrain Government's participation in all aspects of the nation's oil industry. It was in this context that the Supreme Oil Council subsequently addressed the question of the Bahrain National Oil Company's role as envisaged at the time of its incorporation in 1976. On the 25th of March 1981, Legislative Decree No. 3 of 1981 formalized the following amendment:

> "The Object is to engage in the oil industry, inside and outside Bahrain. This shall include undertakings in any stage of the industry from exploration and drilling for oil and natural gas and other hydrocarbon matters, production and storage of the aforesaid materials and to engage in trading, marketing, distribution, sale and export thereof."

This emphasized the Bahrain National Oil Company's role in the management and operation of the Bahrain oilfield and the marketing of the refined petroleum products from the Government of Bahrain's share of the Bahrain refinery.

By the mid 1980s, another important segment of the oil industry had attracted the Government of Bahrain's attention, namely that of aviation fuelling at Bahrain International Airport. The beginning of this service's rapid development goes back to the early 1950s, when international airlines were extending their networks and introducing aircraft capable of long-haul travel. Bahrain became a vital fuelling stop on the routes extending from Europe to the Far East, servicing the new generation of commercial aircraft, including perhaps the most well-known of its time, Concorde. Bahrain was also a hub for air travel within the Arabian Gulf region.

Responding to these fuelling needs, Caltex and BP jointly provided a highly sophisticated service to the international airlines uplifting jet fuel at Bahrain airport on Muharraq island. As the evolution of a national oil industry took shape, and to enhance the Bahrain Government's revenue, it was only natural that these fuelling operations should be provided by a national entity. Therefore, on the 6th of June 1985, the Bahrain Aviation Fuelling Company BSC (closed) was incorporated with the idea that it should

A Bahrain Aviation Fuelling Company bowser at Bahrain International Airport (*Bafco*).

Table 15
OIL RECEIPTS AS A % OF BAHRAIN'S REVENUE 1935 to 1989 **[BD or BD equivalent]**

Year	Oil Receipts	Total Revenue	Oil Receipts as a % of Total Revenue
1935	57,000	134,000	42.54
1936	69,000	165,000	41.82
1937	317,000	434,000	73.05
1938	355,000	475,000	74.74
1939	324,000	458,000	70.75
1940	355,000	490,000	72.45
1941	297,000	443,000	67.05
1942	262,000	398,000	65.83
1943	271,000	470,000	57.66
1944	279,000	550,000	50.73
1945	303,000	640,000	47.35
1946	311,000	667,000	46.63
1947	390,000	868,000	44.93
1948	459,000	1,033,000	44.44
1949	492,000	1,219,000	40.36
1950	938,000	1,701,000	55.15
1951	1,534,000	2,516,000	60.97
1952	2,010,000	3,206,000	62.70
1952/1953*	4,149,000	5,889,000	70.46
1954	5,250,000	6,670,000	78.71
1955	4,256,000	5,898,000	72.16
1956	5,253,000	7,236,000	72.60
1957	5,256,000	7,238,000	72.62
1958	5,933,000	7,839,000	75.69
1959	5,030,000	7,246,000	69.42
1960	5,287,000	7,668,000	68.95
1961	5,211,000	7,222,000	72.16
1962 (budget)	5,410,000	7,500,000	72.13
1963	5,379,000	7,228,000	74.42
1964	5,856,000	8,286,000	70.68
1965	5,903,000	8,468,000	69.71
1966	7,851,000	11,351,000	69.17
1967	7,761,309	12,399,185	62.60
1968	8,070,574	11,970,887	67.42
1969	9,661,489	13,929,441	69.36
1970	9,612,000	19,875,000	48.37
1971	10,943,000	22,304,000	49.07
1972	11,022,000	22,952,000	48.03
1973	13,173,000	38,014,000	34.66
1974	103,456,000	123,747,000	83.61
1975	110,922,000	137,450,000	80.70
1976	156,308,000	201,844,000	77.44
1977	180,717,000	262,777,000	68.78
1978	191,864,000	308,149,000	62.27
1979	213,833,000	311,017,000	68.76
1980	319,767,000	455,200,000	70.25
1981	398,458,000	540,600,000	73.71
1982	401,900,000	554,800,000	72.44
1983	328,607,000	484,100,000	67.88
1984	354,500,000	511,500,000	69.31
1985	374,000,000	531,400,000	70.38
1986	246,100,000	425,000,000	57.91
1987	247,100,000	427,100,000	57.86
1988	209,900,000	401,300,000	52.31
1989	247,400,000	438,000,000	56.55

* The accounting year changed to the Gregorian Calendar in 1953

operate and manage the pipeline laid on the sea bed between the Tank Farm on Sitra island and the Arad depôt on Muharraq island, as well as the hydrant facilities owned by the shareholders. Under the terms of the Articles of Association, Banoco, Caltex and BP are the shareholders of Bafco, owning 60, 27 and 13-per-cent, respectively, of the company's shares.

Positioning for the 1990s

By mid 1985, the Bahrain Government, Caltex technocrats and marketing advisers were beginning to question the ability of the Bapco refinery to meet the petroleum product market demand forecasts for 1995 and beyond. As a result, a joint Caltex-Banoco initiative, a Strategic Plan Study Team, was assigned the task of developing a Strategic Plan so that the refinery might meet market demands to the year 2000.

Until the Government's participation in 1981, the Bahrain refinery had been a swing refinery, the flywheel of the Caltex international refined products supply system. There had always been sufficient demand world wide for Bapco's products as a result of refinery shutdowns, fires, strikes and other unforeseen circumstances. Over the years, as a result of its swing role, the refinery had developed an operational flexibility which enabled it to manufacture a wide range of products of varying specifications to meet the market requirements of much of the world. Whilst this flexibility was an attractive marketing tool, it also made Bapco a high-cost refiner. In addition, with the advances that had taken place in refining technology, Bapco was becoming progressively outmoded, with a diminishing ability to compete successfully against its more modern, lean-staffed competitors in the Arabian Gulf and the Far East.

The Terms of Reference of the Strategic Plan Study Team therefore included four essential requirements. The first was a need to assess future demand for petroleum products, in respect of both volume and quality, in markets best served by the Bahrain refinery. The second was a requirement to assess the likely competitive position of the refinery in meeting future product demand. The third essential was the development of a strategic plan for the refinery which would be operationally flexible and consistent with the overall development plan for the Bahrain economy. The Strategic Plan had to provide for the refinery's modernization and upgrading to ensure that it was brought up to date with the latest process technology. It had to ensure also that the refinery would be economically viable so that it could continue to make a positive contribution to Bahrain's economy. Finally, the study team was required to calculate order-of-magnitude cost estimates, for both short- and long-term refinery capital investments. By mid 1986, the team had developed its work sufficiently to submit a study to a Bapco BSC board meeting held on the 29th of June. The board endorsed the team's work, the foundation of what became the *Strategic Plan*.

Meanwhile, shortly after the Banagas plant was commissioned in late December 1979, it became apparent that the potential gas feed-rate exceeded plant design. Two years later, in December 1981, it was noted that "should the gas rates continue to be significantly greater than those used for design to the plant, consideration should be given to increasing the plant capacity". The outcome was that a Debottlenecking Study was undertaken. On the 4th of December 1982, the Banagas Board agreed to an interim expansion from 110 to 170 million standard cubic feet per day, with all the company's shareholders as participants to the project. When this was completed at the end of March 1986, the project had, in effect, debottlenecked the existing plant to utilize all the gas produced along with crude oil, "associated gas".

During 1986, Banoco foresaw the need to increase substantially the rate of Khuff gas injection into the oil-bearing structures of the producing field in order to maintain the current production of indigenous crude. As a result, the quantity of associated gas was projected to increase from 170 to 280 million standard cubic feet per day early in the 1990s. On the 28th of October 1986, the Banagas Board considered a proposal for further expansion of the plant to process the projected increase in associated gas. Just over a year later, in December 1987, the new project was approved at a cost of US$74 million, with a completion date scheduled for the second half of 1990.

Caltex Nederland and a Bapco tug (*Caltex Bahrain*).

His Highness, Shaikh Khalifa bin Salman Al-Khalifa, Prime Minister of Bahrain, talking to His Excellency Yousuf Shirawi, Minister of Development and Industry, during the Prime Minister's visit to the Bahrain refinery on the 17th of January 1990 (*Bapco archive*).

As 1989 drew to a close, the Bahrain National Oil Company had finalized plans for new offshore exploration in the territorial waters to the north and east of Bahrain.

The Hidden Agenda

In the sixty years since oil was discovered in Bahrain, the industry has made a major contribution directly and indirectly to almost every aspect of the economic and social development of the nation. Over the years since the first oil strike at the foot of Jebel Ad-Dukhan in 1932, the oil industry has developed into an essential component of the economic fabric of the nation.

In the last twenty years, a period of almost constant and frequently intense negotiations, the Bahrain Government never lost sight of the need to keep a balance between maximizing oil revenue and retaining a mutually beneficial relationship with Bapco and Caltex. It also recognized that the oil industry is a high-technology multinational business, requiring a sophisticated level of managerial, technical, administrative and marketing skills. Government negotiators skilfully balanced the need for increasing the nation's oil revenue to meet Bahrain's development objectives against their need for specific foreign involvement. As one corporate observer of the Government and Caltex-Bapco discussions noted:

"It was almost uncanny watching the events of the 1970s and early 80s unfold. Despite all the rhetoric, the almost constant uncertainty of the Government's position and the apparent ambiguity of the debate, it became clear in retrospect that the Bahrain Government always knew precisely where it was going and how it was going to get there. In the final analysis, the Government achieved what we later realized was its hidden agenda. Frankly, it was amazing to watch the negotiators in action. The skilled trader descended from the historic traditions of ancient Dilmun, and the modern corporate negotiator of New York City made for an interesting contrast in styles. The flowing *guttrah* on the one hand and the but-toned-down sharpness on the other, symbolized the two adversaries across the negotiating table – the latter-day 'East meets West' that would have delighted Rudyard Kipling.

"We knew that history was being made. It was enormously exciting to be part of it. Forty years of trust and mutual respect were about to be put to the test, with millions of dollars at stake on the outcome. Both sides were fortunate indeed in that they were represented by men of unique intellectual power, charm and integrity. It was the ultimate test of that special relationship which His Highness Shaikh Salman bin Hamad Al-Khalifa had referred to many years earlier. I sat through many a long meeting in which the esoteric complexities of the oil business were argued to the nth degree of patience and tolerance on both sides. Unequivo-cally, I can attest to the sincerity of the mutual trust and respect which were the bedrock on which negotiations were conducted over those many years – it was indeed a very special relationship."

Perhaps, in reality, as had been seen elsewhere with the multinational oil companies, the negotiations were somewhat weighted in favour of the sovereign power of the host government. But at the end of the long series of negotiations, from which Bahrain's national oil industry evolved, the long-standing "partnership" between the Ruler of Bahrain and the company had survived.

A major factor in the success of this relationship, the first of its kind in the Arabian peninsula region, has been the ongoing commitment of Chevron and Texaco, Caltex's shareholders. For sixty years they have generously encouraged Caltex in the continuing transfer of technical knowledge and support to Bapco which enabled all aspects of Bahrain's oil industry to develop steadily over the years. Patience, skill and enlightened financial self-interest were the new Dilmun "seals" on an enterprise which saw the remarkable sixty-year transition of a foreign corporate subsidiary to a successfully inte-grated national industry. As the decade of the 1980s drew to a close, the citizens of Bahrain had every reason to celebrate their oil industry's Diamond Jubilee.

Epilogue

"One day, the oil wells of Bahrain will become a monument for tourists. After all, oil is a finite source. But in the meantime we hope that the wealth and technology which the industry produces will perpetuate the State of Bahrain, just as the pearl industry did in the past.

"We must be grateful to His Highness Shaikh Isa bin Salman Al-Khalifa, The Amir. Without the wise guidance of His Highness, and the full support of His Highness the Prime Minister, and the commitment of His Highness the Crown Prince who fuelled the enthusiasm of the men who are involved, the nation's oil industry would not be what it is today. As we approach another major celebration next year, the thirtieth anniversary of The Amir's Accession, His Highness Shaikh Isa bin Salman Al-Khalifa has been, and remains, a source of inspiration to us all" [Yousuf A. Shirawi, Minister of Development and Industry, June 1990].

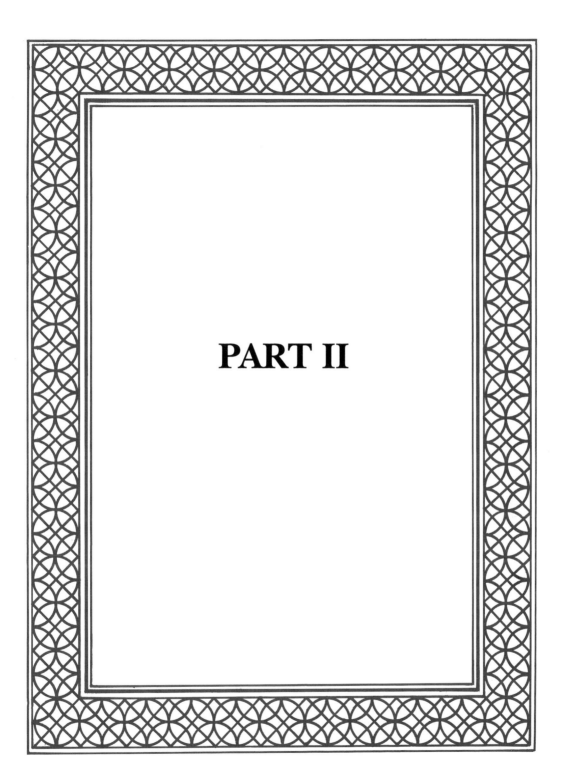

PART II

CHRONOLOGY

BEFORE THE PRESENT

4000 million years ago   The earth was formed.

600 million years ago   The world's first oil is thought to have been formed.

150 million years ago   80-per-cent of the oil found in the Arabian peninsula region, much of it in the Arab zones, was formed.

100 million years ago   The Mauddud oil-producing zone, unique to Bahrain and in which oil was discovered on the 1st June 1932, was formed.

THE PRESENT

1859, 26 August: Edwin L. Drake drilled the world's first oil well in Titusville, Pennsylvania, USA.

1870: Standard Oil Company was founded by John D. Rockefeller and partners in Ohio, USA.

1879: Emil Tietze reviewed the petroleum prospects on the Arabian side of the Gulf, including Bahrain.

1879, 10 September: Pacific Coast Oil Company was formed.

1890, 2 July: The Sherman Antitrust Act became US law.

1904-5: G. E. Pilgrim, of the Geological Survey of India, reconnoitred the Arabian Gulf region, including Bahrain.

1906, 23 July: Pacific Coast Oil Company acquired the business assets of Standard Oil Company (Iowa) and filed a name change to become Standard Oil Company. To distinguish Standard Oil Company, which was incorporated in California, from the other Standard Oil companies, the practice was followed thereafter of adding "(California)" after the name, although this addition was not part of the name as filed.

1908: Pilgrim published a cautious report in which he suggested the possibility of finding oil in Bahrain.

1911, 15 May: The US Supreme Court upheld a Circuit Court decision which had found Standard Oil Company of New Jersey in violation of the 1890 Sherman Antitrust Act. As a result, the Standard Oil Trust was dismantled.

1913: S. Lister James, Chief Geologist of the Anglo-Persian Oil Company, toured Bahrain and concluded that it appeared inadvisable to ignore the area before testing with a fairly deep well.

1914, 14 May: Shaikh Isa bin Ali Al-Khalifa wrote to the British Political Agent that if "there is any prospect of obtaining kerosene oil in my territory of Bahrein" he would not develop it without first referring to him.

1914, 4 August: Britain declared war against Germany.

1919, 28 June: The Treaty of Versailles, the formal end to World War I, was signed, providing for the creation of the League of Nations.

1920, 6 August: The Eastern and General Syndicate Limited was registered as a consortium of British businessmen to acquire and operate oil concessions in the Middle East.

1922, 9 August: Eastern Gulf Oil Company was incorporated as a subsidiary of Gulf Oil Corporation in Pennsylvania, USA.

1924: Dr Arnold Heim wrote in his geological report that drilling on the Arabian coast would have to be classified as a pure gamble.

1924, 8 May: An amendment to the Standard Oil Company's *Articles of Incorporation* officially changed the company's name, which had been filed on 23 July 1906, to Standard Oil Company (California).

1925, 2 December: Shaikh Hamad bin Isa Al-Khalifa, granted the Eastern and General Syndicate Limited an exclusive Exploration Licence for a period not exceeding two years, with a right of renewal for two more years, Schedule I of the *Bahrein Island Concession* sealed on that day. Schedules II and III provided for a Prospecting Licence and Mining Lease, respectively.

1926: Fourteen wells to tap the underground supply of water had been bored by the year end, seven each in Manama and Muharraq.

1926, 27 January: Standard Oil Company of California was incorporated in Delaware, USA.

1926, 26 August: The Texas Company was reincorporated in Delaware as the Texas Corporation and acquired all outstanding stock of the Texas Company, a Texas corporation which had been organized in 1902.

1927, 30 November: Eastern Gulf Oil Company was granted an exclusive option until 1 January 1929 to acquire the Bahrain concession from the Eastern and General Syndicate.

1927, 2 December: The Ruler of Bahrain and Frank Holmes on behalf of the Eastern and General Syndicate, signed an agreement in Bahrain to extend the Exploration Licence for one year until 2 December 1928.

1927, 24 December: Ralph O. Rhoades and his geological party, sailed from New York for Cherbourg and onwards to Bahrain, via Beirut, Baghdad and Basrah.

1928: Fifty-five water wells had been sunk in Bahrain by the year end.

1928, 3 February: The Near East Development Corporation was incorporated in Delaware, USA.

1928, 6 February to 19 March: Ralph Rhoades and his team surveyed Bahrain.

1928, July: The Red Line Agreement was agreed between the partners of the Turkish Petroleum Company, under which each of the participating companies undertook not to conduct independent operations in a large area comprising most of the territory of the old Ottoman Empire.

1928, 30 October: William T. Wallace, having great faith in the oil possibilities in Bahrain, sought the co-operation of the Near East Development Corporation (NEDC), a shareholder of the Iraq Petroleum Company (IPC), to have the IPC buy interests in Bahrain from the Anglo-Dutch-French shareholders for $50,000. The bid failed because Gulf Oil Corporation of Pennsylvania, the parent company of Eastern Gulf Oil Company which held the option on the *Bahrein Island Concession*, was a participant in the NEDC, which itself had entered into the Red Line Agreement. Thus Gulf Oil was precluded from pursuing the Bahrain concession.

1928, 23 November: The Colonial Office agreed to recommend to the Ruler of Bahrain that he might wish to renew the Exploration Licence, due to expire on 2 December 1928, providing that the Eastern and General Syndicate remained a British company and that no land occupied for any of the purposes of the Exploration Licence were, at any time directly or indirectly, controlled or managed by foreigners or a foreign corporation.

1928, 19 December: The Eastern and General Syndicate informed the Colonial Office that its interpretation of Article XIII of the Third Schedule of the 1925 Bahrain concession was that the nationality of the company to which the option on the concession might be assigned, was not specified. The only stipulation was that the country of registration of the ultimate company should be British. The Syndicate added that if some satisfactory arrangement could not be worked out with Gulf Oil, then the latter would resist the new condition and claim a breach of contract, which in turn would cause the liquidation of the syndicate.

1928, 21 December: In Pittsburgh, Pennsylvania, F. A. Leovy for Eastern Gulf Oil Company signed copies of the company's agreement to assign its option on the Bahrain concession to Standard Oil Company of California in return for actual expenses involved: $157,149. On the same day, Thomas Ward reported to William T. Wallace that he could not understand the delay in the discussions with the Colonial Office and was under the impression that the British Government would not raise any objection to the assignment of the option. Meanwhile, he confirmed that Bapco was being organized under the laws of the Dominion of Canada, in readiness for the option to be exercised on 1 January 1929.

1928, 27 December: W. H. Berg authorized Ward to exercise, on behalf of Standard Oil Company of California (Socal), the option dated 30 November 1927 and the letter option of 28 May 1928 covering Bahrain island, entered into between the Eastern and General Syndicate and Eastern Gulf Oil Company, and to designate the Bahrein Petroleum Company Limited of Canada as Socal's nominee. Meanwhile, in the city, county and state of New York, Judge Frank Feuille, Attorney-in-Fact for Socal, signed the duplicate copies of the document which assigned the option on the Bahrain concession from Eastern Gulf Oil Company to Socal.

1928, 28 December: Socal exercised the two options relating to the original 100,000 acres on the Bahrain concession and the negotiations for a concession to the "Additional Area". In so doing, it nominated the Bahrein Petroleum Company Limited as the company to which the concession and option covering any further concession relating to the "Additional Area" should be transferred.

Meanwhile, on the same day and unknown to Socal, representatives of the Colonial Office and the Syndicate met in London. The Colonial Office suggested that it might be more in the interests of the Syndicate for the extension to the Exploration Licence to be refused, which had expired twenty-six days earlier on 2 December.

1929, 2 January; Thomas Ward received a letter from the Syndicate dated 20 December 1928, together with a copy of the Syndicate's 19 December 1928 letter to the Colonial Office which answered the Colonial Office's proposal of 23 November 1928. It became apparent to Ward and Socal, that although the *principle* of a British registered company had been established, the *practice* of it being a wholly owned American subsidiary, financed with American capital, had not yet been approved by the Colonial Office.

1929, 11 January: The Charter of the Bahrein Petroleum Company Limited was sealed in Ottawa, Province of Ontario, Dominion of Canada, by Thomas Mulvey, Under-Secretary of the State of Canada, as a wholly-owned subsidiary of the Standard Oil Company of California. Bapco's registered office was the Trusts Building, 48 Sparks Street, Ottawa, Canada.

1929, 6 February: Socal sought the assistance of the US Department of State in resolving the legal deadlock it faced over the Bahrain concession.

1929, March: Francis B. Loomis (Socal) and William T. Wallace (Gulf Oil) met US Secretary of State Frank B. Kellogg, who, some days later, requested the US Chargé d'Affaires in London to discuss the problem with the British Government.

1929, April: The Colonial Office was informed that the British nationality provisions it required were quite unacceptable to Socal and Gulf Oil.

1929, May: Sir John Cadman, of the Anglo-Persian Oil Company, was asked indirectly by the Colonial Office to use his influence to solve the matter by inducing the Turkish Petroleum Company (TPC) to accept a complicated proposal which would help Eastern Gulf Oil Company out of its difficulty with the Bahrain concession. Meanwhile, Harry G. Davis, Gulf Oil's London representative, discovered that the scheme had not been the Colonial Office's idea, but that of E. W. Janson, one of the Eastern and General Syndicate's directors, who wished to get the syndicate "off the hook".

1929, 3 June: The Colonial Office agreed, in principle, to the participation of US interests in the Bahrain concession. Nevertheless, the deadlock continued because the British Government wished to impose political influence and control on Socal, whilst the latter wished be to free of such restrictions.

1929, 7 June: During a British Government interdepartmental conference it was argued that since the Exploration Licence had not been extended or renewed beyond 1 December 1928, and since the syndicate had not applied either before or upon that date for the grant of a prospecting licence, it was not now entitled to claim such a licence, and its concession had lapsed. This was not a concensus view.

1929, 8 June: The Turkish Petroleum Company became the Iraq Petroleum Company (IPC).

1929, July: The British Government agreed that the "Additional Area" being negotiated for the Bahrain concession would be conceded *if* four conditions were met. By the end of July, the British Government "was inclined to consider favourably" four of the five counter-proposals.

1929, 16 September: The Colonial Office outlined the four conditions, now revised, which when accepted would enable the British Government to advise the Ruler of Bahrain to extend, if he wished, the period of the Exploration Licence granted to the Eastern and General Syndicate on 2 December 1925. The conditions were: that the company should be registered in Canada (which it was); one of the five directors should be nominated by the British Government; the company should maintain a Chief Local Representative in Bahrain, approved by the British Government (Major Frank Holmes to hold this position for the first five years); Bahraini and British subjects should be employed as far as possible.

1929, 18 October: William T. Wallace, James M. Greer and Harry G. Davis (Gulf Oil), Francis B. Loomis and Judge Frank Feuille (Socal), and Major Frank Holmes, E. W. Janson and Thomas

E. Ward (Eastern and General Syndicate) met in New York to discuss the British Government's proposals.

1929, 22 October: Janson attended a Gulf Oil board meeting to outline the proposals. The directors were willing to meet the Colonial Office's conditions, but cautioned that sound business management demanded clarification of certain points. Particularly, for how long would the British Government recommend to the Ruler of Bahrain that the Exploration Licence should be extended?

1929, 24 October: Panic selling caused the New York Stock Exchange to cease functioning temporarily.

1929, 4 December: Socal agreed to the Colonial Office's draft of the form in which the Bahrain concession would be transferred from the Eastern and General Syndicate to the Bahrein Petroleum Company Limited, with the proviso that Socal required definite assurance that the term of the Exploration Licence had been properly extended to 2 December 1930.

1930, 3 January: Gulf Oil wrote a letter of modification to the Colonial Office.

1930, 13 March: The draft agreement, outlining the form of the Bahrain concession transfer, was sent to the Political Resident Persian Gulf, so that it could be formalized between the Ruler of Bahrain and the Eastern and General Syndicate. The document provided for the period of the Exploration Licence to be extended to 2 December 1930.

1930, April: William F. Taylor, General Superintendent of Socal's foreign division, and Fred A. Davies, a Socal geologist, left New York for Bahrain, via Baghdad.

1930, May: Davies and Taylor surveyed Bahrain.

1930, 1 June: Fred A. Davies marked the location of Bahrain's first oil well with a cairn of stones, surmounted by a flag.

1930, 12 June: An Indenture was signed, whereby the Acting Ruler, His Highness Shaikh Hamad bin Isa Al-Khalifa, assented to the transfer by the Eastern and General Syndicate to the Bahrein Petroleum Company Limited of the syndicate's rights under the *Bahrein Island Concession*. On the same day, the Exploration Licence was extended, retroactive to 2 December 1928 and valid until 2 December 1930.

1930, 18 June: Major Frank Holmes filed his application for an "Additional Area" concession.

1930, 30 June: Lt-Col. H. V. Biscoe, Political Resident in the Persian Gulf, informed the Colonial Office that, in his view, Holmes' application was premature.

1930, 1 August: In New York, Cletus Keating, Attorney-in-Fact for the Eastern and General Syndicate Limited, assigned the *Bahrein Island Concession* to the Bahrein Petroleum Company Limited. Judge Frank Feuille, Attorney-in-Fact, executed the transaction on Bapco Limited's behalf and paid the syndicate US$50,000, as agreed.

1930, 18 August: Bapco Limited reimbursed Eastern Gulf Oil Company's expenses, US$57,449.23, as provided for in the 27 December 1928 contract.

1930, 11 September: Major Frank Holmes sent a copy of the 1 August 1930 Deed of Assignment to the Political Agent in Bahrain, asking that it be registered. He also applied for a two-year prospecting licence, to become operative when the Exploration Licence expired on 2 December 1930.

1930, 26 November: Fred A. Davies published his Bahrain report. He recommended that Socal should go ahead and drill a test well.

1930-1: The Egyptian press attacked the Bahrain Government for assigning the Bahrain oil concession. Following these hostile reports, all foreign newspaper.correspondents were ordered to register with the Bahrain Government.

1931: The Anglo-Persian Oil Company became increasingly unpopular throughout the year owing to the much higher rates asked for petrol and kerosene as compared with those elsewhere along the Arabian Gulf coast.

1931, 23 May: A party of four Socal men, headed by Edward A. Skinner, arrived in Bahrain to make preparations for the arrival of the main drilling crew three months later, and the construction of the Jebel Camp.

1931, 16 October: Bahrain's first oil well was "spudded-in" and drilling commenced.

1932, 22 April: Major Frank Holmes applied to the British Political Agent in Bahrain to extend the Prospecting Licence for two years from 2 December 1932.

1932, 1 June: Oil flowed from Bahrain's Well Number One at a depth of 2008 feet (612 metres).

1932, October: The Colonial Office indicated that it was not prepared, at that time, to extend the Prospecting Licence to 2 December 1934. The Additional Area application was stalled.

1932, 9 December: His Highness Shaikh Isa bin Ali Al-Khalifa died. He was succeeded by his son Hamad, who, for almost a decade, had deputized for his elderly father.

1932, 25 December: Well Number Two "came in" with a rush.

1933, 15 February: An Indenture between the Ruler of Bahrain and the Bahrein Petroleum Company was signed, modifying Clause IX of the Prospecting Licence (Schedule II of the 1925 concession). This gave Bapco the right to produce up to 100 tons of oil, free of royalty payment, and further quantities of oil on the payment of a royalty of 3 rupees 8 annas per ton of net crude oil produced and saved.

1933, 4 April: Major Frank Holmes applied, on Bapco's behalf, for a one-year extension to the Prospecting Licence, to be effective from 2 December 1933.

1933, 11 August: Bapco notified the Political Agent in Bahrain, that Socal preferred, at that time, to press for an immediate extension to the Prospecting Licence to 2 December 1934, rather than pursue the Additional Area concession application.

1933, 22 August: H. R. Ballantyne informed Socal that the Ruler of Bahrain had refused to grant the requested extension to the Prospecting Licence.

1933, 14 September: Major Frank Holmes resigned as Bapco's Chief Local Representative. Edward Skinner was appointed in his place.

1934: The construction of the "New Camp" began, renamed Awali in 1938.

1934, 22 February: Standard Oil Company of California's oil tanker, *El Segundo,* used as a freighter, storehouse and floating dormitory during the construction of three fuel oil storage tanks and a jetty, dropped anchor off Sitra island on the anniversary of George Washington's birth, fifty-five days out from San Pedro, USA.

1934, 7 June: *El Segundo* left Sitra jetty with the first shipment of Bahrain crude with a cargo of 25,082 barrels.

1934, 9 July: *El Segundo* docked at Yokohama, Japan, to discharge its cargo, before continuing its round trip back to San Francisco.

1934, 29 December: The Bahrain concession was converted to a *Mining Lease,* provision for which had been made in Schedule III of the 2 December 1925 *Bahrein Island Concession.* This document granted Bapco an exclusive right for a period of fifty-five years from 1 January 1935, to prospect and drill for, extract, treat, refine, manufacture, transport and deal with petroleum products within the area defined in the First Schedule of the 1934 agreement. This new lease formed the basic document to which all amendments and agreements refer up to the 1970s.

1935, October: Construction work began on the Bahrain refinery.

1936: The mounted camel section of the police was formed, part of which patrolled the new refinery.

A new telephone exchange was installed in Manama, linking it with 114 lines including those of the Bahrein Petroleum Company Limited.

1936, 3 June: The first *Supplemental Deed* to the 1934 Mining Lease was signed.

The minimum annual royalty was increased from 75,000 rupees to 150,000 rupees at the rate of 3 rupees and 8 annas per ton.

1936, 26 June: Kenneth R. Kingsbury, President of Socal, and Captain Torkild Rieber, Chairman of the Texas Corporation, issued a press statement which provided for the formation of the California Texas Oil Company Limited.

1936, 30 June: California Texas Oil Company Limited was incorporated as a subsidiary of the Bahrein Petroleum Company Limited.

1936, 2 July: 5000 shares were issued by Bapco Limited to the Texas Corporation in accordance with an agreement dated 2 July, effective 1 July 1936.

1936, 12 July: The first refinery unit, a crude still, was completed with a capacity of 10,000 barrels per calendar day.

1937: Jebel Ad-Dukhan camp was vacated. Air-cooling was installed in the "New Camp".

1937, 30 June: Awali Hospital was opened.

1937, 11 December: His Highness Shaikh Hamad bin Isa Al-Khalifa officially opened the Bahrain refinery. At the year-end it was capable of running 25,000 b.p.c.d. of crude.

1938: By the end of the year, fifty oil wells had been drilled. All but two were "producers".

Bahrain's first service station opened in Government Road (now Government Avenue), Manama.

1938, 23 April: Bapco implemented the suggestion made by His Highness Shaikh Hamad bin Isa Al-Khalifa, Ruler of Bahrain, that the oil community settlement should be named Awali.

1938, 4 September: The first shipment of crude oil from Saudi Arabia was brought to Zallaq terminal on the west coast of Bahrain island in two barges, towed by the steam tug *Arab*.

1939: A polymer plant was added to the refinery.

1939, 3 September: Britain declared war on Germany.

1940: Major improvements were made to the refinery power plant, crude distillation units, SO_2 plant, cracking plant and the pitch pond. The Sitra and Zallaq wharf facilities were altered, pipelines were re-routed and a storage tank was constructed at the Zallaq marine terminal.

1940, 19 June: The *Deed of Further Modification* to the 1934 Mining Lease, was sealed by the Ruler of Bahrain, and signed on behalf of the Bahrein Petroleum Company Limited by M. Weston Thornburg, Vice-President and Attorney-in-Fact.

1940, 31 July: California Texas Oil Company Limited transferred ownership of Balboa Transport Corporation to the Bahrein Petroleum Company Limited.

1940, 19 October: The Italians bombed the Bahrain refinery.

1940, 4 November: Al-Bahrain radio started short-wave broadcasts on 23-m., 50 band with four daily news broadcasts.

1940-1: Bahrain's first census took place: population: 89,970 people, of which 74,040 were Bahraini and 15,930 were other nationalities.

Sitra Wharf was cathodically protected.

1941, 17 May: Bapco's Chief Local Representative suggested to the Bahrain Government saying that Bahrain's geological plans and drilling logs should be taken "to a place of greater safety" in case Bahrain fell into enemy hands.

The India Office informed Bapco's lawyer that the stipulation that Bapco's Chief Local Representative should be a British subject was being waived because of wartime difficulties.

1941, 7 December: Japan attacked Pearl Harbor, Hawaii.

1941, 8 December: The USA declared war on Japan.

1942: With Bapco's assistance, the swing-bridge linking the towns of Manama and Muharraq was completed, opening twice daily to let sailing boats pass through the channel.

Caltex-2 went into operation to make consecutive voyages from Saudi Arabia to Bahrain, carrying crude oil.

Facilities for receiving up to 35,000 barrels of Arabian light crude per day were installed at Zallaq, including a deep-water wharf, steam plant, pumping equipment and a 13,000 barrel tank.

1942, 20 February: The Ruler of Bahrain, H H Shaikh Hamad bin Isa Al-Khalifa died and was succeeded by his son, H H Shaikh Salman bin Hamad Al-Khalifa.

1942, 9 March: Bapco Limited shares held by the Texas Corporation were cancelled and reissued in the name of the Texas Company.

1942, 16 April: At a notorious low-point in World War II, details of a "scorched-earth policy" were outlined in the event of the Allies being forced to abandon the Bahrain islands and Saudi Arabia. The plans included the permanent destruction of all oil wells and blowing up the Bahrain refinery.

1942, 2 May: Following a conference in Baghdad, the Political Resident Persian Gulf informed the Secretary of State for India that a scheme to protect the Bahrain refinery and tanks against attack, had been agreed at a cost of £150,000.

1942, 6 May: The Commander-in-Chief Middle East informed the War Office that Bapco and Casoc representatives did not agree to the permanent junking of oil wells, without covering authority from their directors.

1942, 11 June: Direct radio telephonic communication was established between Bahrain and Dhahran.

1942, 26 July: Ward B. Anderson (Chief Local Representative and Bapco's General Manager) informed A. B. Wakefield, the PRPG, that Stage One of the plan to immobilize and/or destroy the Bahrain oil wells in the event of enemy action had been effected, i.e. all unnecessary wells with a capacity to produce 20,000 b.p.c.d. had been capped with cement. Anderson sought confirmation of his understanding of Stages Two and Three which made provision for the permanent destruction of the wells, including the thirty-three which remained open.

1942, 30 July: The PRPG advised Ward Anderson that the destruction programme in three stages (in the event of it being necessary) had received official approval from the British Government.

1943: With the USA at war, Ward Anderson obtained India Office approval to send and receive telegrams to and from the company's New York office in code or cipher.

Construction of the FCCU (Fluid Catalytic Cracking Unit) started, the US Government having decided to build additional oil refining facilities in Bahrain to produce an estimated 5800 barrels per day of 100-octane aviation fuel. The project included isomerization and alkylation plants.

1944: A third crude still was commissioned.

The drum plant, with a notional maximum capacity of 70,000 drums per month, was placed on a preliminary operational basis.

By the end of the year, a 12-inch (30.48 cm) crude-oil pipeline running from the refinery to Zallaq was completed. Work began on laying a submarine pipeline from Aramco's facilities in Dhahran.

Additional bunkhouse accommodation was constructed in Rafa' and Awali.

1944, 2 September: After negotiations with the Government of the Hijaz and Nejd, the USA opened a consulate in Dhahran, which informally included Bahrain within its jurisdiction. Thereafter, Parker T. Hart, US consular official, regularly visited Bapco Limited in Awali.

1944, 9 October: A four-power meeting at Dumbarton Oaks, Washington DC, USA, announced its decision to form a world-wide organization for preserving peace after the war. The proposals formed the basis of the United Nations Charter.

1945: Bahrain refinery's Fluid Catalytic Cracking Unit (FCCU) went on-stream in the middle of the year. After the end of the war, demand for aviation fuel dropped. Production ceased in October when the aviation gasoline tankage reached capacity. The plant had produced 322,997 barrels of product.

The isomerization and alkylation units were added to the refinery.

1945, February: The Yalta Conference.

1945, 3 March: The last cargo of crude oil to be brought from Saudi Arabia by barge was received at Zallaq terminal, after which the facility was dismantled in favour of the new AB pipeline.

1945, 22 March: The League of Arab States was created. Founding members were Egypt, Iraq, Jordan, Lebanon, Saudi Arabia, Syria and Yemen.

1945, April: A 12-inch (30.48 cm) partially submerged 34-mile (55 km) long pipeline, running from Saudi Arabia to Bahrain, became operational.

1945, 18 April: The League of Nations, after a final meeting in Geneva, dissolved itself and transferred its assets to the United Nations.

1945, 20 April: The Texas Oil Company asked the Ministry of Fuel and Power, on behalf of Bapco, for its assistance in recruiting women stenographers for Bahrain, ten at once and maybe seventy-five over the next year.

1945, 7 May: Victory in Europe Day. General Alfred Jodl, German Army Chief of Staff and emissary, signed the instrument of Germany's unconditional surrender.

1945, 25-26 June: The San Francisco Conference was held. Delegates from 50 states signed the World Security Charter to establish the international peace-keeping organization, the United Nations Organization (UNO).

1945, July: The Potsdam Conference.

1945, 14 August: Victory over Japan Day, following the Japanese unconditional surrender to the Allies.

1945, 24 October: The United Nations Organization (UNO) came into formal existence when the twenty-ninth government ratified the charter. Eypgt, Iran, Saudi Arabia, the UK and the USA were among those countries admitted as members on that day.

On the same day, Ward Anderson received a cable from Bapco's Head Office in New York, indicating that due to world economic circumstances, the Bahrain refinery was threatened with closure.

1946: A four-berth wharf for ocean-going vessels and its 3-mile (4.82 km) long connecting causeway from Sitra island was completed.

The asphalt plant commenced. The polymer plant was reconverted to produce motor polymers and the FCCU was converted to a crude unit.

The Political Resident of the Persian Gulf moved his headquarters from Bushire to Bahrain.

1946, 12 June: The Overseas Tankship Corporation was incorporated in Panama, with California Texas Oil Corporation as the sole owner.

1946, 6 December: The California Texas Corporation was incorporated in Delaware as the successor to the California Texas Oil Company Limited, a subsidiary of the Bahrein Petroleum Company Limited.

1947: Bapco installed an IBM 001 card punch machine, the first in Bahrain.

1947, 4 January: R. M. Brown, Chief Local Representative for Bapco, wrote to the Political Agent informing him that the suggested date for the termination of the moratorium on the Mining Lease extension should be 30 November 1946.

1947, 12 December: Bapco Limited secured permission to drill on the Hawar Islands, the result of a confidential aide-memoire prepared by the British Embassy, Washington DC, USA.

1948: Bapco started full-time industrial study courses and the welding school, installed a telephone system and introduced the Rupee Payroll Thrift Plan.

1948, 1 November: A preparatory school comprising two classrooms opened at Zallaq. Thirty-six Bapco employees attended the first four month course to improve their job-related English vocabulary and mathematical skills.

1949, 4 June: R. M. Brown, pressed by the company's insurers, sought permission to arm the guards transferring Bapco's cash from the Manama banks to Awali, the oil community.

1949, 26 December: The Government of Bahrain informed Bapco's Chief Local Representative that, effective from 1 January 1950, the revised royalty rate would be 10 rupees per ton.

1950: The Al Dar guesthouse was opened.

1950, 1 January: The revised royalty rate of 10 rupees became effective for ten years.

The minimum annual royalty, provided for in Article 7(b) of the 19 June 1940 Deed of Further Modification, reverted to 150,000 rupees, since oil had not been found in the Additional Area.

1951, 28 June: Bapco Limited responded to the Ruler of Bahrain's request to increase its revenue to the Government by way of "income from the importation of oil". The letter agreed to pay, voluntarily and immediately, beginning 1 June 1951, 500,000 rupees per month.

1951, 24 July: The Ruler of Bahrain acknowledged the above arrangement.

1952: *Caltex Bahrain*, a new oil tanker, was launched in England.

An air-conditioned training building was completed at Bapco Limited, allowing for further expansion of the full-time industrial study courses.

1952, 8 December: Bapco agreed to review its preferential terms with the Ruler of Bahrain, should other oil-producing countries bordering the Arabian Gulf receive a better deal than Bahrain's.

The Ruler of Bahrain, sealed the *Supplemental Agreement* to the 19 June 1940 Deed of Further Modification. In summary, this provided for the Bahrain Government to receive:

- A fee of two and one-quarter pence sterling per barrel in respect of all foreign crude oil imported and refined in Bahrain, on and after 1 January 1952
- Royalties paid in pounds sterling at the rate of one shilling and sixpence (1/6) per rupee
- A minimum of £750,000 sterling in royalties and fees during a period of *force majeure*
- For its own internal use, 150,000 imperial gallons of petrol free of charge per year.

1952, 15 December: The Income Tax Decree of the Ruler of Bahrain was made applicable to non-Bahrainis by the Queen's Regulation issued by W. R. Hay, Her Majesty's Political Resident in the Persian Gulf, under Article 85 of the Bahrain Order in Council, 1949. Its effects were retroactive to 31 December 1951.

1952, 16 December: R. M. Brown, Bapco's Chief Local Representative, advised the Ruler of Bahrain, that the company recognized Decree No. 8, which imposed "a 50-50 tax upon income from the sales of crude petroleum and other natural hydrocarbons produced and extracted from the ground and from under the seabed belonging to Bahrain".

1953: Total production of crude oil in Bahrain, from the first discovery in 1932 to the end of 1953, reached 150 million barrels.

Bapco inaugurated a Housing Loan Plan.

The Gregorian calendar replaced the Hijri calendar for official budgets and accounts.

1953, 1 January: The Bahrein Petroleum Company Limited's name was changed officially to the Bahrain Petroleum Company Limited, upon the Government's announcement of the preferred spelling of the country's name.

1954: Bapco instigated a community health service for remote villages, in the form of a mobile dispensary mounted on a six and a half ton truck, capable of being driven on unpaved desert tracks and equipped with essential medical supplies.

1954, 22 February: Bapco agreed to supply the Government of Bahrain with natural gas, without a return on the investment to the company. In effect, C. R. Barkhurst's letter formed the basic document under which natural gas was supplied to the Bahrain Government, on the understanding that it would compensate the company for all expenses and that this reimbursement would not be subject to royalty.

1955: Bapco Limited introduced an Apprenticeship Training Scheme.

No. 5 Vacuum Unit was commissioned at the refinery.

1955, 5 May: *Huna Al-Bahrain* (Bahrain Broadcasting Station) was opened, most of the equipment having been donated by Bapco Limited. Broadcasting times were 8.30-10 a.m. on Friday and 8-9.30 a.m. from Saturday to Thursday.

1955, 28 November: The Ruler of Bahrain's decree extending the 50-50 income tax to income from refined products, was made applicable to non-Bahrainis by Queen's Regulation No. 8,

1955, published by B. A. B. Burrows, British Political Resident in the Persian Gulf. The document was entitled: *Bahrain Income Tax Regulation, 1955*. This repealed the *Bahrain Income Tax Regulation* of 15 December 1952.

1959, 29 November: By agreement between the Ruler of Bahrain and Bapco Limited, the provision of the 1952 *Supplemental Agreement* for payment of fees with respect to imported crude oil was terminated.

1956: A detailed seismographic survey of the land area of Bahrain began during the year.

1957: Bapco launched the Housing Loan Plan which enabled employees to borrow against their thrift plan balances to build new homes or extend old ones. By the end of the year, almost 4 million rupees had been loaned.

A 200,000 pounds-per-hour boiler and a 5000-kilowatt turbo-generator were commissioned. Second-year Craft Apprentices were given the opportunity to join a Technical Apprenticeship.

1957, October: A 11,000 barrels-per-stream-day Unifiner and Platformer went on stream.

1958: Ownership of the Balboa Transport Corporation was transferred from The Bahrain Petroleum Company Limited to California Texas Corporation (re-named Caltex Petroleum Corporation in 1968).

The first 9000-hour run of the new Platformer unit, which upgraded gasoline to a high octane level, was completed.

1958, 7 December: M. H. Lipp, Bapco's Chief Local Representative, informed G. W. R. Smith, Secretary to the Bahrain Government, that as from January 1959 Bapco Limited would provide the Bahrain Government, for its own internal use, with 200,000 imperial gallons of petrol per year free of charge, an increase of 50,000 gallons as provided under the provisions of the *Supplemental Agreement* of 8 December 1952.

1959: The Gulf Rupee was introduced. Its value was the same as the Indian Rupee but the denominations were printed in different colours to prevent smuggling of currency to India.

Muharraq Airport terminal was completed. The foundation stone for an extension was laid.

1959, 1 January: The California Texas Oil Company Limited changed its name to California Texas Oil Corporation.

1959, February: Bapco introduced a 42-hour working week for all employees.

1959, 1 May: The Texas Company changed its name once again to Texaco Inc.

1959, August: Thirty-two Bapco trainees completed the first four-year apprenticeship training course. Their graduation certificates were presented by the then Heir Apparent, His Excellency Shaikh Isa bin Salman Al-Khalifa.

1959, 1 December: M. H. Lipp, on behalf of Bapco Limited, wrote two letters to the Ruler of Bahrain:

- The first indicated the company's wish to relinquish certain marine areas conceded to the company.
- The second sought renewal of the royalty agreement for ten more years at the rate of 10 rupees per ton, which had become 15 shillings (sterling) per ton as a result of the provision.in Clause 2 of the *Supplemental Agreement* dated 8 December 1952.

1960: Fourteen Bahraini Bapco employees started specialist training in the United Kingdom, bringing the total to 46 since the scheme started in 1956.

503 adult Bahrainis attended full-time courses at Bapco's the Vocational Training Centre (VTC) for an average of four months each.

1960, 1 January: A 7.5-per-cent general salary increase was awarded to all Bapco rupee payroll employees.

1960, 7 February: Bapco processed its billionth barrel of crude oil.

1960, 14 September: The Organization of Petroleum Exporting Countries (OPEC) was formed. The founder members were Iran, Iraq, Kuwait, Saudi Arabia and Venezuela.

1961: The Bahrain refinery throughput averaged 218,900 b.p.c.d., reaching a record total of 79.9 million barrels.

Liquefied petroleum gas, a modern domestic fuel, was sold for the first time by Bapco Limited for local distribution.

1961, 8 February: Bapco confirmed to The Ruler the amendment of the 29 November 1955 agreement, namely an increase in the refining profit element from seven pence sterling, to eight and a half pence.

1961, 29 March: Bapco Limited consented to the assignment by Eastern and General Syndicate Limited to Eastern and General Syndicate (Bahamas) Limited of the overriding royalty in respect of Bahrain crude oil, as provided in the 30 November 1927 option agreement. Bapco stated that it would, subject to any British Government limitations on the use of Bapco's London sterling accounts, pay such portions of the royalty as the Bahamas company might request to the Bahamas company in Nassau, Bahamas, in pounds sterling, and any balance to that company in New York City in US dollars. It was agreed that this arrangement was solely an accommodation to Eastern and General Syndicate (Bahamas) Limited, and that Bapco Limited could, at any time, in its sole discretion, revert to full payment in one currency and one sum, and that Bapco retained the right to make all payments out of its sterling balances in London.

1961, 2 November: His Highness Shaikh Salman bin Al-Khalifa, Ruler of Bahrain, died and was succeeded by his son Isa.

1961, 16 December: The accession of His Highness Shaikh Isa bin Salman Al-Khalifa as Ruler of Bahrain and her Dependencies.

1962: A new annual throughput record for the Bahrain refinery of 88 million b.p.c.d., 11-percent more than in 1961.

1962, 1 January: The refinery profit margin agreement no longer applied to Bahrain crude oil, but was retained on imported crude. This arrangement continued until mid-1977.

1962, 17 October: Bapco informed The Ruler that the company had studied the means whereby the Government might obtain additional net income from petroleum products, sold by the company for consumption in Bahrain, without imposing any additional burden on the people of Bahrain, i.e. without increasing the prices.

1962, 2 December: A 50-per-cent tax on local sales of petroleum goods was imposed for the first time.

Bapco confirmed to The Ruler that the agreement made on 8 February 1961 that refinery products consumed in Bahrain should be excluded from the refinery profit margin in respect of all taxable years beginning after 31 December 1961.

1963, 10 February: Bapco confirmed by letter that products destined for consumption in Bahrain would be excluded from the scope of the 29 November 1955 refinery profit margin agreement, as amended, for all taxable years beginning after 31 December 1961.

1963, 25 November: His Highness Shaikh Isa bin Salman Al-Khalifa and Bapco Limited agreed that for taxable years beginning after 31 December 1962, the prices at which the company shall sell finished or semi-finished refined products for export shall be:

- with respect to products made from Bahrain crude oil, the posted prices at which the buyers offered to sell such products f.o.b. Bahrain, less $^3/_7$ pence sterling marketing allowance, except that as to products sold by buyers or their nominees to non-affiliated companies at less than posted prices, then the company's prices shall be the buyers' or their nominees' realization from such products, less a marketing allowance of $^3/_7$ pence sterling per barrel

- with respect to products made from foreign crude oil, the cost to the company of acquiring such crude oil, plus the cost of processing such crude oil into the products sold, plus 8 $^1/_2$ pence sterling per barrel.

It was further agreed that if the company should process foreign crude oil for other persons, its charges would be the costs of processing crude oil, plus 8 $^1/_2$ pence sterling per barrel.

1964: Bapco Limited introduced a Home Ownership Plan for eligible Bahraini employees to obtain home building/purchase loans.

The Bahrain Currency Board was formed.

1964, 11 August: The pricing agreement of 25 November 1963 between His Highness Shaikh Isa bin Salman Al-Khalifa and Bapco, which by its terms was applicable to taxable years beginning after 31 December 1962, was, by further agreement, made applicable to taxable years beginning after 31 December 1961.

1964, 19 August: The British Political Agent, K. Oldfield, confirmed to L. D. Josephson, Bapco's Chief Local Representative, the end of the British Government's role as intermediary between the Government of Bahrain and Bapco Limited.

1965: Bahrain's first traffic lights were connected.

Bapco offered its first secretarial training course for female employees.

1965, 13 February to 12 March: The Bahrain Government conducted a census. Bapco undertook the computer programming and statistical analysis.

1965, 11 March: By agreement, Bapco Limited consented to the reassignment of the Bahrain concession overriding royalty from Eastern and General Syndicate (Bahamas) Limited, back to Eastern and General Syndicate Limited. Furthermore, Bapco would, to the extent from time to time satisfactory to it, make the quarterly overriding royalty payments in London and New York, as may be requested by the Syndicate Limited, in sterling and US dollars respectively. It was understood that all payments would be made from Bapco's London sterling balances, with exchange and costs at the risk and for the account of the Syndicate Limited.

1965, 13 March: Bapco employees went on strike over the Bapco redundancy programme.

1965, 29 March: At the request of the Bahrain Government, the Chief Local Representative of the Bahrain Petroleum Company Limited (L. D. Josephson) submitted an *Amendatory Agreement* to The Ruler of Bahrain which deleted and avoided terminology in the *Mining Lease* (29 December 1934), the first *Supplemental Deed* (3 June 1936) and the *Deed of Further Modification* (19 June 1940) which the Government considered inconsistent with current circumstances.

1965, 9 August: The *Amendatory Agreement*, as proposed above, was signed. This amended the 1934 *Mining Lease*, the 1936 first *Supplemental Deed* and the 1940 *Deed of Further Modification*, but did not alter the basic terms and substance of the agreements. Among the provisions of the *Amendatory Agreement*, in future, Bapco was to refer to "the Sheikh" as "the Ruler", and delete the phrase "acting on the advice of the Political Resident in the Persian Gulf" from all documents.

1965, 27 September: By letter to His Highness Shaikh Isa bin Salman Al-Khalifa, Ruler of Bahrain, signed by L. D. Josephson, the Vice President and General Manager of Bapco Limited advised The Ruler that if a decree were issued, and made applicable to the company, amending the income tax law to expand the definition of an "exploratory well" to include any well drilled for the purpose of discovering a previously unknown petroleum reservoir, or for geological purposes without reasonable assurance of producing economic quantities of petroleum, and further providing that the costs of exploratory wells should be capitalized and amortized at 20-per-cent per year, the company would adhere to the decree.

1965, 16 October: The Bahrain Dinar was created as the unit of currency for the State of Bahrain, equivalent to 10 rupees and a series of bank notes inscribed "Bahrain Currency Board" were placed in circulation.

1966: Bapco Limited introduced a pension plan for Bahraini employees.

To meet increasing world market demands for low-sulphur diesel, the conversion of the Refinery's No. 3 Reformer to a diesel desulphurizer began.

Abu Saa'fa, the offshore oilfield between Saudi Arabia and Bahrain, was brought into production by Aramco at an initial rate of 30,000 barrels per day.

1966, February: The biggest cargo ever handled at Sitra Wharf, 61,000 tons, was loaded aboard the tanker *Chryssi P1 Goulandris*.

New facilities for bulk handling of anti-icing additive for jet fuel were commissioned.

A Study Grant Award Plan was launched for suitably qualified Bahrainis who wished to make a career in the oil industry.

1966, 8 March: Bapco Limited entered into several complex agreements and undertakings with His Highness Shaikh Isa bin Salman Al-Khalifa which may be summarized as:

A When an amendment to the 1955 income tax decree containing the following terms is issued and made applicable to the company, the company would submit to the decree as follows:

Royalties in respect of crude oil produced in Bahrain, on or after January 1964, and sold for export not in excess of $12\,^1/_2$-per-cent of the value of such crude oil, shall be deductible from taxable income. Any other royalties would continue to be a deduction from 50-per-cent of net income to determine the income tax to be paid.

B The prices at which the company sells Bahrain crude oil and Bahrain crude oil products for export in taxable years beginning after 31 December 1963, shall equal (a) the value of Bahrain crude oil sold for export, plus (b) the posted prices of the buyers from the company, of Bahrain crude oil products sold for export f.o.b. Bahrain in such year, minus (c):

(1) a marketing allowance of $^3/_7$ pence sterling per barrel, plus

(2) an allowance equal to the following annual percentages of the aggregate value of Bahrain crude oil and crude equivalents of Bahrain crude oil products sold for export in the year: $8\,^1/_2$ per-cent for 1964; $7\,^1/_2$ per-cent for 1965; $6\,^1/_2$ per-cent for 1966 and thereafter; plus any allowance in excess of the above percentages of value of crude equivalents in respect of Bahrain crude oil products sold by the company, or an affiliate to non-affiliates; plus

(3) US\$0.0026470 per barrel by which the API gravity of such crude and equivalent exceeds 27;

(4)   all such allowances are to be reduced or eliminated for years after 1966 when the company deems this justified by changes in competitive marketing situations compared with 1964;

The pricing agreement of 29 November 1955, between the Ruler and the company, as amended, shall remain in effect in respect of products processed in Bahrain from foreign crude oil, but shall not be applicable to Bahrain crude oil products. Exported Bahrain crude oil products shall be deemed to be that proportion of products processed by the company in Bahrain and exported corresponding to the proportion of Bahrain crude oil to total crude oil so processed and exported. Products and crude oil destined for consumption in Bahrain are excluded from the agreement. The value of Bahrain crude oil shall be:

(a)   if there is a published price for such crude oil, such published price, less adjustment for terminaling costs saved when such crude oil is refined in Bahrain;

(b)   if there is no published price for Bahrain crude oil, a value equal to the weighted average of posted prices of foreign crude oil brought into Bahrain from Arabian Gulf countries and refined in the year concerned, adjusted for gravity differences, it being understood that crude oil brought in Bahrain via the AB pipeline shall be the published price of similar crude f.o.b. Ras Tannurah, adjusted for stabilization. The company's income tax reports shall conform with the above pricing provisions.

C No payment to the Government shall be required, related to production, manufacture, sale, export, transport, shipment of or dealings in crude oil or products produced or processed by the company, or profits therefrom or their distribution, other than as provided in the agreements of 8 March 1966 and prior agreements and laws, except with respect to

(1)   charges which are for services by the Government rendered to others on request, or to the public generally, and are reasonable and of general application, and

(2)   payments by others than the company, pursuant to laws of general application, in respect of income from services or goods provided to the company.

D The company shall never be required to make total payments to the Government under terms less favourable than those applicable to the most favoured or any other enterprises producing and exporting crude oil anywhere within His Highness' domains, provided that total company payments shall never be less than they would have been under arrangements prior to the 8 March 1966 agreements.

The Ruler of Bahrain also informed L. D. Josephson, General Manager of Bapco, that regarding the marketing allowance, effective from the 1st January 1968, the expression of half an American cent would replace $^3/_7$ sterling pence when calculating income tax.

1966, 21 March: The Eastern and General Investment Company Limited was formed as the successor to the Eastern and General Syndicate Limited.

1966, 17 April: L. D. Josephson, Vice President and Director of Bapco, wrote to His Excellency Shaikh Khalifa bin Salman Al-Khalifa, Head of Finance of the Government of Bahrain, proposing an accelerated schedule for the payment of income tax.

1966, 11 June: Bapco confirmed its agreement to pay royalties on a monthly basis, fifteen days in arrears, by 15 February 1967. Tax was to be paid monthly during the taxable year on the basis of estimated income tax declarations, in twelve equal instalments, with estimates revised quarterly.

1966, 13 June: *The Bahrain Income Tax (Amendment) Decree* was published, due to take effect on 15 June 1966. In summary, this lengthy document comprised redefinitions of the term "exploratory well", the accounting treatment of the costs related thereto for tax calculation purposes, redefined the accounting treatment, for tax purposes, of tangible and intangible oilfield assets, and dealt with the expensing of royalty payments. Much of this document's content had been detailed in the 8 March 1966 agreements. This decree reflected OPEC's first big "bite" and the first breach of the previously sacrosanct fifty-fifty principle.

1967: A new Bahrain refinery annual throughput record was achieved of 87.8 million barrels per year.

A two-year secretarial training course by Bapco Limited, the first of its kind for Bahraini girl employees, was successfully completed in the middle of the year.

Corporate income tax and royalties, payable by Bapco, began to be computed in US dollars, but continued to be paid in pounds sterling.

Bapco Limited created a new position of Health Educator to be awarded to a Bahraini member of the Medical Department following a year's training at the American University of Beirut.

1967, 24 April: L. D. Josephson, Vice-President and Director of Bapco Limited, confirmed to the Ruler of Bahrain, a royalty revision, "an amount per barrel equal to twelve and one-half (12.5) per cent of the value of such crude oil including casing-head petroleum spirit". In respect of natural gas, the rate of royalty would be 12.5-per-cent of the field price received by the company.

1967, May: The Bahrain Economic Development Study Report was published.

1967, June: The Arab-Israeli Six-Day War.

1967, July: The Bahrain Government created a Development Bureau under the auspices of the Finance Department, headed by His Excellency Shaikh Khalifa bin Salman Al-Khalifa. The Bureau was attached to the Petroleum Affairs Bureau, led by Mr Yousuf Shirawi (later appointed Minister of Development and Industry). The officers assigned to the Bureau were Mr Habib Qassim, Economic Adviser (later appointed Minister of Commerce and Agriculture), and Mr Denis Jones, a process engineer seconded from Bapco as Technical Adviser.

1967, September: The 100 fil note was issued in place of the original coin.

1967, October: The UK devalued the pound sterling from US$2.8 to US$2.4.

1967, 17 November: Bahrain driving convention was changed from the left to the right side of the road. Bapco assisted with the preparation of this move by setting up a "Go Right" committee to co-operate with the Bahrain Government Traffic Department.

1968: The National Guard (now the Bahrain Defence Force) was formed.

Isa Town was inaugurated.

Delmon Hotel, Bahrain's first international hotel, was built with Bapco Limited's technical assistance.

1968, 1 January: California Texas Oil Corporation was renamed Caltex Petroleum Corporation (CPC).

The marketing allowance of three-sevenths pence was substituted by the words "one half ($\frac{1}{2}$) US cent"

1968, 9 January: The Organization of Arab Petroleum Exporting Countries (OAPEC) was formed by Kuwait, Libya and Saudi Arabia in Kuwait.

1968, 1 February: The eight and a half (8 ½) pence sterling refining profit margin was substituted by ten (10) US cents.

1968, 24 June: Eastern and General Holdings Limited was registered as the successor to Eastern and General Investment Company Limited.

1968, 3 August By agreements between the Ruler of Bahrain and Bapco, the crude oil pricing agreement of 8 March 1966 was amended:

- Moving from pounds sterling to US dollars, (1) effective 1 January 1968, the marketing allowance was changed from $^3/_7$ pence sterling to half a US cent per barrel; (2) effective 1 February 1968, the company's refining profit element was changed from 8 ½ pence sterling to ten US cents.

- Effective 1 January 1968, the aggregate of allowances based on the value of Bahrain crude oil and crude oil equivalent shall be the sum of (1) a percentage of the value of such Bahrain crude oil, decreasing from 5 ½ per-cent in 1968 to zero per-cent in 1972 and thereafter, and (2) an amount per barrel by which API gravity exceeds 27° (with all crude in excess of 37° API being treated as 37° API), specified for each year through 1974, after which the gravity allowance will terminate.

1968, 9 August: Aluminium Bahrain (ALBA) was incorporated.

1969, 1 April: The Ruler of Bahrain confirmed to Bapco's General Manager that ten American cents would replace 8 ½ pence sterling for tax purposes.

1969, 14 July: Bapco Limited donated a language laboratory to the Gulf Technical College.

1969, September: The first gas development well was drilled into the sweet Khuff gas zone underlying the oil reservoir, in preparation for supplying gas-turbine fuel to the new aluminium plant, Alba.

The directors of the Bahrain Petroleum Company Limited and the Caltex Petroleum Corporation received His Highness Shaikh Isa bin Salman Al-Khalifa at an informal meeting in the Caltex Board Room, New York, on the occasion of His Highness' historic visit to the USA.

1969, October: The first full meeting of Bapco's Board of Directors ever to be convened in Bahrain took place in Awali.

A Health Education Centre was established at the Bahrain refinery.

1969: At the close of the decade, Bahrain had 213 oil-producing wells and seven gas-producing wells.

1970: The Bahrain refinery throughput reached 254,000 b.p.c.d., 6% above the 1967 record.

91.4 million barrels of crude had been processed, of which 63.5 million had come from Saudi Arabia and 27.9 million from the Bahrain field.

Bahrain's first museum opened in Muharraq with Bapco's financial assistance.

Bapco hosted the Third International Conference on Asian Archaeology.

A Computer Users Association was formed with Bapco's initiative.

1970, March: Bahrain became a member of OAPEC.

1970, 27 November: The corporate income tax rate was increased to 55-per-cent.

The refining profit margin was reduced to 9.0909 cents.

The marketing allowance of half a US cent was discontinued.

1971, 15 February: The UK decimalized sterling.

1971, 14 August: The declaration of the creation of the State of Bahrain was broadcast by The Ruler, His Highness Shaikh Isa bin Salman Al-Khalifa, following the formation of the United Arab Emirates.

1971, 15 August: A new Treaty of Friendship was signed by the State of Bahrain and the United Kingdom of Great Britain and Northern Ireland.

1971, 11 September: Bahrain became a member of the Arab League.

1971, 21 September: The State of Bahrain was admitted to the United Nations.

1971, November: Bahrain joined the World Health Organization (WHO) and the Food and Agriculture Organization (FAO).

1971, 27 November: An agreement to implement the amendments to the Income Tax Law was issued, namely the price of finished products for income tax purposes. The deletion of marketing allowances was made official.

1971, 16 December: The State of Bahrain's first National Day.

1972: Natural gas output rose above 100 million cubic feet per day. A revenue-sharing agreement was signed between Bahrain and Saudi Arabia, whereby Bahrain received a 50-per-cent share of revenue from the Aramco-operated Abu Saa'fa offshore field which lies between the two countries.

Work started on a US$65 million project so that the Bahrain Refinery could manufacture 50,000 US barrels daily of low-sulphur fuel oil.

1972, January: Bahrain became a member of the United Nations Educational, Scientific and Cultural Organization (UNESCO).

1972, July: The Gulf Area Oil Companies Mutual Aid Organization (GAOCMAO) was formed to provide an effective response mechanism to a major oil spill in the Arabian Gulf region, greater than any one member company could expect to handle alone. Bapco BSC(c) is one of the ten member companies.

1972, September: The value of the pound sterling fell to 942 fils (1,333 fils at the time of the Bahrain Dinar's creation in 1965), caused by US action in 1971 which precipitated the collapse of the fixed-exchange-rate structure.

Bahrain became a member of the International Monetary Fund (IMF) and the International Bank for Reconstruction and Development (IBRD), the World Bank.

1972, 29 September: Berry Wiggins and Company Limited purchased from Cavendish Land Company Limited, an overriding royalty of five pence (sterling) per ton on all oil produced in excess of 750 tons per day by Bapco Limited.

1972, 24 November: Tank 474 in the Bahrain refinery, containing about 19,000 barrels of naphtha base stock for jet fuel production, ruptured without warning, and spilled two thirds of its contents. The fire which followed took six days to put out and completely destroyed four other tanks, damaged three others beyond economic repair and destroyed 1500 feet (457.2 m) of main pipeline. It also forced the shutdown of the Bahrain refinery and the AB pipeline for approximately twelve and sixteen days respectively, and caused the Bahrain oilfield to be shut down for five days.

1973: Two new gas wells "came in", bringing the total number to thirteen and increasing production to an average of 227 mm.c.f.d.

Kanoo Nursing School was opened with the participation of Bapco's medical department.

1973, 1 January: The Bahrain Government acquired 25-per-cent of the Bahrain producing field and its related assets.

1973, 6 January: The Arab Marine Petroleum Transport Company (AMPTC), in which the Bahrain Government owns 3.5-per-cent equity, was formed in Kuwait.

1973, 1 September: The refinery profit margin on imported crude oil was increased from 9.0909 cents per barrel to 11 cents per barrel (less 55-per-cent tax).

1973, 6 October: Egypt and Syria launch a joint military campaign to repossess part of the Arab territory which Israel had captured and occupied in 1967.

1973, 16 October: OPEC increased the posted price of Arabian light crude, on a "take it or leave it basis", by 70-per-cent to US$5.119 per barrel.

1973, 17 October: The oil ministers of the OAPEC members, except Iraq, agreed to cut their crude oil production by a minimum of 5-per-cent with immediate effect, precipitating a sense of crisis as the Arab oil embargo began to be felt.

1973, 5 December: The Monetary Agency Law was signed, replacing the Currency Board with the Bahrain Monetary Agency, which has wide central-banking powers over the monetary and banking system.

1974: The effect of the Abu Saa'fa oilfield revenue was reflected in the Government of Bahrain's budget for the first time.

1974, 1 January: The Bahrain Government acquired a further 35-per-cent of the Bahrain producing field and its assets.

1974, 11 March: Bapco Limited notified Berry Wiggins that following the Government of Bahrain's participation in the production and ownership of Bahrain crude oil, effective 1 January 1973, the overriding royalty payable to Berry Wiggins should end.

1974, July-September: Royalties were paid at the rate of 14.5-per-cent of the deemed posted price of the value of crude oil.

1974, October: The royalty rate was increased to 16.6-per-cent of the deemed posted price of crude.

1974, November: The royalty rate increased again to 20-per-cent of the deemed posted price of crude.

1974, 23 November: The combined effect of the Government of Bahrain's incremental acquisition of the Bahrain oilfield to 60-per-cent, was expressed in an *Agreement* signed on this date.

1975, January: The Bahrain Monetary Agency became fully operational and established control of the Dinar exchange rate, which it set slightly below parity at 395.80 fils per US$.

1975, 25 March: Bapco's Board of Directors in New York ratified the 11 March 1974 notification to Berry Wiggins regarding the cessation of the overriding royalty payment.

1975, May: The Bahrain Monetary Agency took over the function of purchasing dollar oil revenues from the Government and established full control over the Dinar exchange rate.

1975, May: The royalty rate on *exports* was retained at 20-per-cent on the deemed posted price of the value of crude. The royalty rate on *local* sales was reduced to 12.5-per-cent of the deemed posted value of crude.

1975, 5 June: The Suez Canal was reopened to commercial shipping after an eight-year closure.

1975, October: An offshore banking sector was launched in Bahrain.

1975, 23 November: Arab Petroleum Investment Corporation (APICORP) was incorporated. APICORP owns 12.5-per-cent of the Bahrain National Gas Company.

1976, 23 February: Bahrain National Oil Company BSC (Banoco) was incorporated, wholly-owned by the Government of Bahrain, with the intention of assuming responsibility for managing the State's 60-per-cent share in the producing field assets of Bapco Limited.

1976, 21 August: Arab Petroleum Services Company (APSC) was formed in Libya as an OAPEC holding company, in which the Bahrain Government owns 3-per-cent of the equity.

1977: The Bahrain Refinery broke its 1971 daily average record of 260,000 b.p.c.d. The run of 95.2 million US barrels was more than 1 million above the previous record high in 1971, including 21.2 million barrels of Bahrain crude and 73.2 million from Saudi Arabia. The production of natural gas rose to 332 mm.c.f.d., of which 35-per-cent went as fuel to the aluminium smelter, Alba.

299 oil wells and seventeen gas wells were in production.

The Sitra-Manama Causeway was completed.

1977, 15 December: The inauguration ceremony of the Arab Shipbuilding and Repair Yard (ASRY).

1978, 26 January: The Bahrain Dinar was revalued by 2-per-cent to 388 fils per US$ with the declaration that the par value was now to be described in terms of the IMF "Special Drawing Rights" at a rate of 476.19 fils per SDR – the original 1966 dollar parity.

1978, 9 May: The Arab Petroleum Training Institute (APTI) was formed in Baghdad by the OAPEC Council of Ministers. The Bahrain Government has a 1-per-cent shareholding.

1978, July: Bahrain issued the BD20 bank note. The replacement of 100 fil notes by coins was announced, owing to the unacceptable cost of maintaining the quality of notes of small denomination.

1979: Royalty payments ended.

1979, 22 March: Bahrain National Gas Company BSC(c) was incorporated.

1979, December: The Amir opened the US$95 million Bahrain associated gas project at Jebel Ad-Dukhan.

Bahrain issued new 500 fils, BD1, BD5 and BD10 notes.

1979, 5 December: Gulf Petrochemical Industries Company (GPIC) was incorporated as Bahrain's first petrochemical venture to utilize the State's natural Khuff gas in basic petrochemical products such as ammonia and methanol.

1979, 15 December: An *Agreement* was signed, providing for the Bahrain Government to acquire Bapco's 40-per-cent remaining interest in the exploration and producing right, operations and facilities and the related production of the Bahrain oilfield, effective from 1 January 1980.

A number of other significant changes were made by, or concurrently with, this document. In particular, all previously existing agreements between Bapco Limited and the Government of Bahrain were cancelled, except for a short list of "Agreements to be Preserved". Notable among the cancelled agreements was the 1934 *Mining Lease*.

At the same time, Banoco took over the local marketing of refined products but left the management and operation of the producing field in the care of Bapco Limited.

An *Interim Operating Agreement* was signed simultaneously.

1980, 29 May: GPIC was re-incorporated to include Saudi Basic Industries Corporation (SABIC) representing the Government of Saudi Arabia.

1980, 26 June: The Bahrain Government and Bapco Limited executed a document entitled

Agreed Principles which set forth the basic terms of the Bahrain Government's acquisition of a 60-per-cent interest in the Bahrain refinery.

1980, July: The Bahrain Monetary Agency announced that oil still accounted for 70-per-cent of total government revenue, even though Bahrain's actual crude oil production was dropping.

1980, 19 July: The *Bahrain Refinery Participation Agreement* was signed which effectively enabled the Bahrain Petroleum Company BSC(c) to commence operations, although the company was not incorporated until 21 May 1981 as the successor to the Bahrain Petroleum Company Limited. The company was formed to operate the Bahrain refinery and to administer Awali town, together with its related facilities, following a formal change in the company's shareholding: the Bahrain Government (60-per-cent) and Caltex Bahrain (40-per-cent).

By the year end, the Petroleum Marketing Unit (PMU) had been created within the Ministry of Development and Industry to provide the Bahrain Government with international petroleum marketing capability for its recently acquired shareholding in the refinery.

1980, September: Bapco Limited introduced a giant oil and water separator for use in the fight against sea pollution.

1980, 3 November: The Supreme Oil Council was created in the State of Bahrain under the Chairmanship of the Prime Minister, His Highness Shaikh Khalifa bin Salman Al-Khalifa, and affiliated to the Council of Ministers. Its function was to formulate Bahrain's general oil policy to ensure the conservation of petroleum resources and to develop alternatives to them.

1980, December: The Bahrain Dinar was revalued: 377 fils = US$. This rate has held since.

1980, 9 December: Caltex Bahrain Limited was incorporated in Bermuda as a wholly-owned subsidiary of the Caltex Petroleum Corporation.

1981: The throughput of the Bahrain refinery reached 259,300 b.p.c.d., a new record.

Bapco BSC(c) launched a new house journal, *Bapco News*, to replace *Awali Weekend*.

1981, 4 May: The *Participants' and Operating Agreement* was signed, effective on 1 July 1980.

1981, 21 May: The Bahrain Petroleum Company BSC(c) was incorporated as the successor to the Bahrain Petroleum Company Limited to operate the refinery and protect the shareholders' assets. The Bahrain Government and Caltex Bahrain Limited owned 60-per-cent and 40-per-cent respectively.

1981, 25 May: The Gulf Co-operation Council Charter was signed in Abu Dhabi.

1981, 8 July: The Bahrain-Saudi Arabia Causeway construction contract was signed with the Dutch consortium, Ballast Nedam.

1981, November: The Banoco Phase-In Committee was formed with the Government's intention that "Banoco should take over the operational responsibilities of the Bahrain producing field from Caltex Bahrain Limited, with a transfer date for accounting purposes of 1 January 1982."

1981, 8 December: The Bahrain Petroleum Company Limited was dissolved.

1982: The Bapco refinery operated at a reduced level compared with 1981 owing to the general world economic recession.

A US$7.8 million revamp of the Fluid Catalytic Cracking Unit was completed.

Two of the three Arabia-Bahrain (AB) pipelines carrying Saudi Arabian light crude to the Refinery were relocated in the East Rafa' area to make way for a new dual-carriageway (divided highway).

1982, 1 January: Banoco assumed complete management responsibility for the exploration and production of the Bahrain oil and natural gas field.

1982, February: No 1. Power Plant was shut down after 47 years of service.

1982, March: Bapco's communications achieved a major breakthrough with the first successful transmission of text between the company and Caltex in the US.

1982, April: The 500th loan as part of the Home Ownership Plan was signed. The scheme, started in 1964, had granted BD6,279,633 (US$16.6 million).

1982, July: Bapco BSC(c) was given the go-ahead by the Ministry of Labour to form a Joint-Labour Committee.

1982, September: Drilling operations for the first of six new gas wells began in the Khuff field as part of a BD15 million (US$40 million) programme to increase the island's gas production.

1983: The Bapco workforce was 3,985, of whom 82-per-cent were Bahraini nationals.

1983, February: The Arabian-Bahrain pipeline was shut down and the throughput of Bahrain indigenous crude cut back to offset the refinery's inability to sell its products economically.

1983, April: Parking meters appeared for the first time in Bahrain.

Bapco's Light Isomate Production Project was completed, the largest single investment in Bapco for many years.

1983, May: The Fluid Catalytic Cracking Unit's Process Computer, costing US$3.79 million, was installed.

An explosion and blaze at Bapco's hydrogen plant claimed the lives of two workers and caused damage in excess of BD1 million.

1983, 26 May: Caltex Bahrain Limited was delimited to become a division of Caltex Trading and Transport Corporation (CTTC).

1983, October: Petrol prices rose between 11 and 14-per-cent.

1984: A new sulphur pelletizing plant, owned and operated by Universal Chemicals, was opened adjacent to the Bahrain refinery.

1984, 1 July: Standard Oil Company of California was renamed Chevron Corporation.

1985, January: An Agreement was signed between Bapco, its shareholders and the Bahrain National Oil company (Banoco) for the transfer of the Sitra marketing terminal equity to Banoco.

1985, 6 June: The Bahrain Aviation Fuelling Company BSC(c) was incorporated as an affiliate of Banoco to operate the jet fuel facilities at Bahrain International Airport.

1986, November: King Fahad of Saudi Arabia arrived in Bahrain to the biggest welcome ever for a visiting Head of State on the occasion of the official opening of the BD212 million (US$562 million) 25 km (15.5 mile) long causeway which links the two countries. "The Bridge to the Future" was named the King Fahad Causeway.

1986, December: The Amir celebrated the Silver Jubilee of his accession as Ruler of Bahrain.

1987, 19 October: Black Monday, when Wall Street experienced its worst trading day ever during which the Dow Jones industrial average fell 508 points to 1739. Thus 22.5-per-cent was wiped off share values, almost double the drop experienced in the great "crash" of October 1929.

1988: Bahrain refinery operated at more than 95-per-cent its design capacity for the third successive year.

The average daily crude run for the year was 242,000 barrels of which 199,000 barrels came from Saudi Arabia.

A new US$1.4 million Pressure Swing Adsorption Unit, part of the Low Sulphur Fuel Oil (LSFO) complex at the refinery, was commissioned.

1988, August: Bapco's new Finance and Administrative building was commissioned at the refinery, including the installation of a new IBM 4381-Q14 computer.

1988, 22 September: Caltex Services Corporation was incorporated in Delaware, to provide services to the Caltex group of operating companies world wide.

1988, 20 October: It was announced that Bahrain was finalizing plans to build a petrochemical plant with an annual capacity of 100,000 tonnes each of propylene and polypropylene. The Bahrain National Gas Company (Banagas) would supply propane for processing the materials.

1988, December: The FCCU achieved a record throughput of 44,000 barrels per day, following extensive modifications.

1989, 1 January: Caltex Petroleum Corporation commenced operations as a holding company for its numerous regional subsidiary operating companies, following the group's restructuring in October 1988.

1989, 29 March: The foundation stone was laid for the US$74 million Banagas expansion project, due to be completed in late-1990. The boosted production of liquefied petroleum gas (LPG), from 9000 to about 15,000 barrels per day, was to be used as feedstock for the BD75 million polypropylene plant, part of the islands' developing petrochemical industry.

1990, 17 January: The Prime Minister, His Highness Shaikh Khalifa bin Salman Al-Khalifa visited the Bahrain refinery to mark Bapco's Diamond Jubilee.

DIRECTORY
Companies and Organizations
Associated with the Evolution
of the Oil Industry in Bahrain

The main purpose of this directory is to help readers track the mosaic of nomenclature and corporate relationships, the result of antitrust legislation, acquisition, joint ventures, restructuring, place-name changes or government participation. Each entry relates to the life-cycle of a company's entity. "Before and after" progressions are cross-referenced when appropriate.

company name usage: to maintain historical accuracy, particularly when referring to legal documents or quoting directly from sources, incorporated company names have been retained within their contemporary context. In particular:

The Bahrein Petroleum Company Limited (11 January 1929 to 31 December 1952)
The Bahrain Petroleum Company Limited (1 January 1953 to 8 December 1981)
The Bahrain Petroleum Company BSC(c) (since 21 May 1981)

Information contained under each entry relates only to the period in which the company was known by that name, that is, the life-cycle of the incorporated name. To avoid confusion, previous or subsequent lineages are indicated as cross-references.

ownership: equity, shares and stock may be interpreted as synonymous expressions in the context of this guide. Participation refers to host-government ownership, with the distinct proviso that such shares/stocks are not divisible and may not be offered for sale in a public subscription.

American Overseas Petroleum Limited – AMOSEAS

The company was incorporated in the Bahamas on 9 January 1952 as a non-profit service company owned equally by Standard Oil Company of California (now Chevron) and the Texas Company (now Texaco). It provides by contract a broad line of services to various petroleum exploration and producing ventures in the eastern hemisphere, ownership of which resides half in Chevron and half in Texaco, or their subsidiaries. From 1 February 1957 until 1 January 1969, it provided extensive services to the Bahrain Petroleum Company Limited, on a non-profit basis. On 16 December 1970, Amoseas Indonesia Inc. was incorporated in Delaware, as a wholly owned subsidiary of American Overseas Petroleum Limited, to provide, under non-profit contracts, services to jointly owned exploration and production enterprises of Chevron and Texaco, or their subsidiaries, in areas of Indonesia outside Sumatra. Caltex Petroleum Corporation and Amoseas are sister companies, each owned 50-per-cent by Chevron and 50-per-cent by Texaco. Until 1 January 1989 Caltex performed, on a non-profit basis, some services for Amoseas. Their principal operating relationship has been through the sale of crude oil by Amoseas and its client producing companies to Caltex and its client marketing companies. (*See also* Caltex Petroleum Corporation, Chevron Corporation *and* Texaco)

Anglo-Iranian Oil Company Limited – AIOC

When Persia was renamed Iran on 21 May 1935, the Anglo-Persian Oil Company Limited correspondingly changed its name. On December 1954 AIOC became the British Petroleum Company Limited (BP). (*See also* Anglo-Persian Oil Company Limited)

Anglo-Persian Oil Company Limited – APOC

After the discovery of oil in Persia in 1908, the concession granted to William Knox d'Arcy on 28 May 1901 was taken over by the Anglo-Persian Oil Company Limited, registered in England on 14 April 1909. The rights of Anglo-Persian Oil Company Limited in the "transferred territories" were confirmed by the Government of Iraq on 30 August 1925, supplemented by further agreement on 25 May 1926. On 27 November 1932 the Persian Government attempted to cancel the 1901 concession. An appeal to the Council of the League of Nations by the UK Government led to a new concession agreement in 1933, covering a small area of Persia. Anglo-Persian Oil Company existed until 21 May 1935, after which the name was changed to Anglo-Iranian Oil Company (AIOC) because of the official change in the country name. (*See also* Anglo-Iranian Oil Company Limited)

Arab Marine Petroleum Transport Company – AMPTC

This was formed on 6 January 1973 in the State of Kuwait by three members of the Organization of Arab Petroleum Exporting Countries (OAPEC), Bahrain, Kuwait and Saudi Arabia, with a share capital of $500 million, of which the Government of Bahrain owns 3.5-per-cent.

Arab Petroleum Investment Corporation – APICORP

APICORP was incorporated on 23 November 1975. Shareholders are OAPEC members: Kuwait*, Saudi Arabia* and the UAE*, 17-per-cent each; Libya, 15-per-cent; Iraq and Qatar*, 10-per-cent each; Algeria, 5-per-cent; Bahrain*, Egypt and Syria, 3-per-cent each. (The five starred countries are members of the Gulf Co-operation Council.) Now based in Dammam, Eastern Province, Saudi Arabia, the corporation finances investments in petroleum and petrochemical projects and related industries in the Arab world and in developing countries, with emphasis placed on Arab joint ventures. APICORP owns 12.5-per-cent of the Bahrain National Gas Company.

Arab Petroleum Services Company – APSC

An OAPEC holding company, formed on 21 August 1976 in Libya, in which the Bahrain Government has a 3-per-cent shareholding. Two subsidiary companies, based in Tripoli, operate in drilling and seismic exploration, whilst a third subsidiary, based in Baghdad and affiliated to the American Santa Fe company, operates oil-well logging.

Arab Petroleum Training Institute – APTI

APTI was formed on 9 May 1978 by the OAPEC Council of Ministers in Baghdad, Iraq, for the purpose of raising all levels of oil refining skills in the Arab world. The Bahrain Government owns 1-per-cent of the Institute's equity. APTI's new headquarters were inaugurated on 27 April 1987.

Arab Shipbuilding and Repair Yard Company – ASRY

The foundation stone for the dry-dock was laid in Bahrain on 30 November 1974 and the inauguration ceremony took place on 15 December 1977. The ASRY shareholding is divided between seven of the nine OAPEC member countries, five of which are Gulf Co-operation Council members: Bahrain, Kuwait, Qatar, Saudi Arabia and the UAE each hold 18.8358-per-cent of the equity (totalling 94.179-per-cent); Iraq 4.7180-per-cent and Libya 1.1030-per-cent. (Oman, the sixth member of the GCC, is not a member of OAPEC. Algeria and Syria, the other two members of OAPEC, are not equity holders of ASRY.)

Arabian American Oil Company – ARAMCO

Formerly and formally, the company was known as California Arabian Standard Oil Company (Casoc), incorporated on 8 November 1933, whilst the Arabian American Oil Company (Aramco) evolved as the company's operating name during World War II. During negotiations for US Government participation in Casoc, eventually abandoned, it was planned to create the American Arabian Oil Company. This latter name was adopted officially on 31 January 1944, after which a series of complex manoeuvres occurred.

By the end of the Second World War, Secretary of the Interior Ickes' plan that the US Government should build a pipeline from Aramco's eastern Saudi Arabian fields to the Mediterranean port of Sidon had been rejected. In 1945, the Trans-Arabian Pipe Line Company was incorporated, the plan being that the Standard Oil Company of California (Socal) and the Texas Company should build and operate the pipeline instead. In order to raise funds for the venture, Socal and the Texas Company decided to sell 40-per-cent of their shares in Aramco. Therefore, in 1948, the Standard Oil Company (New Jersey) (now Exxon) and Socony-Vacuum (now Mobil), having denounced the Red Line Agreement, bought these available shares, which gave them 30-per-cent and 10-per-cent ownership of Aramco respectively. This left Standard Oil Company of California (now Chevron) and the Texas

Company (now Texaco) with 30-per-cent each. Thus, the shareholding of the Trans-Arabian Pipe Line Company (Tapline) became the same percentage split.

The assets of the Aramco operating company were purchased by the Saudi Arabian Government in 1980 and the name of that entity was changed in 1988 to Saudi Aramco. A separate entity, the Arabian American Oil Company, still exists, but only as a supplier of management and technical services. Saudi Aramco headquarters are in Dhahran, Eastern Province, Kingdom of Saudi Arabia.

Bahrain Aviation Fuelling Company BSC(c) – BAFCO

Incorporated on 6 June 1985 as an affiliate of Banoco to operate the aviation fuelling facilities at Bahrain International Airport. Equity owners are the Bahrain National Oil Company (60-per-cent); Caltex Bahrain (27-per-cent) and BP Arabian Agencies Limited (13-per-cent).

Bahrain National Gas Company BSC(c) – BANAGAS

Incorporated on 22 March 1979, Banagas was created with the object of maximizing utilization of Bahrain's gas resources by processing associated gas into marketable products. Banagas is owned by the Bahrain Government (75-per-cent), Caltex Bahrain (12.5-per-cent), and the Arab Petroleum Investment Corporation (12.5-per-cent). Corporate headquarters are located in the heart of the Bahrain producing field, close to *Jebel Ad-Dukhan* (Mountain of Smoke) and the site of the first producing oil well in the Arabian peninsula area.

Bahrain National Oil Company BSC – BANOCO

Incorporated on 23 February 1976 by an Amiri Decree, Banoco was formed with the objective to become a fully integrated oil company, wholly-owned by the Government of the State of Bahrain. Between 1 January 1973 and 15 December 1979, the Government of Bahrain acquired 100-per-cent ownership of Bahrain's producing field and related assets, as defined in the 1934 *Mining Lease.* Thus, in the new agreement signed on the 15 December 1979, effective on 1 January 1980, the Bahrain producing field land, which had previously been leased to the Bahrain Petroleum Company Limited, reverted to the Bahrain Government. The agreement also contained terms that made the financial effect precisely the same as it would have been if the agreement had been effected on 1 January 1979.

At the same time, Banoco took over the local marketing operation for refined products, but left the management and operation of the producing field in the care of Bapco Limited. In November 1981, a Banoco Phase-In Committee was formed to phase into Banoco the Bapco Limited employees who had been operating the Bahrain producing field on behalf of Banoco since the 15 December 1979 agreement. This secondment lasted until 1 January 1982. On this date Banoco assumed complete management and operational responsibility for the exploration and production of the Bahrain oil and natural gas field. Titles for the natural gas distribution network and the local marketing terminal at Sitra Terminal were acquired by the Bahrain Government and later transferred as assets to Banoco in 1983 and 1985 respectively. Corporate headquarters are located in Awali.

Bahrain Petroleum Company BSC(c) – BAPCO

In its new form, Bapco BSC(c) effectively commenced operations on 1 July 1980, but the company was not incorporated until 21 May 1981 as the successor to the Bahrain Petroleum Company Limited, following complex negotiations to facilitate Bahrain Government participation. The company was formed to operate the Bahrain refinery and to administer Awali town, together with its related facilities.

Bapco BSC(c) does not own land or fixed assets. Instead, its two shareholders, the Bahrain Government (60-per-cent) and Caltex Bahrain (40-per-cent), have granted Power-of-Attorney to the Chairman and Chief Executive of the company to manage and operate the refinery and to protect and control the shareholders' assets. Bapco BSC(c) operations focus on refining crude oil supplied from the Bahrain oilfield and Arabian light crude imported along the Arabian-Bahrain (AB) pipeline from Saudi Arabia. The refinery manufactures a full petroleum product range to meet the needs of both the domestic and export markets of Bahrain, with increasing emphasis on product yield improvements for its shareholders. (*See also* Bahrain Petroleum Company Limited, Bahrein Petroleum Company Limited, Caltex Bahrain, Caltex Bahrain Limited *and* Caltex Trading and Transport Corporation)

Bahrain Petroleum Company Limited – BAPCO

On 1 January 1953, the spelling of Bahrein changed officially to Bahrain, as did the company name. Thereafter and until 1971, the Bahrain Petroleum Company Limited was vertically-integrated, responsible for the exploration, production, refining and marketing of Bahrain indigenous crude, together with that supplied from the Arabian fields. On 1 January 1973, the Bahrain Government acquired 25-per-cent of the Bahrain producing field's assets. Acquisition of a further 35-per-cent occurred on 1 January 1974. Effective from 1 January 1980, Bapco Limited no longer owned any of the Bahrain oilfield or related assets, although it still managed them. At the same time, the Government took over 100-per-cent of the local marketing operations.

On the following day, the Bahrain Government indicated its wish to acquire an equity interest in the refinery. This was effected on 1 July 1980, with the Bahrain Government acquiring 60-per-cent and Caltex retaining 40-per-cent, although the procedure for formalizing the documentation took a further ten months to complete. To this end, the Bahrain Petroleum Company Limited (registered in Canada) was dissolved on 8 December 1981, following the incorporation of the Bahrain Petroleum Company BSC(c) on 21 May 1981 (registered in Bahrain). (*See also* Bahrain Petroleum Company BSC(c) *and* Bahrein Petroleum Company Limited)

Bahrein Petroleum Company Limited – BAPCO

The Charter of the Bahrein Petroleum Company Limited Charter was sealed on 11 January 1929 in Ottawa, Province of Ontario, Dominion of Canada. The company was a wholly owned subsidiary of the Standard Oil Company of California and maintained its registered office in the Trusts Building, 48 Sparks Street, Ottawa, Canada. The Bahrain concession was granted on 2 December 1925 to Eastern and General Syndicate Limited, which later granted an option to acquire the concession to Eastern Gulf Oil Company Inc. on 30 November 1927. On 21 December 1928, Eastern Gulf Oil transferred its rights to Standard Oil of California. On 1 August 1930, Eastern and General Syndicate Limited assigned the concession to Bapco Limited, after exercising the option which Bapco held at that time. (*See also* Bahrain Petroleum Company BSC(c) *and* Bahrain Petroleum Company Limited)

Balboa Transport Corporation

This tanker company was incorporated on 24 February 1937 in the Republic of Panama. On 31 July 1940, California Texas Oil Company Limited transferred ownership of Balboa Transport Corporation to the Bahrein Petroleum Company Limited. Ownership was transferred from Bapco Limited to California Texas Corporation in 1958. Effective on 13 May 1976, Balboa Transport Corporation changed its name to Caltex Trading and Transport Corporation. (*See also* Caltex Trading and Transport Corporation)

Berry Wiggins and Company Limited

The company was registered in Rochester, Kent, England, on 6 January 1922 to operate an oil refinery at Kingsnorth, Kent, and oil storage facilities at Ellesmere Port, Cheshire. Following a letter written to the company on 29 September 1972 by Cavendish Land Company Limited, Berry Wiggins and Company Limited purchased in November 1972 an overriding royalty of five pence sterling (one shilling prior to decimalization) per ton on all oil produced in excess of 750 tons per day by the Bahrain Petroleum Company Limited. On 11 March 1974, Bapco Limited notified Berry Wiggins that following the Government of Bahrain's participation in the production and ownership of Bahrain crude oil, effective as of 1 January 1973, the overriding royalty payable on the Bapco Limited share of production should end. An agreement to this effect was ratified by the Bahrain Petroleum Company Limited's Board of Directors in New York on 25 March 1975. (*See also* Cavendish Land Company Limited, Eastern and General Holdings Limited, Eastern and General Investment Company Limited *and* Eastern and General Syndicate Limited)

(The) British Petroleum Company plc – BP

Registered in England as Anglo-Persian Oil Company Limited on 14 April 1909, the name was changed on 21 May 1935 to Anglo-Iranian Oil Company Limited, conforming to the official change of country name proclaimed on that date. On 17 December 1954, the name was changed again to the British Petroleum Company Limited. Effective on 1 January 1955, the company became a holding company. Trading activities were taken over by a new, wholly owned subsidiary, BP Trading Limited, formed for the purpose. On 4 January 1982, the British Petroleum Company became a public limited company, quoted on the London Stock Exchange. The BP group has extensive operations throughout the Middle East, including a 13-per-cent shareholding in the Bahrain Aviation Fuelling Company BSC(c).

California Arabian Standard Oil Company – CASOC

Incorporated in Delaware on 8 November 1933 as California Arabian Standard Oil Company (Casoc). In 1934 the company acquired the exclusive concession granted by Saudi Arabia, on 29 May 1933, to Standard Oil of California (Socal), following the discovery of oil in

Bahrain on 1 June 1932. On 21 December 1936, the Texas Corporation obtained a 50-per-cent interest in Casoc. Until 1940, Casoc existed, in effect, as a legal entity only, whilst the practical development of the Saudi Arabian field had been conducted by Socal. During World War II, a Petroleum Reserves Corporation was created by the US Government which attempted to buy a controlling interest in Casoc, based in New York. Protracted negotiations followed, but led nowhere. Casoc ceased to exist on 31 January 1944, following a formal name change to Arabian American Oil Company. (*See also* Arabian American Oil Company, Standard Oil Company of California *and* The Texas Company)

California Texas Corporation – CALTEX

This was incorporated on 6 December 1946 in Delaware as the successor to the California Texas Oil Company Limited. Its name was changed to California Texas Oil Corporation on 1 January 1959, and to Caltex Petroleum Corporation on 1 January 1968. Together with its Bahamas predecessor, Caltex has at all times been owned 50-per-cent by Texaco (or its predecessors) and 50-per-cent by Chevron Corporation (formerly Standard Oil Company of California) either directly or through subsidiaries of those companies. (*See also* California Texas Oil Company Limited, Caltex Bahrain *and* Caltex Petroleum Corporation)

California Texas Oil Company Limited – CALTEX

This joint venture was incorporated on 30 June 1936 in Nassau, Bahamas, as a subsidiary of Bapco Limited. This deal provided the Texas Corporation with a source of petroleum to supply its extensive markets east of Suez, as well as its markets in Europe following the outbreak of World War II, when they became harder to reach from the USA. On 6 December 1946 the company's place of incorporation changed to Delaware, and its name became the California Texas Corporation. (*See also* California Texas Corporation, Caltex Bahrain *and* Caltex Petroleum Corporation)

Caltex

See California Texas Corporation, California Texas Oil Company Limited, *the* Caltex Bahrain *entries,* Caltex Petroleum Corporation *and* Caltex Trading and Transport Corporation

Caltex Bahrain

This entity was established as a division of the Caltex Trading and Transport Corporation on 26 May 1983. Caltex Bahrain owns 40-per-cent of the Bahrain Petroleum Company BSC(c), 12.5-per-cent of the Bahrain National Gas Company and 27-per-cent of the Bahrain Aviation Fuelling Company. The division is based in Awali, State of Bahrain. (*See also* Caltex Bahrain Limited *and* Caltex Trading and Transport Corporation)

Caltex Bahrain Limited

Caltex Bahrain Limited was incorporated in Bermuda on 9 December 1980 in response to the participation transition taking place in Bahrain, as well as revised Canadian legislation, the Canadian Companies Act to which the Bahrain Petroleum Company Limited, being a Canadian registered company, was subject. The amendment required that the majority of the directors of all Canadian corporations should be Canadian nationals. Under the participation agreement of 15 December 1979, effective from 1 January 1980, when the

Bahrain field became 100-per-cent Government-owned, this condition could not have been met. Thus, Caltex Bahrain Limited was created as a new wholly-owned subsidiary of Caltex Petroleum Corporation (CPC). The parent company assigned the Bahrain Petroleum Company Limited's shares to Caltex Bahrain Limited, after which on 8 December 1981, the old Bapco Limited was dissolved.

On 1 July 1980, the Bahrain Government acquired 60-per-cent of the assets of what remained of The Bahrain Petroleum Company Limited. This arrangement was not formalized until the incorporation of the Bahrain Petroleum Company BSC(c) on 21 May 1981. Caltex Bahrain Limited became a 40-per-cent shareholder of this new company. On 26 May 1983, Caltex Bahrain Limited was dissolved and became a division of Caltex Trading and Transport Corporation (CTTC), a wholly-owned subsidiary of Caltex Petroleum Corporation in Dallas, USA. (*See also* Caltex Bahrain, Caltex Petroleum Corporation *and* Caltex Trading and Transport Corporation)

Caltex Petroleum Corporation – CPC

CPC was incorporated as California Texas Corporation in Delaware, USA, on 6 December 1946, owned equally by the Standard Oil Company of California (San Francisco) and the Texas Oil Company (New York). On 1 January 1959, the company changed its name to California Texas Oil Corporation. Within a decade, on 1 January 1968, it changed its name once more to become Caltex Petroleum Corporation. In 1989, Chevron Corporation (formerly Standard Oil Company of California) and Texaco Overseas Holdings Inc. owned 580,000 shares each in Caltex Petroleum Corporation. Since 1 January 1989, CPC has operated primarily as a holding company for its numerous regional subsidiary operating companies. Caltex Services Corporation, incorporated on 22 September 1988 in Delaware, provides services to the Caltex group of operating companies, as can best be handled centrally. Since 1 January 1989, the corporate headquarters of both CPC and Caltex Services Corporation have been in Dallas, Texas, USA.

Caltex Trading and Transport Corporation – CTTC

CTTC was incorporated as Balboa Transport Corporation on 24 February 1937 in the Republic of Panama. On 31 July 1940, Balboa Transport Corporation's ownership was transferred to Bapco Limited, and in 1958 its ownership was transferred once more to California Texas Corporation (renamed Caltex Petroleum Corporation in 1968). Effective on 13 May 1976, Balboa Transport Corporation changed its name to Caltex Trading and Transport Corporation (CTTC), a wholly-owned subsidiary of Caltex Petroleum Corporation (CPC), of which Caltex Bahrain is now an affiliate and a division of CTTC.

Cavendish Land Company Limited

This is the name to which Eastern and General Holdings Limited changed on 29 August 1972. Within a month the company had sold its overriding royalty on the Bahrain oil concession to Berry Wiggins and Company Limited. (*See also* Berry Wiggins and Company Limited, Eastern and General Holdings Limited, Eastern and General Investment Company Limited *and* Eastern and General Syndicate Limited)

Chevron Corporation

The Chevron Corporation can trace its origins back to the Pacific Coast Oil Company (PCO), incorporated in California on 19 February 1879 and to John Rockefeller's Standard Oil Company formed in Ohio, USA, in 1870. On 15 May 1911, the US Supreme Court upheld a Circuit Court decision which found Standard Oil Company (New Jersey) in violation of the Sherman Act, and caused the dissolution of Standard Oil Company within six months. On 30 September 1911, Standard Oil Company (California) became a separate company, although it was not until 27 January 1926 that the Standard Oil Company of California was incorporated. This situation continued until 1 July 1984, when the company became known officially as the Chevron Corporation so that it could be associated more closely with its marketing symbol, the chevron.

Chevron and Texaco jointly own the equity of the Caltex Petroleum Corporation (CPC), one subsidiary of which is the Caltex Trading and Transport Corporation (CTTC). Caltex Bahrain (a division of CTTC) has various equity and operational interests in Bahrain, principally a 40-per-cent share in the Bahrain Petroleum Company BSC(c). (*See also the* Aramco, Caltex, Casoc *and* Standard Oil Company of California *entries*)

d'Arcy Exploration Company Limited

An English company, registered on 12 June 1914, as a wholly owned subsidiary of Anglo-Persian Oil Company Limited. It held a 23.75-per-cent equity interest in the Iraq Petroleum Company (IPC).

Eastern and General Holdings Limited

The successor to Eastern and General Investment Company Limited, registered in London on 24 June 1968. The company was formed to effect the merger of Eastern and General Investment Company Limited, Swithin's Investment Company Limited and Tarbutt Management Limited. Eastern and General Investment Company Limited's major asset was its 90-per-cent interest in a royalty agreement with the Bahrain Petroleum Company Limited. On 29 August 1972, the company changed its name to Cavendish Land Company Limited. Within a month, on 25 September, Cavendish Land agreed to sell the Bahrain overriding royalty to Berry Wiggins and Company Limited (confirmed in a letter of 29 September 1972). (*See also* Berry Wiggins and Company Limited, Eastern and General Investment Company Limited *and* Eastern and General Syndicate Limited)

Eastern and General Investment Company Limited

The successor to the Eastern and General Syndicate, formed on 21 March 1966. The group associated with the company in its oil interests exercised the option held over the Bahrain oil concession. As a result, Eastern and General Investment Company Limited was entitled to receive a royalty of one shilling per ton on all oil produced by the Bahrain Petroleum Company Limited under the terms of the 1934 *Mining Lease*. In 1968, the company merged into Eastern and General Holdings Limited, which in turn was disposed of in a letter to Berry Wiggins and Company on 29 September 1972. (*See also* Berry Wiggins and Company Limited, Cavendish Land Company Limited, Eastern and General Holdings Limited *and* Eastern and General Syndicate Limited)

Eastern and General Syndicate Limited – EGS

This was a consortium of British businessmen, who registered the company on 6 August 1920 to acquire and operate oil concessions in the Middle East. In May 1923 it acquired from Ibn Saud, then Sultan of Nejd, a concession in the Al-Hasa province of Saudi Arabia. In May 1924, it was granted a concession in the Kuwait-Nejd neutral zone. On 2 December 1925, Shaikh Hamad bin Isa Al-Khalifa, Regent of Bahrain, granted the company an exclusive oil exploration licence for a period not exceeding two years. EGS transferred the option to Eastern Gulf Oil on 30 November 1927. On 1 August 1930, the concession was assigned by EGS in New York to Bapco Limited. (*See also* Eastern and General Holdings Limited *and* Eastern and General Investment Company Limited)

Eastern Gulf Oil Company Inc.

This was a subsidiary of Gulf Oil Corporation which itself had been incorporated on 9 August 1922 in Pennsylvania, USA. Eastern Gulf Oil Company was granted an option by Eastern and General Syndicate Limited on 30 November 1927 to exercise oil exploration rights in Bahrain up to 1 January 1929. Eastern Gulf Oil transferred the option to Socal on 21 December 1928. (*See also* California Texas Oil Company Limited, Eastern and General Syndicate Limited, Gulf Oil Corporation *and* Standard Oil Company of California)

Exxon

Incorporated in New Jersey, USA, in 1882 as Standard Oil Company (New Jersey), the company was often referred to as Standard Jersey, or simply Jersey, until November 1972, when it became known as Exxon Corporation. In some parts of the world, affiliates are still known by the company's trade name, Esso, which was derived from the company name Eastern States Standard Oil Company. Exxon is the trade name in the USA and much of the world, thereby permitting the company to sell petroleum in all the US states by avoiding any use of the "Standard" name or derivatives. This limitation was placed on it by the 15 May 1911 US Supreme Court Order that broke up John D. Rockefeller's Standard Oil Trust.

As one of the "majors", Exxon interests in the Middle East focused on Aramco, the Iraq Petroleum Company, Mosul Petroleum Company and Tapline. Exxon is by far the largest US oil company and has long vied with Royal Dutch/Shell as the largest non-government oil company in the world.

Gulf Area Oil Companies Mutual Aid Organization – GAOCMAO

Formed in July 1972, GAOCMAO aims to provide an effective response mechanism to a major oil spill in the Arabian Gulf region, greater than any one of the ten member companies could expect to handle alone. The members: Abu Dhabi National Oil Company, Arabian Oil Company (divided zone, formerly the neutral zone), Saudi Aramco, Bapco BSC(c), Dubai Petroleum Company, Getty Oil Company, Iraq National Oil Company, Kuwait Oil Company KSC, Petroleum Development Oman and Qatar General Petroleum Corporation. The Secretariat, based in Bahrain, is funded by the member companies on an equally-shared basis. The organization's chairmanship rotates on an alphabetical basis according to the company name.

Gulf Oil Corporation

On 10 January 1901, Spindletop, reputedly the most famous oil well in the world, established Texas as a major oil source. The well was financed with cash borrowed from the Pittsburgh banking house of T. Mellon & Sons. In May, Andrew W. and Richard B. Mellon organized a partnership which later became the Gulf Oil Corporation, incorporated in January 1907 in New Jersey, with A. W. Mellon as President. On 9 August 1922, Gulf Oil Corporation of Pennsylvania was incorporated to acquire the undertaking of Gulf Oil's New Jersey corporation. Following this reorganization, the corporation owned a controlling interest in several subsidiaries, including Eastern Gulf Oil Company Inc., incorporated in Kentucky, and South American Gulf Oil Company, registered in Pittsburgh. Confusingly, using both letterheads at random, William T. Wallace (Vice-President of both subsidiaries) corresponded with Thomas Ward, the New York representative of the Eastern and General Syndicate (EGS).

On 30 November 1927, Eastern Gulf Oil Company was granted an option by Eastern and General Syndicate to exercise oil exploration rights in Bahrain up to 1 January 1929. On 21 December 1928, Eastern Gulf Oil transferred the option to Standard Oil Company of California (Socal). Many years later, Gulf Oil Corporation merged with Chevron Corporation, formerly Standard Oil Company of California and a 50-per-cent shareholder of Caltex Petroleum Corporation (CPC). On 18 January 1984, Gulf Oil Corporation, a Pennsylvania corporation, became a wholly owned subsidiary of Gulf Corporation, a Delaware corporation. On 15 June 1984, Gulf Corporation stock holders approved a merger agreement whereby the company became a wholly owned subsidiary of Socal. (*See also* Chevron Corporation, Eastern Gulf Oil Company Inc. *and* Standard Oil Company of California)

Gulf Petrochemical Industries Company – GPIC

Created as the country's first petrochemical venture, GPIC utilized Khuff gas as feedstock in the production of two basic petrochemical products, ammonia and methanol. Initially, GPIC was incorporated on 5 December 1979 with the Bahrain National Oil Company and Petrochemical Industries Company as shareholders on behalf of the Governments of Bahrain and Kuwait respectively. On 29 May 1980, GPIC was reincorporated to include Saudi Basic Industries Corporation, representing the Government of Saudi Arabia. Each participant acquired one third of the company's equity. The project plan required GPIC's output to reach 1000 tonnes of ammonia and 1000 tonnes of methanol per day. The US$450 million complex went on stream on 19 May 1985 with the production of methanol, followed by the production of ammonia on 18 July 1985.

Iraq Petroleum Company – IPC

Formerly the Turkish Petroleum Company, the Iraq Petroleum Company came into being on 8 June 1929 as the first of several oil consortia to develop in the Middle East. Its shareholders included five of the eight companies that were to become the international "majors". BP, Shell and Compagnie Française des Pétroles (CFP) each had a 23.75-per-cent share. Standard Oil Company (New Jersey) and Socony-Vacuum equally shared a further 23.75-per-cent, (11.875-per-cent each). The remaining 5-per-cent was allocated to Participations and Investments Limited (Partex), an organization which took care of Calouste

S. Gulbenkian's interests. Originally, the IPC's concession embraced only part of Iraqi territory, east of the Tigris River, comprising the Kirkuk oilfield.

"Majors"

A term frequently applied to the eight major oil companies: British Petroleum, Chevron, Compagnie Française des Pétroles, Exxon, Gulf Oil, Mobil, Shell and Texaco. (*See also* "Seven Sisters")

Mobil Oil Corporation

First known as Standard Oil Company of New York, hence Socony, the company amalgamated with Vacuum Oil in 1931 to become Socony-Vacuum. A further name change in 1955 created Socony-Mobil, which finally became Mobil Oil Corporation in 1966. In April 1975, Mobil announced its decision to purchase additional shares in the Arabian American Oil Company (Aramco), thus increasing the company's holdings from 10-per-cent to 15-per-cent.

Near East Development Corporation – NEDC

When NEDC was incorporated in Delaware, USA, on 3 February 1928, its shareholders were: Standard Oil Company (New Jersey) (25-per-cent), Socony (25-per-cent), Gulf Oil (16.6-per-cent), Arco (16.6-per-cent) and Amoco (16.6-per-cent). The latter two were bought out in 1931 by Standard Oil Company (New Jersey) and Socony (by then merged with Vacuum Oil to become Socony-Vacuum). Gulf Oil withdrew in 1934, leaving Standard Oil (New Jersey) and Socony-Vacuum each owning 50-per-cent of NEDC, whilst NEDC itself owned 23.75-per-cent of Petroleum Concessions Limited and the Iraq Petroleum Company.

Oilfield Equipment Company Incorporated

Founded on 27 December 1923 at 30 Church Street, New York, as an oilfield equipment supply company. Thomas E. Ward, founder and President of the company until his retirement in 1958, was the New York representative of Eastern and General Syndicate Limited. In this capacity he handled the communications and negotiations between EGS and Eastern Gulf Oil Company relating to the Bahrain oil concession. (*See also* Eastern and General Syndicate Limited *and* Eastern Gulf Oil Company Inc.)

Organization of Arab Petroleum Exporting Countries – OAPEC

Formed on 9 January 1968 with headquarters in Kuwait, the nine member countries are: Algeria, Bahrain, Iraq, Kuwait, Libya, Qatar, Saudi Arabia, Syria and the United Arab Emirates. Egypt was suspended temporarily on 17 April 1979; Tunisia was suspended on 1 January 1987.

Organization of Petroleum Exporting Countries – OPEC

Formed in September 1960, with headquarters in Vienna, Austria. The five founder member countries were: Iran, Iraq, Kuwait, Saudi Arabia and Venezuela. The subsequent eight members (making a total of thirteen) are: Algeria, Ecuador, Gabon, Indonesia, Libya, Nigeria, Qatar and the United Arab Emirates.

Overseas Tankship Corporation – OTC

Incorporated in Panama on 12 June 1946 to handle the tanker requirements of the expanded Caltex Group, OTC bought forty surplus T2 tankers from the US Government, drydocked them, repaired and stored them, before putting them into service. Like its predecessor, the Balboa Transport Corporation, OTC was administered from the Caltex New York office. The first "run" was from New York to Bahrain to pick up crude, sail to Shanghai, deposit the cargo and return to Bahrain. Office space for the four Caltex marine personnel based in Bahrain was established with Bapco Limited. The Bahrain/Shanghai "run" ended in 1949, after China's largest city fell to the troops of Mao Tse-tung's People's Liberation Army on 26 May. By 1950, when it became clear that OTC could no longer command dollar freights, tanker companies were established outside the US and the T2 tankers were sold to them: Nederlandsche Pacific Tankvaart Mij, Outremer Navigaçion de Pétrole, Tokyo Tanker Company and Overseas Tankship (UK) Limited. In 1967, the European properties of Caltex, including NPTM, ONP and OTUK, reverted to Caltex shareholders: Standard Oil Company of California and Texaco.

Pacific Coast Oil Company

Incorporated on 10 September 1879, following a meeting between D. G. Scofield, F. B. Taylor, Charles Felton and others. This historic meeting and hence this company, began the corporate evolution through which the Bahrain Petroleum Company BSC(c) can trace its origins. Pacific Coast Oil Company was bought by Standard Oil Company in 1900. In 1906 Standard Oil Company of California was formed. On 11 January 1929, the Bahrein Petroleum Company Limited formally became a wholly owned subsidiary of Socal. (*See also* Standard Oil Company of California)

Petroleum Concessions Limited – PCL

An English company, registered on 12 October 1935, its stock being held by the owners of the Iraq Petroleum Company Limited: Turkish Petroleum Company Limited, 23.75-per-cent; Compagnie Française des Pétroles, 23.75-per-cent; d'Arcy Exploration Company Limited, 23.75-per-cent; Near East Development Corporation, 23.75-per-cent and Calouste Gulbenkian, 5-per-cent (through Participations and Investments Limited). PCL's many subsidiaries held concessions, exploration or prospecting licences throughout the Middle East.

Saudi Aramco

See Arabian American Oil Company

"Seven Sisters"

A phrase which evolved many years ago recognizing that the world's oil was more or less controlled by seven huge corporations: British Petroleum, Exxon, Gulf Oil, Mobil, Shell, Standard Oil Company of California (now Chevron) and Texaco. The eighth "major", Compagnie Française des Pétroles, was not considered to be one of the "Seven Sisters".

Socony

See Mobil Oil Corporation

Socony-Vacuum
See Mobil Oil Corporation

Standard Oil Company of California – SOCAL
The company owes its origins to the Pacific Coast Oil Company, formed on 10 September 1879 and to the Standard Oil Company founded by John D. Rockefeller in 1870. In 1900, the San Francisco marketing operation of Standard Oil Company bought Pacific Coast Oil. On 23 July 1906, PCO acquired the business assets of Standard Oil Company (Iowa) and filed a name change to become Standard Oil Company. To distinguish Standard Oil Company, which was incorporated in California, from the other Standard Oil companies, the practice was followed whereby "(California)" was added after the name. However, this addition was not part of the name as filed.

On 15 May 1911, a US Supreme Court upheld a Circuit Court decision finding Standard Oil Company (New Jersey) in violation of the 1890 Sherman Antitrust Act, and caused the dissolution of Standard Oil Company (Iowa) on 18 November 1911. Meanwhile, on 30 September 1911, 249,995 shares held by Standard Oil Company (New Jersey) in 34 subsidiaries were distributed to individual stockholders of the New Jersey corporation.

On 8 May 1924, an amendment to the Standard Oil Company's *Articles of Incorporation* officially changed the company's name, which had been filed on the 23 July 1906, to Standard Oil Company (California). Then, on 27 January 1926, Standard Oil Company of California was incorporated in Delaware, USA, and became known in an abbreviated form as Socal. On 1 July 1984 that Standard Oil Company of California changed its name officially to Chevron Corporation. Socal's main pioneering interests in Arabia are vested with the Arabian American Oil Company, the Bahrein Petroleum Company Limited and Caltex. (*See* Arabian American Oil Company *and the Caltex entries*)

Standard Oil Company (New Jersey)
See Exxon

Standard Oil Company of New York (Socony)
See Mobil Oil Corporation

Standard Oil Trust
This was formed in 1882 as part of the John D. Rockefeller empire. As a result of the Sherman Antitrust Act of 2 July 1890, a US Supreme Court Order issued on 15 May 1911 decreed that the empire's companies should be dismantled into independent operations. (*See also* Standard Oil Company of California, Standard Oil Company (New Jersey) *and* Standard Oil Company of New York)

Supreme Oil Council, State of Bahrain
Established in the State of Bahrain on 3 November 1980 under the Chairmanship of The Prime Minister, HH Shaikh Khalifa bin Salman Al-Khalifa, and affiliated to the Council of Ministers, the council's function is to formulate the general oil policy for the country. The

other members of the council are Their Excellencies the Ministers of Foreign Affairs, Development and Industry, Finance and National Economy, Works, Power and Water, and Labour and Social Affairs.

Texaco
See The Texas Company

(The) Texas Company

The original Texas Company was reincorporated in Delaware on 26 August 1926 as the Texas Corporation and acquired, by exchange of shares, substantially all outstanding stock of the Texas Company, a Texas corporation which had been organized in 1902. Thereafter, the company incorporated in Texas was dissolved. In 1936, the Bahrein Petroleum Company Limited, in the process of constructing a refinery and without adequate retail marketing facilities, doubled its stock, so that the Texas Corporation and Standard Oil Company of California could own 50-per-cent each of Bapco Limited's subsidiary, California Texas Oil Company Limited (Caltex).

Effective on 1 November 1941, the Texas Corporation merged the Texas Company (Delaware) and caused the Texas Company (California) to be dissolved. The Texas Corporation acquired all assets and assumed all liabilities of both these companies. Following this reorganization, the company became known as the Texas Company once again. Eighteen years later, on 1 May 1959, the company became Texaco Incorporated.

As one of the "Seven Sisters", Texaco (as the company became known later) owned a 30-per-cent interest in Aramco, one of the four major oil consortia operating in the Middle East from the 1950s to the 1970s. During the late 1980s, Texaco attracted much publicity during a lengthy legal contest with Pennzoil, which was settled in December 1987 in Pennzoil's favour for US$3 billion.

Trans-Arabian Pipe Line Company Inc. – TAPLINE

Incorporated in Delaware in 1945 to build and operate a pipeline from the eastern province of Saudi Arabia to Sidon, Lebanon, on the Mediterranean coast. (*See also* Arabian American Oil Company)

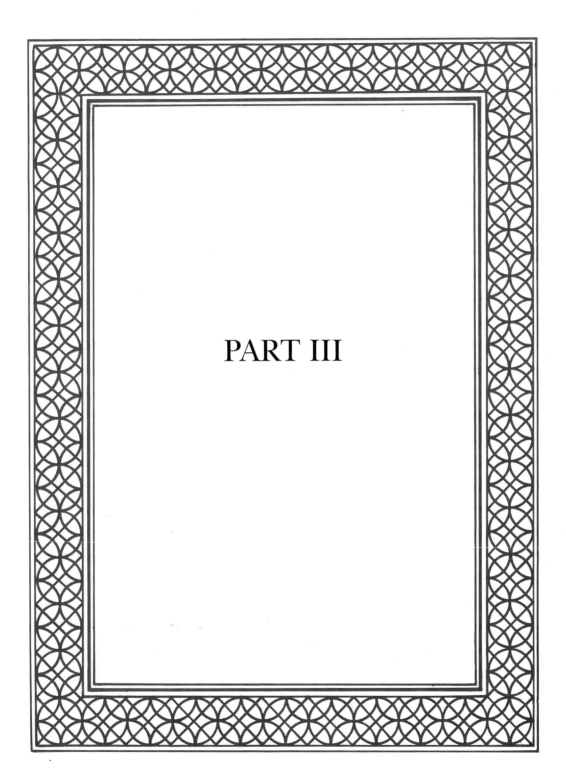

PART III

ABBREVIATIONS and ACRONYMS

Acronyms are listed in upper case with no punctuation.
Abbreviations are listed in upper and lower case with punctuation, when required.

AB Arabian-Bahrain (pipeline)

AD Anno Domini (in the year of the Lord), as used in the Gregorian Calendar established by Pope Gregory XIII in 1582

ADNOC Abu Dhabi National Oil Company

ADPC Abu Dhabi Petroleum Company Limited

AH Anno Hijira (in the year of Mohammed's departure from Mecca to Medina), as used in the Hijri Calendar which began that year, corresponding to AD 622

AIOC Anglo-Iranian Oil Company

ALBA Aluminium Bahrain BSC

AMOSEAS American Overseas Petroleum

AMPTC Arab Marine Petroleum Transport Company

API American Petroleum Institute

APICORP Arab Petroleum Investment Corporation

APOC Anglo-Persian Oil Company Limited

APSC Arab Petroleum Services Company

APTI Arab Petroleum Training Institute

ARAMCO Arabian American Oil Company

ARCO Atlantic Refining Company (until 3 May 1966), now Atlantic Richfield Company

ARPGPR Administration Report on the Persian Gulf Political Residency and Maskat Political Agency

ASRY Arab Shipbuilding and Repair Yard

AVTUR aviation turbine fuel

BAFCO Bahrain Aviation Fuelling Company BSC(c)

BALEXCO Bahrain Aluminium Extrusion Company BSC

BANAGAS Bahrain National Gas Company BSC(c)

BANOCO Bahrain National Oil Company BSC

BAPCO Bahrein Petroleum Company Limited (from 11 January 1929) Bahrain Petroleum Company Limited (from 1 January 1953) Bahrain Petroleum Company BSC(c) (from 21 May 1981)

BATELCO Bahrain Telecommunications Company

BBK Bank of Bahrain and Kuwait BSC

bbl barrel

BBME British Bank of the Middle East

BC before Christ, as used to express ancient historical time

BD Bahrain Dinar

BMA Bahrain Monetary Agency

BOAC British Overseas Airways Corporation

BP British Petroleum Company plc; *or* before the present, as used to express geological time

b.p.c.d. barrels per calendar day

BSC(c) Bahrain Stock Company (closed)

C Celsius and/or Centigrade scale of temperature

CALTEX California Texas Corporation (from 6 December 1946) California Texas Oil Corporation (1 January 1959/1 January 1968)

Capt. Captain

CASOC California Arabian Standard Oil Company

CCL Chevron Corporate Library

c.d. calendar day
Cdr Commander
CFP Compagnie Française des Pétroles
Ch. chapter
CHEVRON Chevron Corporation
CIE Companion (of the Order of the) Indian Empire
cm centimetre(s)
CMG Companion (of the Order of St) Michael (and St) George
CO Colonial Office (United Kingdom)
Co. company; county
CPC Caltex Petroleum Corporation
CTTC Caltex Trading and Transport Corporation
d. penny, a unit of sterling currency before decimalization
DC District of Columbia, as in 'Washington DC', the federal capital of the USA
Dept Department
Dr Doctor
EEC European Economic Community
EGS Eastern and General Syndicate Limited
ESSO Eastern States Standard Oil Company
F Fahrenheit scale of temperature
FAO Food and Agricultural Organization (of the United Nations)
FCCU Fluid Catalytic Cracking Unit (often referred to colloquially as a "cat cracker")
FO Foreign Office (United Kingdom)
GAOCMAO Gulf Area Oil Companies Mutual Aid Organization
GCC Gulf Co-operation Council
GIB Gulf International Bank
GPIC Gulf Petrochemical Industries Company
HE His Excellency
HH His Highness
HM Her/His Majesty
IBRD International Bank for Reconstruction and Development (also known as the World Bank)
IEDC International Energy Development Corporation
IMF International Monetary Fund
IMO International Maritime Organization
Inc. Incorporated
incl. inclusive
IO India Office
IPC Iraq Petroleum Company
Jr junior
km kilometre(s)

k.p.h. kilometres per hour
£ s.d. pounds, shillings and pence (sterling prior to decimalization)
LNG liquefied natural gas
LPG liquefied petroleum gas
Lt Lieutenant
Lt-Col. Lieutenant Colonel
Ltd Limited
m metre(s)
m. thousand
MC Military Cross
mm millimetre
mm. million
mm.b.p.c.d. million barrels per calender day
mm.c.f.d. million cubic feet per day
MP Member of Parliament (refers to the British House of Commons)
m.p.h. miles per hour
NB nota bene (Latin, meaning: note well)
NEDC Near East Development Corporation
NGL natural gas liquids
no. number
n.p. new paisa
NPTM Nederlandsche Pacific Tankvaart Mij
NYMEX New York Mercantile Exchange
OAPEC Organization of Arab Petroleum Exporting Countries
OBE Officer (of the Order) of the British Empire
OBU offshore banking unit
OECD Organization for Economic Co-operation and Development
ONP Outremer Navigaçion de Pétrole
OPEC Organization of Petroleum Exporting Countries
OTC Overseas Tankship Corporation
OTUK Overseas Tankship (UK) Limited
OUP Oxford University Press
p penny *or* pence, post-decimalization of sterling
p. page *or* paisa
PA Political Agent
PARTEX Participations and Investments Limited
PCL Petroleum Concessions Limited
per bbl. per barrel of crude run in the refinery
PIC Petroleum Industries Company
PIW *Petroleum Intelligence Weekly*
pl. plural
plc public limited company
PMU Petroleum Marketing Unit
pp. pages

PRPG Political Resident Persian Gulf

RPM refinery profit margin

RGIFD Records of the Government of India Foreign Department

Rps. (Indian *or* Gulf) rupees, as often abbreviated today to avoid confusion with Riyals (Rys)

Rs. rupees, as the abbreviation appears in most historical documents

s. shilling

SABIC Saudi Basic Industries Corporation

sing. singular

SOCAL Standard Oil Company of California

SOCONY Standard Oil Company of New York

SS Steam Ship

St Saint

TAPLINE Trans-Arabian Pipeline Company

TEXACO Texas Company (26 August 1926 to 1 November 1941) Texas Corporation (since 1 November 1941)

TPC Turkish Petroleum Company

TTKK Tokyo Tanker Company

UAE United Arab Emirates

UK United Kingdom of Great Britain and Northern Ireland

UNO United Nations Organization

UNESCO United Nations Educational, Scientific and Cultural Organization

USA United States of America

VE Day Victory in Europe Day, World War II

VJ Day Victory over Japan Day, World War II

VLCC very large crude carrier

vol(s) volume(s)

GLOSSARIES

──────── ECONOMIC AND FINANCIAL ────────

anna, a former coin of Indian and Pakistani currencies. Prior to the decimalization of India's currency in 1957, 12 paise = one anna; 16 annas = one rupee. Thus, for example, ten rupees, fifteen annas and eleven paise (sing. paisa) were expressed: Rs.10/15/11.

After India decimalized its currency, 100 new paise (n.p.) = one rupee, and the anna became obsolete. (*See also* lakh, paisa *and* rupee)

Bahrain Dinar, created on 16 October 1965 as the unit of currency in Bahrain. The management of the Bahrain Dinar was entrusted to a Currency Board, whose main responsibility was to hold foreign exchange and gold backing, equal to the value of notes in circulation, thus ensuring that the Government could not print bank notes in excess of gold holdings to finance its operations and so devalue the currency.

At the time of its creation, the Bahrain Dinar was stated to be equal to 1.86221 grammes of fine gold and by reference to the declared gold value of other currencies, "par" exchange rates could be calculated. In the case of sterling this gave the value of fifteen shillings (15/- or 1333 fils per £). With the US$ declared to be convertible at the rate of US$35 per ounce of gold (a rate fixed in 1933) the Dinar/dollar exchange rate was US$2.10 per Dinar or 476.19 fils per US$. Much more important in the local suuq was the fact that 1 dinar equalled 10 rupees.

Initially, BD7.8 million Dinar notes were issued in exchange for Gulf rupee notes; the latter were returned to the Reserve Bank of India, which paid their equivalent value in sterling to the account of the Currency Board in London over a period of years.

Bahrain Monetary Agency. Partly in response to the turbulent exchange markets, and more particularly to meet a need for more sophisticated management of the financial sector, the Bahrain Government decided to replace the Currency Board with a Monetary Agency, having full central-banking powers over the banking and monetary sectors, although technically subordinate to the Ministry of Finance. The Agency Law was signed on 5 December 1973, and the BMA became fully operational in January 1975 following the appointment of its first Director General.

Bretton Woods Agreement, an international monetary system established following the United Nations Monetary and Financial Conference held at Bretton Woods, New Hampshire, USA, in 1944. This system was originally based upon a policy of fixed exchange rates, the elimination of exchange restrictions, currency convertibility and the development of a multilateral system of international payments. (*See also* International Monetary Fund)

cartel, a combination of independent businesses

formed to regulate production, pricing and marketing of goods by the members. In the context of the oil industry, OPEC is the best known example.

exchange rate, the price of a currency in terms of another currency. Exchange rates are quoted regularly between all major currencies, but frequently the US$ is used as a standard in which to express and compare all rates. The exchange rate of all fully convertible currencies is determined, like any price, by the supply-and-demand conditions of the foreign exchange market, but in countries where the supply of foreign exchange is largely controlled by the government, the central bank can, in practice, set the rate in terms of a benchmark, usually US$. The Bahrain Dinar is effectively "tied" to the US$ by the Bahrain Monetary Agency's readiness to sell $ at 377 fils.

foreign tax credit. The USA considers all world wide income of companies having their principal place of business in the USA to be subject to domestic taxation, for example the Caltex Petroleum Corporation, which is headquartered in Dallas. To avoid or minimize double taxation of the same income, the USA (with numerous exceptions and limitations) grants a credit against US taxes for those paid to a foreign country with respect to income generated in that country. This matter is often covered by treaties. This has applied to the Bahrain Petroleum Company and, *de facto*, its American shareholders for many years.

forward market, *see* futures market

futures market, or forward market, is an operation in which traders exchange forward contracts, that is, individually tailored contracts which state a future date and price (the forward price) at which a stated quantity of a commodity will be exchanged for money. NYMEX, the New York Mercantile Exchange, is where much of the crude-oil futures market operates.

Gulf rupee was introduced in Bahrain during 1959, having the same value as the Indian rupee but printed in different colours. Gulf rupee notes benefited from the freer exchange system in the Gulf, while the smuggling of Indian rupee bank notes out of India effectively ceased. (*See also* Indian rupee)

Indian rupee had been used as a local means of exchange in the Gulf from the time when the British political presence in the area was directed from India. When India became independent in 1947, the Reserve Bank of India continued to supply notes and coins to the Arabian Gulf countries. Both India and the Gulf countries were part of the Sterling Area, within which these currencies were freely exchangeable. However, exchange controls were in force to prevent rupee movements into nonsterling Area currencies, notably the USA (except for approved transactions). It was partly because of this restriction that the Gulf rupee came into being to allow greater freedom for Arabian Gulf residents compared with Indian citizens. Prior to the decimalization of India's currency in April 1957, one rupee = 16 annas. (*See also* anna, lakh *and* paisa)

International Monetary Fund (IMF), established in December 1945 following ratification of the Articles of Agreement of the Fund which were formulated at the United Nations Monetary and Financial Conference held at BrettonWoods, New Hampshire, USA, in 1944. The IMF became a specialized agency of the United Nations in 1947. Its purposes are to encourage international monetary co-operation, facilitate the expansion and balanced growth of international trade, assist member countries in correcting balance-of-payments deficits and promote foreign exchange stability.

lakh, a financial term expressing 100,000 rupees. During the first decades of the twentieth century in Bahrain, government and oil industry budgets and accounts were expressed in *lakhs*, *rupees*, *annas* and *paise*. Thus, for example:

Rps. 1,22,000/15/11 = one lakh and twenty-two thousand rupees (i.e. one hundred and twenty-two thousand rupees), fifteen annas, and eleven paise. Or, expressed another way, 1 x 100,000 rupees *plus* 22,000 rupees, 15 annas and 11 paise. (*See also* anna, rupee *and* paisa)

NYMEX, the New York Mercantile Exchange, where much of the crude oil futures market, or forward market, operates.

paisa, an "old" paisa (sing.) was a coin of Indian currency. 12 paise (pl.) = one anna; 16 annas = one Indian and/or Gulf rupee.

After the Indian currency was decimalized in April 1957, the anna became obsolete and "new" paise (n.p.) were created: 100 n.p. = one rupee. Today, new paise are used in India, Pakistan, Nepal and Bangladesh.

In historical documents it is easy to confuse the currency symbol p. for paisa (paise) with p sterling, which refers to a penny or pence post-decimalization. For the sake of clarity, particularly when reading historical documents, it is helpful to note:

Pre-decimalization: p. = paisa (paise); d. = penny (pence).

Post-decimalization: n.p. = new paisa (new paise); p = penny (pence) (*See also* anna *and* penny)

paper barrel, crude oil or products traded on the forward or futures markets which are closed out by subsequent sale or settlement without physical or "wet" delivery. (*See also* wet barrel)

penny, an old unit of sterling. 12 pence = one shilling, 20 shillings = £1.

After decimalization on 15 February 1971, the shilling ceased to be legal tender although the coins remained in circulation as values of five new pence (5p) for several years. Thus 100 new pence = £1. (*See also* sterling)

petrodollars, revenue generated from the sale of oil used by a national of an oil-producing country, which generally refers to oil-exporting countries. The monetary designation is derived from the fact that oil sales are transacted in US$. The term came into fashion after the first oil shock in 1973.

royalty, in the context of the oil industry, a payment to the owner of a natural resource, made by the explorer/producer/marketer. In the case of Bahrain's oil and gas reserves, the owner is the Government.

rupee, *see* Gulf rupee *and* Indian rupee

sterling, British monetary unit. Before decimalization in February 1971, sterling was divided into pounds (£), shillings (s.) and pennies/pence (d.). Twelve pence = one shilling; twenty shillings = one pound. Thus, ten pounds, nineteen shillings and eleven pence were expressed as £10 19s.11d.

Following decimalization, one hundred pence (100p) = one pound. Now, ten pounds and ninety-nine pence is expressed as £10.99.

wet barrel, a term used in the futures market in association with paper barrels. A paper barrel is crude oil or products with no specified delivery date. For example, an April paper barrel refers to a barrel for delivery some time in April. Once the delivery date has been set, the paper barrel becomes a wet barrel. (*See also* paper barrel)

——————— LANGUAGE ———————

For those readers unfamiliar with the Arabic language who may wish to pronounce the transliteration in this book as closely as possible to the sounds heard spoken in Bahrain, many of which are not heard in the English language, explanatory notes follow the vocabulary. To help this process, emphatic Arabic consonants H ح (haa), S ص (saad) and T ط (taa) are capitalized in the following glossary so that they may be differentiated from the softer valued consonants h ه (haa), s س (siin) and t ط (taa). However, as precise pronunciation is not essential to a flow of reading, and to avoid confusion with the English convention of capitalizing proper names and the first letters of sentences, the values of these Arabic consonants are not identified in the narrative.

Vocabulary

a'ain (pl. oyuun), fresh water spring

aba, a variant of *abba* or *abbaya*

abbaya or **abba**, a black ankle-length cotton cape, trimmed with black lace round the bottom edge, worn by a Bahraini woman or mature girl to cover her head (not face), shoulders and body in public places, or in her home when any male who is not an immediate relative is present.

abu, father, as in Abu Dhabi, Abu Saa'fa (*See also* Place-Name Glossary)

abzu, interpreted as a symbolic underground shrine, as deciphered from cuneiform texts in Mesopotamia

adhari, virgin, as in A'ain Adhari (Virgin's Spring or Pool)

al-gahwa, the coffee shop

al-ghoos, the pearl diving season

ameer, a senior prince (a title often reserved for a Ruler) (*See also* amir)

amir, a variant of *ameer*, (as in the Amir of Bahrain)

awali, high place, as in the name of Bahrain's oil town, Awali

bab, gateway or door, as in Bab Al-Bahrain (Gateway of Bahrain)

baghla, jar (usually a clay pot for storing water)

bait, house, as in Bait Al-Jasra

baHar, sea

BaHrain, the State and main island name, which means two seas

bakhuur, scent or perfume

barasti, a traditional Bahraini house constructed with palm fronds

bilad, town or country, as in Bilad Al-Qadim (the Old Town)

bin, son of (**see also** ibn), as in HH Shaikh Isa bin Salman Al-Khalifa

chai, tea

cuneiform, writing in wedge-shaped characters, usually in clay, in ancient Assyrian and Mesopotamian texts

dhabi, gazelle, as in Abu Dhabi (an Emirate name meaning father gazelle)

dhow, a wooden sailing vessel, usually associated with the Arabian Gulf, the Red Sea, the India Ocean and east African coastal waters. Today, most dhows are powered by engines

dukhan, smoke, as in Jebel Ad-Dukhan

faroush, a building material, usually comprising coral stone

ftaam, pearl diver's nose clip

gahwa coffee, or a traditional coffee-shop

ghaiS, a pearl diver

gutch, a term applied to a type of house built in the "upper camp" of Awali

guttrah, the large kerchief comprising part of the traditional male head-dress worn by many Gulf Arabs, with the exception of Omanis who tend to wear small embroidered pill box hats. (*See also* ighal)

Halat, a sand island

HaSiir, palm mat

Hdhuur, reed fish traps

helow, sweet water

ibn, son of (*see also* bin), as in the ibn Saud dynasty

ighal, black braiding forming part of the traditional male head-dress worn by many Gulf Arabs. (*See also* guttrah)

jalbout, a type of dhow

jebel, mountain or hill, as in Jebel Ad-Dukhan

khaliij, gulf (as in sea)

khamis, Thursday, as in Masjid Al-Khamis (Thursday mosque) and Suuq Al-Khamis (Thursday market)

khariij, a mixture of salt and fresh water as found in some springs

khor, bay, as in Khor Qala'aia

khuff, camel's hoof, as in Khuff gas

madina, town, as in Madinat Hamad

majlis, traditional Arabic meeting room

manama, place of sleeping, as in Manama, the capital of Bahrain

masjid, mosque, as in Masjid Al-Khamis

mina, port, as in Mina Manama and Mina Salman

mughaidrat, place of little dust storms

muharraq, place of burning, as in Muharraq, the nineteenth century capital of Bahrain which, according to legend, was attributed to the place on the island of the same name where the Hindus cremated their deceased

nabiih, a well informed person, not a prophet

nahaam, singer who sails with the pearl divers

nawkhadha, pearl dhow captain (sing.)

nefT, crude oil, from the which the name naphtha, a crude oil fraction, is derived

nowaakhdha, pearling dhow captains (plural)

oyuun (sing. a'ain), water springs

qadim, old, as in Bilad Al-Qadim (the Old Town)

qala', fort or castle, as in Qala'at Al-Bahrain (the Fort of Bahrain)

qala'aia, little fort or little castle, as in Khor Qala'aia (Bay of the Little Fort)

qasr, a palace or large house

radm, bridge, as in Radm Al-Kawara

Rafa', a name for two towns in Bahrain, East Rafa' and West Rafa'

ras, headland, as in Ras Al-Qala' (Headland of the Fort)

ruTab, dates (to eat)

Saa'af, palm frond, as in Abu Saa'fa oilfield

Sala, prayer
Salat Al-Fajr (dawn prayers)
Salat Al-Thuhr (noon prayers)
Salat Al-A'asr (mid-afternoon prayers)
Salat Al-Maghrib (sunset prayers)
Salat Al-A'isha (evening prayers)

Sanduug, safe or chest for clothes or precious items, e.g. pearls

sharika, company, as in Ash-Sharika, literally meaning the Company. In Bahrain, this expression is traditionally associated with The Bahrain Petroleum Company.

Sharikat NefT Al-Bahrain, the Oil or Petroleum Company of Bahrain

Sharikat NefT Al-MaHduuda, Bahrain Petroleum Company Limited

Sharikat NefT Al-Bahrain (Muqfalah), Bahrain Petroleum Company BSC (closed)

Sharikat NefT Al-WaTaniiya, National Oil Company, known as the Bahrain National Oil Company

sharqi, east, as in Rafa' Ash-Sharqi (East Rafa')

shaT, bank of a river, as in ShaT Al-Arab

SubH, morning

suuq, market, as in Suuq Al-Khamis

Tawaash (pl. Tawawiish), pearl merchant

umm, mother, as in *Umm Al-Qaiwain* (one of the seven United Arab Emirates) and *Umm An-Nassan* (one of Bahrain's islands near which the customs post for the King Fahad Causeway is stationed)

zawiia, corner or angle

Transliteration Notes

Sometimes it is necessary to use a combination of English letters and/or punctuation to obtain an acceptable equivalent to *one* Arabic consonant. Also, these sounds may appear at the beginning, in the middle and at the end of words, so the following examples have been chosen to show how this works.

Consonants

a' (the Arabic letter *ayn*) is a glottal sound produced from the back of the throat and is pronounced as if one is being strangled. It is not a vowel and therefore is not pronounced like an /a/.

Most commonly in this book, it appears in the village name of *A'ali*, the names of the springs, such as *A'ain Adhari*, and the towns of *Rafa'*.

dh (the Arabic letter *dhaad*) is pronounced with the tongue behind the teeth, as heard in *Riyadh* or *nowaakhdha* (pearling dhow captains). *Dhaad* is not the letter /d/ heard in English.

gh (the Arabic letter *ghayn*) is pronounced like a gargling sound from the back of the throat, as heard in *Al-Maghrib* (sunset prayers). *Ghayn* is not a soft-sounding /g/, spoken in English.

H (the Arabic letter *haa*) is pronounced from further down the throat than a soft /h/ which has a different value. The following examples demonstrate these sounds at the beginning, in the middle and at the end of words: *Hidd* (town), *BaHar* (sea) and *SubH* (morning).

kh (the Arabic letter *khaa*) is pronounced like the Gaelic sound in loch, that is, a soft sound from the back of the throat, as heard also in *nowaakhdha*. In this book, *kh* is found most frequently in the names *Al-Khalifa* (the Ruling Family name) and *Jebel Ad-Dukhan* (Mountain of Smoke). *Khaa* is not the hard /k/ heard in English.

q (the Arabic letter *qaaf*) is pronounced like a letter /G/. The precise sound does not exist in English and should not be confused with the English /q/ which, usually, is followed by the vowel /u/. Examples of the sound in this book arise in *qala'* (fort) and *suuq* (market).

S (the Arabic letter *saad*) is pronounced like a hissing sound. The sound does not exist in English and should not be associated with the softer consonant /s/ (siin) which has a different value in Arabic. For example, the letter *saad* is found in *ghaiS* (pearl diver).

T (the Arabic letter *taa*) is pronounced with the tongue held in the roof of the mouth and then let go! This sound does not exist in English either and should not be compared with an Arabic soft /t/ (*taa*). Most often, this letter appears in this book as *Tawawiish* (pearl merchants) and *nefT* (oil).

Vowels

In Arabic, there are long and short vowels. Short vowels are pronounced more or less as they are heard in English.

aa is the long form of /a/ as heard in the English word ram. Examples in this book appear in *nowaakhdha* (pearling dhow captains) or *Tawaash* (pearl merchant).

ii is the long form of /i/ heard in English, but pronounced ee (as in bee), for example, as in *Al-Khaliij* (the Gulf) and *Tawawiish* (pearl merchants).

oo is the long form of the English vowel /o/. In Arabic it sounds like /or/, as heard in *fooq* (up[stairs]) or *jooz* (walnuts).

uu is the long form of /u/ which in English is pronunciated /oo/ as in room. Examples of the sound in this book are heard in *oyuun* (watersprings (sing. *a'ain*) and *suuq* (market).

However:

ai, as heard in *Bahrain* and *A'ain* (water spring), is the long form of the vowel /e/ heard in English and is pronounced like ay, as in the month of May. It is not a diphthong, a union of two vowels, as heard in English.

Definite Article

As the definite article in Arabic has a complex grammatical construction, a little guidance in recognizing certain features may help and interest readers.

al means the

Al-Khamis means the place named Khamis

In Arabic, there are two types of letters, known as the moon and sun letters. For example, kh, m, j are *moon* letters. In the example above, *al* is written correctly, since *kh*, being a moon letter, has not affected the word sound and spelling construction.

However, when the definite article, *al*, precedes a *sun* letter, such as s, n, d, sh, normally the letter /l/ is dropped and replaced by the appropriate sun letter. For example:

Ash-Sharika (the Company, usually meaning Bapco in Bahrain)

Jebel Ad-Dukhan (Mountain of Smoke)

Halat An-Naim (the community of Naim)

Rafa' Ash-Sharqi (Rafaa of the East (East Rafa'), a town which today is bisected by the AB pipeline)

LEGAL

antitrust, opposition to or the regulation of trusts, cartels and business monopolies.

The Sherman Antitrust Act was enacted and became US law on 2 July 1890. The Clayton Act, which supplements it, was enacted and became US law on 14 October 1914. The two laws, although amended many times since, remain the basic laws of the USA concerning civil and criminal antitrust matters.

assignment, a term which, in the petroleum industry, refers to the transfer of certain rights and obligations (e.g. as a party to a mining contract, or to a mineral concession) from one party to another. Thus, the rights and obligations of Eastern and General Syndicate, under its contract with the Ruler of Bahrain (prior to independence in 1971), were assigned to Bapco. In a broader, less defined sense, one may also say they were transferred to Bapco. (*See also* transfer)

Attorney-in-Fact, an agent, appointed in accordance with the laws of a particular jurisdiction and given authority to bind the appointing person or company. A company will often appoint an attorney-in-fact so that it has a legally authorized signatory who may legally bind the company to the terms of an agreement. In both the USA and the UK, the appointment of an attorney-in-fact is effected by a notary public witnessing and certifying the appointing signature. This occurred in the case of most of the Bahrain Petroleum Company's legal agreements

up until Bahrain Petroleum Company BSC(c) was incorporated in 1981.

concession, as used in the petroleum industry, is a general term that generally gives the holder broad rights (within a defined period of time) to do whatever may be necessary to search for, find, produce, transfer from and sell petroleum in a specified area. Thus, the 1934 Bapco Mining Lease is the equivalent of a concession, *lease* being considered a more precise term.

A *concession*, in petroleum industry parlance, is a contract, pursuant to which the owner of certain land (and of any petroleum underlying the land) gives to the other party (the concession holder) broad rights to explore for, produce and sell petroleum, in return for the performance of numerous specified obligations, including payments which may involve initial payments, plus rentals, royalties, certain (or all) taxes, specified environmental duties, delivery of free oil to the landowner, and other listed duties.

The term *concession* has been used much less in the last thirty years or so than previously. Perhaps the reason is that the most recent contracts of this nature now more clearly retain the legal ownership of the petroleum for the land-owning party, usually a state, as is the case in Bahrain. This concept may create home country tax credit problems for the company acquiring the petroleum rights. Thus, terminology is often important to the company, even when the practical effects are almost the same.

income tax, income tax was established in Bahrain by decree, that is, by legislation and not by contract. This distinction was important for the Bahrein Petroleum Company Limited to ensure that it was entitled to US foreign tax credit. Only a tax imposed on the company by the Government of Bahrain could qualify for foreign tax credit.

licence (English)/**license** (American), gives the holder the right to do certain specified things, such as conduct geological surveys seeking evidence of the possibility of oil being under a certain ground area. Usually the licence is for a specified period. Similar to an option, a licence gives the holder a right to do something but not an obligation to do it. Whilst an option alone gives only the right to enter into another transaction, a licence gives the right to perform other specified acts. In the petroleum industry, a licence to undertake geological work frequently is coupled with an option to acquire additional working rights (e.g. to drill for oil) if the licence holder's surveys, or other allowed prospecting activities, give him reason to believe oil may be available in the specified area.

nationalization, conversion into a national undertaking, e.g. the oil industry in Bahrain

option, a word which, in the petroleum industry, has the same meaning as in common parlance. The options held by Gulf Oil and Eastern and General Syndicate were the right, but not the obligation, to enter into a specified contract by which they could acquire the right to seek and, if successful, produce petroleum in a certain area of Bahrain. Usually, the option agreement states that it terminates unless the right to exercise the option, that is, to enter into the next agreement (e.g. a purchase), is performed within a certain period.

participation, in Bahrain, a procedure whereby the Government acquired, in stages, equity ownership of the country's indigenous oilfield and natural gas reserves. Now, this includes Bahrain Government participation in the refinery and other companies related to the hydrocarbons industry.

transfer is a broader term than *assignment*, as used within the petroleum industry. *Transfer* may be used to refer to the movement of money or other goods from one currency or country to another, without any change of ownership. (*See also* assignment)

PLACE NAMES

A'ain, a water spring. (For specific spring names, see the place-names in this glossary.)

A'ali, a village in the northern region of Bahrain island where the largest of the ancient tumuli are situated. The village is well-known for its range of local pottery.

Abadan, the most western coastal town in modern Iran, lying to the east of the Iraqi border and the Shat Al-Arab. The town was well-known for its vast refinery, particularly during World War II, when it supplied essential fuel to the Allies.

Abdan, a water spring on Sitra island.

Abu Dhabi, formerly a Shaikhdom within the Trucial States, adjacent to the northern shores of the Arabian peninsula. Joined the federation of United Arab Emirates in 1971.

Abu Saa'fa, an offshore oilfield between the eastern province of the Kingdom of Saudi Arabia and the State of Bahrain.

Abu Zeidan, formerly a sweet water spring near the old medieval capital of Bahrain, Bilad Al-Qadim.

Adari, *see* Adhari.

Adhari, a sweet water spring in the northern sector of Bahrain island. In old documents it is occasionally spelt Adari or Avari.

Al-Hasa, a province in Arabia.

Al-Raha, a fresh water spring on Sitra island.

Anatolia, a plain in central Turkey.

Arabia, a peninsula which embraced the regions of Asir, Hasa, Hijaz and Nejd. In 1932 they were incorporated into the Kingdom of Saudi Arabia. Other countries which occupy the peninsula are the States of Kuwait and Qatar, the United Arab Emirates, the Sultanate of Oman and the Republic of Yemen.

Arabian Gulf, the official name used in Bahrain and other Arab states for that body of water sometimes called the Persian Gulf.

Avari, *see* Adhari.

Awali, the oil town created in the centre of Bahrain island during the 1930s by the Bahrein Petroleum Company Limited to accommodate company offices, personnel and ancillary facilities.

Baghdad, the capital of Iraq.

Bahrain, an archipelago in the Arabian Gulf, which was declared a State on 14 August 1971 by His Highness Shaikh Isa bin Salman Al-Khalifa.

Bahrein, until 31 December 1952, the official spelling of Bahrain.

Baku, an oilfield and port, within Russia, located on the western shore of the Caspian Sea.

Bandar Abbas, a port on the southern coast of Iran, due north of the Strait of Hormuz and the Emirate of Ras Al-Khaimah.

Barbar Temple, located on the northern shore of Bahrain island, where the Dilmuns commemorated Enki, God of Wisdom and all the Sweet Waters under the Earth (meaning fresh water), at a subterranean shrine within the temple complex.

Basrah, the most southerly port of modern Iraq, adjacent to the Shat Al-Arab.

Beirut, the capital of the Republic of Lebanon.

Bellad-i-Kadim or **Billad-i-Kadim**, early variants of the modern spelling Bilad Al-Qadim.

Bilad Al-Qadim, the medieval capital of Bahrain, located in the north-eastern region of Bahrain island.

Buri, a village in the north-western sector of Bahrain island.

Bushire, a port on the Iranian coast, north-east of Bahrain.

Diraz, in ancient times, the site of a temple complex attributed to various phases of the Dilmun civilization, a trading station circa 2400 BC and now a village located on the northern coast of Bahrain island.

Dilmun, a region believed to have incorporated the islands of Bahrain, also known in cuneiform texts as Ni-Tukki, Kur-Ni-Tuk and Niduk-ki. The Formative Dilmun Period is now thought to date from 3200 to 2200 BC.

Eridu, the most southerly, and reputedly the oldest, Sumerian city.

Euphrates, a river which flows south into the Shat Al-Arab waterway through ancient Sumer, later part of Mesopotamia and now known as Iraq.

Hamad Town (Madinat Hamad), Bahrain's newest and most extensive town, founded in the 1980s and dedicated to the Crown Prince and Commander-in-Chief of the Bahrain Defence Force, His Excellency Shaikh Hamad bin Isa Al-Khalifa.

Hanaini, in old documents, a variant of Hunainiiyah.

Hawar Islands, an archipelago of approximately sixteen small islands which lies to the south-east of Bahrain island and forms part of the State of Bahrain.

Hedd, in old documents, a variant of Hidd.

Hidd, originally a pearl-diving and fishing community on the eastern shore of the most northern island of the State of Bahrain, Muharraq. Now it is a modern town.

Hormuz, a small island lying in the northern sector of the Strait of Hormuz and just south of the Iranian port of Bandar Abbas.

Hunainiiyah, a spring lying at the foot of the Rafa' rimrock which, at one time, was reputed to be 20 fathoms (35.57 m) deep.

Iran, now an Islamic republic, following the exile of the Shah of Iran in 1979. Until 21 May 1935 Iran was known as Persia.

Iraq, prior to 1921, when the monarchy was declared, Iraq had been the province known as Mesopotamia, part of the Ottoman Empire. In 1958 the monarchy ended and Iraq became a republic.

Isa Town, a modern town located on Bahrain island, founded soon after the accession in 1961 of His Highness Shaikh Isa bin Salman Al-Khalifa, to whom it is dedicated.

Jasra, a former pearl-diving community, on the north-west coast of Bahrain island, where His Highness Shaikh Isa bin Salman Al-Khalifa was born. His birthplace, Bait Al-Jasra (Jasra House), is now a museum.

Jebel Ad-Dukhan, "the mountain of smoke" below which the first oil camp in Bahrain was established, where the first oil well was spudded in, and from which oil flowed on 1 June 1932.

Karzakhan, a small village lying near the west coast of Bahrain island.

Khor Qala'aia, Bay or creek of the little fort or castle, which lies between the islands of Bahrain and Muharraq. The little fort is Qala' Abu Mahiir, now incorporated into the Muharraq coastguard station.

Khor Tuubli, Toobli Bay, which lies north-west of Nabiih Saleh island and off the north-east coast of Bahrain island.

Kur-Ni-Tuk, "the mountain of Dilmun", as deciphered from the Sumerian language.

Kuwait, a modern city-state, major oil producer, member of the GCC and OPEC, located in the north of the Arabian peninsula.

Madinat Hamad, *see* Hamad Town.

Magan, a region contemporary with the Dilmun civilization, now known as Oman.

Mahazza, a fresh water spring on Sitra island.

Makan, a variant of Magan.

Manama, the modern capital of the State of Bahrain. In old documents it is occasionally spelt Manameh.

Mascat, the capital of the Sultanate of Oman.

Masjid Al-Khamis, the oldest mosque in Bahrain, the foundations of which date back to the 11th century AD.

Masjid Sulaiman, a major refinery town in Iran, located north-east of Abadan, which is also a refinery complex and town located at the northern end of the Arabian Gulf.

Meluhha, a region, contemporary with the Dilmun civilization, which incorporated much of the Indus Valley.

Mesopotamia, a region in the Near East later known as the Republic of Iraq.

Mughaidrat, the place of little dust storms, the first name given to Bahrain's oil town, later renamed Awali.

Muharrak, an early spelling of Muharraq.

Muharraq, the nineteenth century capital of Bahrain and the most northern island in the archipelago State of Bahrain.

Nabiih Saleh, a small island lying to the west of the

Sitra Causeway, State of Bahrain.

Niduk-ki, an alternative place-name for Dilmun which appears in the Babylonian *Myth of the Creation.*

Ni-Tukki, a variant in the Sumerian language of Dilmun and sometimes deciphered as "the place of the bringing of oil".

Oman, an area in the southern portion of the Arabian peninsula, with an Arabian coastline second in extent only to that of Saudi Arabia. Following independence, the regions of Mascat and Oman were declared one country, the Sultanate of Oman.

Persia, a kingdom in the Near East until it was renamed Iran on 21 May 1935.

Radm Al-Kawara, Kawara Bridge, crossed a water channel lying north of modern Isa Town. A six-lane flyover has since replaced the bridge, known also as the Wiggly Bridge.

Rafa', two towns located in the centre of Bahrain island. West Rafa' is the home of the Ruling Family. East Rafa', a traditional Bahraini community with a fort and a market as the focal points; the AB pipeline bisects the modern town.

Ras Al-Qala', the headland of the fort. This area incorporates an ancient trading site and palace attributed to the Dilmun civilization, as well as several fortress occupation levels, which represent the Tylos, Islamic and Portuguese periods.

Saffahiiyah, a fresh water spring on Sitra island.

Salmabad, a village south of the former Kawara Bridge and west of the northern sector of Isa Town, Bahrain island.

Saudi Arabia, a modern Kingdom created by the House of Saud, embracing much of the Arabian peninsula, including the provinces of Asir, Hasa, Hijaz and Nejd. Saudi Arabia is a major oil producer, member of the GCC, OAPEC and OPEC.

Shanghai, a port on the east coast of mainland China.

Shat Al-Arab, a waterway at the mouth of the Euphrates and Tigris Rivers in southern Iraq.

Sitra, an island close to the north-eastern shore of Bahrain island. In the 1930s the west coast of Sitra was linked to Bahrain island by a trestle bridge. Later the bridge was replaced by a six-lane highway constructed on reclaimed land; the north shore of Sitra island was linked to

Bahrain island by a four-lane causeway which itself provides a link road to the small island of Nabiih Saleh.

Sumer, the regional homeland of the Sumerians in antiquity, located in an area adjacent to the Tigris and Euphrates Rivers.

Suuq Al-Khamis, the former Thursday market, located beside the mosque of the same name. The market was also the donkey market.

Tigris, a river which flows south through Iraq into the Shat Al-Arab waterway.

Titusville, a town in the state of Pennsylvania, USA, where Edwin Drake drilled the world's first oil well in 1859.

Toobli Bay, *see* Khor Tuubli.

Trucial States, the seven emirates (Abu Dhabi, Ajman, Dubai, Fujairah, Ras Al-Khaimah, Sharjah and Umm Al-Qaiwain) which took this designation from a truce imposed by Britain in the late nineteenth century, later to become the United Arab Emirates in December 1971.

Tylos, the classical name for Bahrain.

United Arab Emirates (UAE), proclaimed as a new state on 2 December 1971 by the Rulers of Abu Dhabi, Dubai, Sharjah, Ajman, Umm Al-Qaiwain and Fujairah [Heard-Bey, 1982, p. 367], the opportunity for Ras Al-Khaimah to join them being left open, together with the expressed wish that Bahrain and Qatar would join in due course. Ras Al-Khaimah subsequently joined the UAE on 10 February 1972, whilst Bahrain and Qatar remained independent states.

Ur, an ancient city of Mesopotamia.

Uruk, an ancient city-state, north-west of Ur and south-east of Nippur, beside the Euphrates River.

Yokohama, a port in Japan.

Zagros Mountains, a range lying adjacent to the northern shore of the Arabian Gulf in modern Iran.

——— TECHNICAL GLOSSARY ———

The terms in this glossary refer to usage in Bahrain and are not intended to be applied universally.

alkylation, a refining process whereby two four-carbon atom molecules are combined into a single eight-carbon atom molecule. The product of the process is a high octane gasoline blendstock known as alkylate.

ammonia, a nitrogen-based chemical primarily used either as, or in the manufacture of, fertilizers. In Bahrain, ammonia is produced from Khuff gas, a form of natural gas.

anticlinal crest, an alternative term for the apex of a geological dome structure, often used in early geological reports relating to Bahrain.

anticline, a geological term to define the consequence of colossal pressure having been exerted on each side of a sedimentary rock structure to cause it to fold into a dome-like structure. This is particularly evident along the Arabian Platform, parallel to the Zagros mountains. In this type of fold or bend, the layers normally dip or slope away from the crest of the dome in opposite directions. After this dramatic movement has taken place up and down, and from side to side, folds, cracks and faults occur to form a rearranged stratigraphy in which the oilfields of the Arabian peninsula region have been found. In Bahrain, five periods of such change, or migration, took place. (*See also* Arabian Platform *and* steep dip)

API Scale, an American Petroleum Institute index used to classify crude oil according to specific gravity or density. The higher the index number, the lighter the crude. Bahrain's indigenous crude oil is 30.5°-31° gravity, but the gravity of the crude oil discovered in Bahrain's test well on 1 June 1932 was 33.8°.

Arab Zone, a geological formation in the Riyadh Group of the Jurassic Period, Mesozoic Era, which comprises limestone and anhydrite rocks. It is an oil and associated gas bearing stratum. Much of the oil found in the Arabian peninsula region and Bahrain was formed in this zone about 150 million years ago. (*See also* Mesozoic)

Arabian heavy, an export grade of Saudi Arabian

crude oil, with an API specific gravity of between 24° and 26°.

Arabian light, the most plentiful export grade of Saudi Arabian crude oil, with an API specific gravity of between 33° and 34°. Arabian light is a blend of crudes extracted from various fields in the Eastern Province, a proportion of which is pumped along the Arabian-Bahrain (AB) pipeline to the Bahrain refinery.

Arabian medium, an export grade of Saudi Arabian crude oil with an API specific gravity of 29°-30°. It is extracted from various fields in eastern Saudi Arabia, including the offshore Abu Saa'fa field owned and operated by Saudi Aramco. Bahrain derives a 50-per-cent benefit from the field's net revenue through a crude oil arrangement with Saudi Arabia.

Arabian Plate, a geological province, defining Arabia and part of East Africa, the motion of which is shifting to narrow the Arabian Gulf by a few centimetres annually, as indicated by the number of earthquakes along the Zagros mountain zone. In association with this movement, a process of tilting and subduction is causing the northern end of the Musandam peninsula, northern Oman, to sink. If this pattern continues, in several million years' time the Strait of Hormuz may close and the Arabian Gulf will become an enclosed, narrower sea.

Arabian Platform, a geological province, comprising a thick basin of sedimentary rocks in which rich oil deposits were found. The Arabian Platform incorporates much of the eastern region of the Arabian peninsula and the Arabian Gulf, including the islands of Bahrain. It is now believed that the Arabian Platform and the Arabian Shelf were formed during a succession of geological movements throughout the Palaeozoic and the Mesozoic Eras.

Arabian Shelf, a geological province, comprising thin sedimentary rock resting on a bed of igneous and metamorphic rocks, in which some oil is present. The Arabian Shelf occupies a central north/south band of the Arabian peninsula and part of the peninsula's south-east region.

Arabian Shield, a geological province comprising igneous and metamorphic rocks in which no oil is present. This province occupies the western sector of the Arabian peninsula, running north/south beside the Red Sea and along the bottom edge of the peninsula.

asphalt, a heavy, normally solid, petroleum product used predominantly in road surfacing. Also known as bitumen.

associated gas, a gas similar in chemical composition to natural gas but found in association with crude oil rather than in a separate geological stratum. (*See also* Khuff gas *and* natural gas)

aviation spirit is the fuel used in spark-ignition aviation engines to power propellor driven aircraft. During the early years of the oil industry in Bahrain, this was commonly referred to as AVGAS. (*See also* jet fuel *and* aviation turbine fuel)

aviation turbine fuel, often abbreviated to AVTUR and also known as jet fuel. (*See also* jet fuel)

barrel, the basic unit of volume used in relation to crude or refined oil. One barrel is equivalent to 35 imperial gallons, 42 US gallons or 159 litres.

bitumen, a naturally occurring tarry substance. In Bahrain it is produced by the surface weathering of crude oil which has seeped to the surface.

blow-out, a sudden expulsion of fluid under pressure, caused by a loss of pressure containment during the drilling process. This is very hazardous because, if the expelled hydrocarbons ignite, the entire rig can be destroyed. (*See also* hydrocarbons *and* rig)

butane, four-carbon atom hydrocarbons found in both crude oil and natural gas, and a component of LPG. (*See also* liquefied petroleum gas)

cable tool, *see* rig.

cat cracker, or more precisely Fluid Catalytic Cracking Unit (FCCU), a refinery process unit which converts or "cracks" high-boiling molecules into lower-boiling molecules, primarily high-octane gasoline blendstock.

catalyst, a substance used to cause or hasten a chemical reaction without undergoing a chemical change itself during the process. In the case of an oil refinery, various catalysts are used to change molecular structures. (*See also* cat cracker)

Celsius, a scale of temperature on which water freezes at 0° and boils at 100°. Its origin is attributed to the Swiss astronomer of the same name.

Cenozoic, the most recent geological Era, which occurred some 56 million years ago, during which the rocks of the Eocene Epoch found in Bahrain were deposited.

centigrade, a scale derived from the Latin *centum* (hundred) + *gradus* (step), applied to the Celsius scale of temperature. (*See also* Celsius)

cracking, an oil industry term used to describe refining processes which convert or "crack" large high-boiling hydrocarbon molecules into smaller lower-boiling molecules. The desired product of cracking processes is usually gasoline blendstock.

Cretaceous, a geological Period of the Mesozoic Era, during which the oil of the Mauddud Zone, unique to Bahrain, was formed about 100 million years ago.

crude oil, a mixture of naturally occurring liquid hydrocarbons found in the rock strata of certain geological formations. (*See* hydrocarbons *and* petroleum)

derrick, a term which usually refers to the structure portion of the oil drilling apparatus. (*See also* rig)

derrick man, normally the second-in-command in a drilling crew of four. (*See also* driller *and* roughneck)

diesel, a refined crude oil product in the middle distillate range.

diesel cut, a mid-boiling range cut, taken after kerosene in crude oil fractionation.

distillate, the product obtained from the condensation of lighter boiling components in the distillation process.

distillation, a physical separation process which involves evaporating the lower boiling components of a mixture and separating the resulting vapours by selective condensation. The distillation of crude oil is the basic refining process. This involves the separation of the crude oil into a number of fractions, or intermediate components, from which finished products may be made. (*See also* fractionation)

distillation column, a vertical cylindrical vessel in which materials are vaporized and condensed in stages to effect their fractionation.

driller, in the 1930s in Arabia and Bahrain, normally the man in charge of a crew of four who handled the drilling machinery. (*See also* derrick man *and* roughneck)

drilling rig, *see* rig.

Epoch, a geological subdivision of a Period, which is a subdivision of an Era.

Era, the principal division of geological time, subdivided first into Periods, then into Epochs and finally into Units.

Fahrenheit, a scale of temperature on which water freezes at 32° and boils at 212°. Its origin is attributed to the German physicist of the same name.

feedstock, that primary material which enters any process and is converted to a final product. The following industries in Bahrain use specific feedstocks: Bapco (crude oil); Banagas (associated gas); GPIC (Khuff gas); ALBA (bauxite); GARMCO (aluminium ingots); and Rafa' power station (Khuff gas for conversion to electricity).

flaring, burning off excess gas.

Fluid Catalytic Cracking Unit (FCCU), *see* cat cracker.

fossil fuel, any naturally occurring hydrocarbon formed by the deposition of prehistoric plant or animal life. Fossil fuels include coal, crude oil, natural gas and peat.

fractionating column, *see* distillation column.

fractionation, a distillation process in which crude oil is separated into fractions by boiling in broad temperature ranges. The heavier the product, the higher its boiling point. (*See also* distillation)

fuel oil, a general term applied to describe high boiling liquid petroleum products used for the production of heat and power, particularly as a fuel for boilers or industrial furnaces.

gas, *see* gasoline *and* natural gas.

gasoline, a refined light petroleum distillate. Gasoline is known internationally as motor spirit, in the UK as petrol, in France as pétrole and in the USA as gas or mogas (motor gasoline).

geosyncline, a large sedimentary basin where very deep layers of sediment have accumulated.

hydrocarbons, a class of chemical compounds consisting of carbon atoms with hydrogen atoms attached to each of the carbon atoms. Other atoms, such as sulphur, nitrogen, nickel and vanadium, may also be present in hydrocarbons. Oil, gas, petroleum, fossil fuels and natural gas are all composed of mixtures of hydrocarbons.

jet fuel, the liquid petroleum products burned as fuel in jet engines. Commercial jet fuels are essentially kerosene. Military jet fuels may be a naphtha kerosene mixture or a special grade of kerosene.

Jurassic, a Period of the Mesozoic Era, during which 80-per-cent of the oil found in the Arabian peninsula region was formed between about 195 and 140 million years ago. (*See also* Arab Zone *and* Mesozoic)

kerosene, a refined product derived from the "middle of the crude oil barrel". Its boiling range is between those of gasoline and diesel oil. Originally, kerosene was used as lamp oil and cooking fuel. Now, its main use is as aviation turbine fuel (jet fuel). (*See also* naphtha)

Khuff gas, natural gas produced from the Khuff zone in Bahrain at a depth of about 10,500 feet (3200 metres).

light end of the barrel products, a common industry term for transportation fuels such as gasoline, kerosene and diesel oil.

liquefied natural gas (LNG) is the result of chilling natural gas to very low temperatures, so that it may be stored and transported as a liquid. At present, this is not produced in Bahrain and should not be confused with NGL (natural gas liquids). (*See also* natural gas liquids)

liquefied petroleum gas (LPG) consists mainly of liquified butane and propane.

long ton, 2240 pounds and equal approximately to 7.4 barrels of crude oil, 32° API gravity. (*See also* metric ton, short ton *and* tonne)

lubes, an abbreviated term for lubricating oils.

Mauddud, an oil-producing limestone zone, unique to Bahrain, in which oil was discovered on 1 June 1932 at a depth of 2,008 feet (612 m). This zone was formed in the Cretaceous Period of the Mesozoic Era, about 90 million years ago. (*See also* Mesozoic)

Mesozoic, the geological Era which occurred after the Palaeozoic and before the Cenozoic Eras, between about 195 and 56 million years BP (before the present). The formation of oil in the Middle East seems to have intensified mostly during this time. It was also during this Era that, in Bahrain, both the oil-producing Mauddud zone (Cretaceous Period, circa 140 to 56 million years BP) and the oil and gas producing Arab Zone (Jurassic Period, circa 195 to 140 million years BP) were formed.

methanol, an industrial alcohol, used as a chemical feedstock. In Bahrain, methanol is produced from natural gas.

metric ton, 1000 kg, equal to approximately 7.2 barrels of crude oil of 32°API gravity. (*See also* long ton, short ton *and* tonne)

middle distillate, describes distillates in the boiling point range of kerosene and diesel oil.

Miocene, a geological Epoch of the Tertiary Period of the Cenozoic Era, which occurred about 35 million years ago.

naphtha, historically, in refining parlance, the first crude oil fraction removed prior to the kerosene (paraffin oil (UK), lamp oil, petroleum spirit) cut. More precisely, naphtha comprises any petroleum hydrocarbons which boil in the range between LPG and kerosene, whether derived from crude oil fractionation or gas processing. Naphtha may be used as a petrochemical feedstock or as a feedstock for gasoline production. (*See also* kerosene *and* paraffin)

natural gas, historically, any gas which comes out of the ground and replaces gas manufactured from coal gasification. More precisely, natural gas is a light hydrocarbon gas, primarily methane, which occurs naturally in subterranean geological structures. (*See also* hydrocarbons)

natural gas liquids (NGL), liquid hydrocarbons recovered from gas processing.

nodding donkey, in Bahrain, an oilfield term for a beam pump used to bring crude oil to the surface from a well in which there is insufficient pressure to force the crude oil to the surface.

octane number (also called octane) is a measure of the performance of gasolines, as determined in a laboratory test engine. The reference fuel for the test is "iso octane", a pure hydrocarbon compound which is defined to have a rating or octane number of 100.

oilfield, a geological expression to define the place where crude oil is hosted (stored) beneath the earth's surface in the cracks, fissures, joints and porous zones of carbonate rocks, mostly limestone, dolomite and shale in Bahrain.

Palaeozoic, the geological Era prior to the Mesozoic Era, between approximately 600 and 230 million years ago, during which oil formation in the Middle East seems to have started and the Khuff natural gas zone of Bahrain was formed.

paraffin, the name of a class of hydrocarbons, but also commonly used in the UK to describe kerosene. (*See also* kerosene *and* naphtha)

Period, *see* Era.

petrochemical, any chemical derived from a petroleum feedstock. Present examples in Bahrain include ammonia and methanol.

petroleum, naturally occurring mixtures of liquid hydrocarbons found in subterranean structures. 'Petroleum' can also be used as an adjective to describe products refined from petroleum, as in "liquified petroleum gas".

pitch, in Bahrain, the black residue obtained from the thermal cracking process. The dump for this residue near the Bahrain refinery is known locally as the "pitch pond".

plane table, a portable surveying instrument consisting primarily of a drawing board and a ruler mounted on a tripod and used to sight and map topographical details, hence "plane tabling" (referred to by Fred Davies in his 1930 Supplementary Report on Bahrain).

plugging, a process of filling a well with concrete, normally when it has ceased production. During World War II provision was made to plug the wells of Arabia in the event of the Axis powers occupying the region.

propane, a three-carbon atom hydrocarbon found in crude oil and natural gas, and also a component of liquefied petroleum gas.

province, an expression to define a geological area, as in the Arabian Shelf, the Arabian Shield and the Arabian Platform.

residue gas, in Bahrain, the gas remaining after the extraction of natural gas liquids.

rig (or drill rig). A term used to describe collectively all the equipment employed in drilling an oil or gas well. The main components of the present day rig are a derrick structure above the well (used to raise and lower the string of pipe connected to the rotary drilling tool); a pressure-containing device or blow-out preventer; a turntable to rotate the drilling apparatus and a mud system which circulates mud down through the drill pipe to flush cuttings away from the face of the drill bit at the bottom of the hole. In Bahrain, all the drill rigs in use at present are truck mounted mobile rigs.

The early wells in Bahrain were drilled with a cable tool rig which raised and dropped a pointed drilling bit by means of a long cable. The dropped bit chipped out the well hole once it struck the well bottom. The early drill rigs were called standard rigs, devices whereby the derrick was either dismantled and re-erected at another site after the completion of drilling, or left standing over the well hole.

rotary drilling, a process in which an entire "string" of pipe is rotated to turn the drill bit. In Bahrain, this later replaced cable tool drilling, used for the first oil wells, where the bit was repeatedly raised and dropped to chip a hole through the rock strata.

roughneck(s), two men on a drilling crew of four who work "on the rig floor", manipulating the pipe of the drill string. (*See also* derrick man *and* driller)

seepage, the leaking of oil or gas to the surface.

short ton, US parlance, 2000 pounds, equivalent to approximately 6.7 barrels of crude oil. (*See also* long ton, metric ton *and* tonne)

shut-in pressure, the pressure at the bottom of an oil well when it is not flowing.

specific gravity, the ratio of the density of a liquid substance to the density of water.

steep dip, a geological expression to describe a particularly steep domed structure (anticline) where the sides slope (dip) sharply from the apex or crest of the structure. (*See also* anticline)

tail gas, *see* residue gas.

terminal, a facility that receives oil for storage and onward transportation. In Bahrain, this is the Sitra Terminal.

thermal cracking, breaking large molecules into smaller molecules using heat as the medium.

tonne, *see* metric ton.

tun (tunne), a large cask or vat.

venting, in Bahrain, the release of excess gas from oil production into the atmosphere.

wildcat, an exploration term used to denote an exploration well drilled on a previously undrilled prospect.

NOTES

Chapter One

1. First Assistant Resident, who reported to the Political Resident of the Persian Gulf (PRPG), in turn, reported to the Government of India.
2. Found in the Saar tumuli complex during the 1989 excavation season conducted by the Bahrain Department of Antiquities, led by Abdulaziz Sowaileh.
3. Found during the 1989 excavation season conducted by the French Archaeological Mission to Bahrain, led by Dr Pierre Lombard.
4. Calcott Gaskin, 1904, p. 60.
5. Namely, the British India, the Bombay and Persian, the British and Colonial, and the Anglo-Arabian and Persian Steam Navigation Companies.
6. Calcott Gaskin, 1902.
7. The Gulf rupee was introduced into Bahrain in 1959. It had the same value as the Indian rupee but was printed in different colours. Only Gulf rupee notes were able to benefit from the freer exchange system in the Gulf, and the smuggling of Indian rupee banknotes out of India effectively ceased.
8. Lorimer, p. 937.
9. Adel A'ali.
10. Haji Abdul Nabi Al-Fardan.
11. Ali Hassan.
12. Aisha Yateem.
13. After independence in August 1971, the Ruler of Bahrain adopted the title The Amir of the State of Bahrain.
14. HE Yousuf A. Shirawi, Minister of Development and Industry.

Chapter Two

1. Tuson, 1979, pp. xiii-xv.
2. Bidwell, 1986.
3. On 27 January 1926 the Pacific Oil Company and Standard Oil Company (California) merged to become the Standard Oil Company of California (Socal). On 28 December 1928 the company exercised its option on the Bahrain concession, which it had just acquired from Eastern Gulf Oil Company. In so doing, it nominated the Bahrein Petroleum Company as the company to which it should be transferred. On 11 January 1929 the Charter of the Bahrein Petroleum Company Limited as a wholly-owned subsidiary of the Standard Oil Company of California was signed.
4. ARCO – Atlantic Refining Company until 3 May 1966, when it merged with Richfield Oil Company to become Atlantic Richfield.
5. Hewins, 1958, p. 141.
6. *Qasr*: a palatial home or palace
7. *Saiyed Hashem*: a holy or religious man
8. *Ameer*: a prince, often a Ruler. In this context, the name of the Ameer to whom Rihani was referring is not mentioned, but almost certainly he would have been an eminent person in the Ojair community. (Of interest also, the Ruler of Bahrain adopted the same title in 1971 following his country's independence. Today, *Amir* is the most usual transliteration.)
9. Ward, 1965, p. 28.

Chapter Three

1. Beatty, 1939, p. 110.
2. Moore, 1948, p. 24.
3. *Eocene* : a geological epoch in the Tertiary period of the Cenozoic Era.
4. Research note, Thomas E. Ward Jr, 9 December 1988.
5. Owen, 1975, p. 1323.
6. Ward, 1965, p. 36.
7. The cable acronym for the Mellon family, which controlled Gulf Oil Corporation. In 1928 Andrew W. Mellon was Secretary of the United States Treasury. Subsequently he was appointed United States Ambassador to London.
8. Eastern Gulf Oil Company and South American Gulf Oil Company, both subsidiaries of Gulf Oil Corporation, had offices in Battery Park Building, 21 State Street, New York City. Confusion often arose when communications concerning Bahrain, which should have originated from or been destined for Eastern Gulf Oil, appeared on the letterheads of, or were signed in the name of, South American Gulf Oil. This letter of the 27th of May to T. E. Ward, signed by J. Volney Lewis, is an example.
9. 3 R. 8A. = 3 rupees 8 annas (or three and a half rupees); 5/3 = 5/3d. (five shillings and threepence).
10. Quoted from the Encyclopaedia Britannica.
11. Holmes was referring to the Political Resident Persian Gulf (PRPG).
12. A reference to Gulf Oil's negotiations for the Al-Hasa and Neutral Zone concessions with Ibn Saud, being undertaken on its behalf by Major Holmes.
13. Rhoades, 1928, p. 9.
14. Wallace, letter to Holmes, 22 May 1928.
15. The concession agreement consisted of a firm, exclusive Exploration Licence, plus conditional rights to move to a Prospecting Licence, and then to a *Mining Lease*, the principal terms of which were specified in three Schedules attached to the 1925 *Bahrein Island Concession*.
16. Moore, 1948, p. 31.
17. The *Mining Lease* is the Third Schedule of the *Bahrein Island Concession* granted by HE Shaikh Hamad bin Isa Al-Khalifa, Deputy Ruler of Bahrain, to the Eastern and General Syndicate Limited on 2 December 1925. The Secretary of State was referring to Article XIII of the *Mining Lease*.
18. Moore, 1948, p. 32.

Chapter Four

1. Adams, 19 December 1928.
2. Williams, 12 January 1929.
3. Williams, 12 January 1929.
4. Williams, 12 January 1929, No. 27.
5. Williams, memorandum, 23 November 1928.
6. Kellogg, cable to Atherton, 28 March 1929.
7. Seymour, memorandum, 6 April 1929.
8. Williams, memorandum to the Admiralty, 20 March 1929.
9. Adams, letter from the Eastern and General Syndicate to the Colonial Office, 6 April 1929.
10. Memorandum of a Conference at the Colonial Office, 10 May 1929.
11. Davis, letter to Wallace, 27 May 1929.
12. Secretary of State, letter to the US embassy, 29 May 1929.
13. Davis, letter to Wallace, 3 June 1929.
14. Davis, letter to Janson, 3 June 1929.
15. Report of an Interdepartmental Conference, 7 June 1929.
16. Davis, telegram to South American Gulf Oil, 19 July 1929.
17. Record of a discussion between the Colonial Office and the Eastern and General Syndicate, 19 July 1929.
18. Davis, telegram to South American Gulf Oil, 22 July 1929.
19. Hall, letter from the Colonial Office to the Foreign Office and India Office, 3 August 1929.
20. Patrick, letter from the India Office to the Colonial Office, 23 August 1929.
21. Williams, letter from the Colonial Office to the Eastern and General Syndicate, 16 September 1929.
22. Leovy, 24 October 1929.
23. Adams, letter to Wallace, 15 November 1929.
24. Lombardi, night letter to Loomis, 4 December 1929.
25. Feuille, letter to Loomis, 28 February 1930.

Chapter Five

1. The Iraq Petroleum Company succeeded the Turkish Petroleum Company on 8 June 1929.
2. Presumed, but not stated, to be Major Harry G. Davis.
3. Also on 12 June 1930, the Acting Ruler of Bahrain signed an Indenture which provided for the transfer by the Eastern and General Syndicate to the Bahrein Petroleum Company of the syndicate's rights under the Bahrain concession.
4. Standard Oil Company of California (Chevron since July 1984) and Pillsbury, Madison and Sutro occupy the same premises today: 225 Bush Street, San Francisco.
5. The Mining Lease, a separate issue, formed Schedule 3 of the 2 December 1925 agreement. It set forth the terms on which the lease could be obtained during, or at the expiration of, the Prospecting Licence.
6. C. G. Prior, *ARPGPR*, 1932, Ch. VIII, p. 44. Prior gave another likely reason for the rapid drop in the water table: he reported that water was being wasted in incredible quantities, as if the supply were inexhaustible.
7. *Steep dip* = a geological expression to describe a particularly steep domed structure, or anticline, where the sides slope (dip) sharply from the apex or crest of the structure.
8. *Nissen hut* = a tunnel-shaped hut of corrugated iron with a cement floor, so named after its British inventor.
9. Gornall, 1965, p. 7.
10. *The Depression* = a phrase usually reserved for the severe downturn in trade which occurred in the USA and Europe following the "crash" in share values on the New York Stock Exchange in October 1929.
11. Almost certainly, Ward was referring to the Eastern and General Syndicate, the Bahrein Petroleum Company (Bapco) and the Standard Oil Company of California (Socal).
12. Holmes was referring to Bapco and Socal.
13. 34 gravity oil refers to the American Petroleum Institute's index which is used to classify crude oil according to specific gravity or density. The higher the index number, the lighter the crude. On 1 June 1932, the index number for Bahrain's indigenous crude was 33.8°, whereas today it is 30.5-31°.
14. A relationship between barrels of oil per day, and tons of oil, both referred to by Lombardi in his letter to Loomis on 2 June 1932, may be calculated by assuming an *average* of 7 barrels of oil corresponding to one ton, derived from:
    short ton (US parlance) = 6.7 barrels of oil per ton
    metric ton = 7.4 barrels of oil per ton.
15. Lombardi was referring to the *Bahrein Island Concession*, dated 2 December 1925, which contained three schedules. The required change concerned Clause IX of the Second Schedule.
16. On 16 October 1965, the Bahrain Dinar (BD) was created as the unit of currency in Bahrain. One BD comprises 1000 fils.

Chapter Six

1. Vane, memo to Stoner, 15 February 1933.
2. Bahrain *Annual Report for the Year* 1351, p. 33.
3. H. R. Ballantyne, telegram to Loomis, 20 March 1933.
4. H. R. Ballantyne, telegram to Vane, 23 March 1933.
5. Political Resident, telegram, 27 April 1933, 18135/33 (No. 84).
6. Holmes and Skinner, cable to Berg, 8 May 1933.
7. Holmes, letter to the Political Agent, Bahrain, 11 August 1933.
8. Holmes, cable to Bapco/London, 18 August 1933.
9. Lombardi, cable to Ballantyne, 8 September 1933.
10. Memorandum of a Discussion at the India Office on 26 October 1934, Public Record Office classification 33/194, reference P.Z. 6687/34.
11. Petroleum coke: the residue left following a high-pressure, high-temperature "cracking" operation. After a prolonged period, the residual forms a solidified mass of hard granular carbon.
12. "Nonsense milk": a reference to powdered milk.

Chapter Seven

1. India Office record from classification L/P & S/12/382.
2. Article 10 of the *Deed of Further Modification* prohibits drilling within the municipal boundaries of Manama and Muharraq. It is believed that the reference applied to the city of Muharraq, not to the entire island.
3. Secret Telegram from the PRPG to the Under Secretary of State for India, ref. 11720, 2 May 1942.
4. The last of these "stone blast" walls were dismantled in 1986.
5. India Office records from classification R/15/2/423.
6. Foreign Office records from classifications FO 371/31317 and 31318.
7. Secret telegram from the Secretary of State for India to the Government of India, classification 2404/796/91, 30 April 1943.
8. Formerly and formally, Arabian American Oil Company (Aramco) was known as California Arabian Standard Oil Company (Casoc). 'Aramco' evolved as Casoc's operating name during World War II and was adopted officially on 31 January 1944.

Chapter Eight

1. The Yalta Conference took place between the 4th and 11th of February 1945.
2. On the day following the German surrender, Victory in Europe Day [VE Day] was celebrated on 8 May 1945.
3. The Potsdam Conference took place between 17 July and 2 August 1945, during which time the Labour Party won the British General Election. Thus, in late July, Prime Minister Winston Churchill, leader of the Conservative Party in the British wartime coalition government, was replaced at the Potsdam Conference by Clement Attlee.
4. Victory over Japan Day [VJ Day] was celebrated on 15 August 1945.
5. The State of Bahrain was admitted as a member of the UNO on 21 September 1971.
6. Standard Oil Company (New Jersey) became the Exxon Corporation in November 1972.
7. Socony-Vacuum was the result of Standard Oil Company of New York (Socony) and Vacuum Oil (Vacuum) amalgamating in 1931. In 1955, Socony-Vacuum became Socony-Mobil, which in turn became Mobil Oil Corporation in 1966.
8. Arabian American Oil Company (Aramco) was renamed Saudi Aramco in 1988.
9. Standard Oil Company of California (Socal) changed its named to the Chevron Corporation on 1 July 1984.
10. The Texas Company, a Texas corporation, was reincorporated as the Texas Corporation in August 1926. Effective on 1 November 1941, the Texas Corporation merged the Texas Company (Delaware) and caused the Texas Company (California) to be dissolved. Following this reorganization, the company became known as the Texas Company, which in turn on 1 May 1959 changed its name to Texaco Inc.
11. The pipeline referred to was built and operated by Trans-Arabian Pipe Line Company Inc., usually abbreviated to Tapline.
12. The 1948 sale changed the Aramco shareholding to: Standard Oil Company (New Jersey), 30-per-cent; Socony-Vacuum, 10-per-cent; Standard Oil Company of California, 30-per-cent; Texas Company, 30-per-cent. Thus, three offshoots of the original Standard Oil Trust had become equity owners of the Arabian American Oil Company, which officially adopted its name on 31 January 1944, having been known formerly and formally as California Arabian Standard Oil Company.
13. The Marshall Plan took its name from General George Catlett Marshall, who was US Secretary of State at the time. As a result of World War II, European countries had built up large balance of payments deficits with regard to the United States. The loans made by the USA and Canada were thus designed to deal with a shortage of dollars in Europe. The initial loans were made as early as 1946, but by 1948 further massive loans were necessary to avoid a liquidity crisis.
14. J. P. Tripp, CMG, Economic Secretary, Bahrain Residency, 1961-63, and Political Agent, Bahrain, 1963-5.

15. The Abadan refinery was shut down on 15 April 1951 when negotiations between the Iranian Government, and the British and American oil companies party to the Anglo Iranian Oil Company (AIOC), foundered over the nationalization issue. On 28 April, the Majlis voted unanimously for the immediate seizure of AIOC's properties in Iran and elected Dr Muhammad Mossadegh as Prime Minister. On 5 September, after having obtained a 26-0 vote of confidence in the Senate on his oil policy, Premier Mossadegh threatened to expel from Iran all British oil refinery technicians unless the British Government resumed negotiations within fifteen days. The threat was implemented on 4 October, following which the British made representation to the United Nations to mediate in the dispute. By the end of 1951, the matter remained unresolved.

16. In 1947 India gained her independence from the British. Gradually, Indian Political Service officials, who had been responsible for maintaining relations with the Rulers of the Arabian Gulf nations, were replaced by Foreign Service officials.

17. J. P. Tripp, unpublished *Lecture No. 4*, "Political Progress and Development in the Persian Gulf since 1947", p. 3.

18. Between July 1955 and December 1964, 928 apprentices were enrolled by Bapco. At the end of those nine years, 718 (about 77-percent) were still working with the company (1964 *Annual Report*). In 1989, at the request of the Ministry of Labour, Bapco resurrected its apprenticeship programme to provide Bahraini students with essential craft training so that they might find employment outside the company. The Government of Bahrain pays each trainee a stipend and Bapco provides the training free of charge.

19. The *Bahrain Government Annual Report for the Year 1348 (1929-30)*, pp. 25-6 in the original document, now in the British Library, London. Also published in *The Bahrain Government Annual Reports, 1924-1956*, Volume I, 1924-1937, pp. 144-5, Archive Editions, 1986.

20. *Swing refinery:* capable of changing its product slate at short notice.

21. *Sprint refining capacity:* the ability to process crude oil at increased production rates for a limited period.

22. In a *spot* market, commodities or currency (in this case oil products) are traded for one-off sales purchase, as distinguished from short- or long-term contracts.

23. The League of Arab States was created on 22 March 1945. Founder members were Egypt, Iraq, Jordan, Lebanon, Saudi Arabia, Syria and Yemen.

Chapter Nine

1. The *Bahrein Island Concession* from Sheikh Hamad bin Sheikh Issa Al Khalifah, Sheikh of Bahrein ("The Sheikh") to the Eastern and General Syndicate Limited ("The Company"), dated 2 December 1925.

   The basic document, containing ten Articles, comprised the Agreement. This was accompanied by three Schedules. The First Schedule (seven clauses) defined the privileges to be enjoyed by the company under the Exploration Licence. The Second Schedule (ten clauses) defined the privileges to be enjoyed by the company under the Prospecting Licence. The Third Schedule, the Mining Lease, comprised fourteen Articles.

2. Article I of the Agreement, *Bahrein Island Concession*, 2 December 1925.

3. Article II of the above Agreement.

4. Article VI of the above Agreement.

5. Clause I, First Schedule, *Bahrein Island Concession*.

6. Clauses II-V of the above Schedule.

7. Clause VI of the above Schedule.

8. Article III of the Agreement, *Bahrein Island Concession*.

9. Second Schedule, *Bahrein Island Concession*.

10. Clause II of the above Schedule.

11. Also Clause II of the above Schedule.

12. Article VI of the Agreement, *Bahrein Island Concession*.

13. The quotation from the Prospecting Licence, refers to Clause IX, Second Schedule, *Bahrein Island Concession*. The Mining Lease comprised the Third Schedule of the concession.

14. Article VIII, Third Schedule (Mining Lease) of the *Bahrein Island Concession*, 2 December 1925.
15. Clause X, *Option Covering Bahrein Concession from Eastern and General Syndicate Limited to Eastern Gulf Oil Company*, dated 30 November 1927.
16. The Third Schedule, *Bahrein Island Concession*, dated 2 December 1925.
17. Article I, *Mining Lease*, 29 December 1934.
18. Articles I and II of the *Mining Lease*, 29 December 1934.
19. Second Schedule to the *Mining Lease*, 29 December 1934.
20. Article I, *Mining Lease*, 29 December 1934.
21. Article IV, *Mining Lease*, 29 December 1934.
22. Article VIII, *Mining Lease*, 29 December 1934.
23. Clause C of the first *Supplemental Deed* to the Mining Lease, dated 3 June 1936.
24. Clause B (4), of the above first *Supplemental Deed*.
25. Clause A, of the above first *Supplemental Deed*.
26. Clause B (2) of the above first *Supplemental Deed* which was substituted for the original Article III of the 29 December 1934 *Mining Lease*.
27. Article 3, *Deed of Further Modification* to the Mining Lease, 19 June 1940.
28. Article 10, of the above *Deed*.
29. After the signing of the *Deed of Further Modification* (19 June 1940) the minimum royalty provisions in Article 7 applied to both the original 100,000 acres and the "Additional Area". Thus, Article 7 of the 1940 *Deed* replaced the minimum royalty provisions of Article VIII of the 1934 *Mining Lease*, as altered by Article B-4 of the 1936 first *Supplemental Deed*.
30. Article 7(a), *Deed of Further Modification*, 19 June 1940.
31. Article 7(b) of the above *Deed*.
32. Article 8 of the above *Deed*.
33. No relation of Thomas Ward, the Eastern and General Syndicate's New York Representative.
34. J. E. Chadwick, *Minutes*, dated 25 November 1948, of an undated call (unstated as to whether it was a record of a meeting or a telephone conversation) with R. J. Ward, classification FO 371/68333.
35. It must be assumed that Nuttall erroneously referred to the concession, instead of Article VII (a) of the 1934 *Mining Lease*.
36. Section 1, *Supplemental Agreement*, dated 8 December 1952.
37. Section 3, *Supplemental Agreement* as above.
38. Section 4, *Supplemental Agreement* as above.
39. Section 2, *Supplemental Agreement* as above.
40. Section 5, *Supplemental Agreement* as above.
41. *Article 85 of the Bahrain Order in Council*, 1949.
42. Queen's Regulation made under Article 82 of the Bahrain Orders, 1952 and 1953, the *Bahrain Income Tax Regulation, 1955*, comprises the Bahrain Income Tax Decree 1955.
43. *Decree No. 80*, issued by His Highness Shaikh Salman bin Hamad Al-Khalifa, Ruler of Bahrain, on 28 November 1955, superseding Decree No. 8, issued by His Highness on 6 December 1952. C. R. Barkhurst, Chief Local Representative, confirmed that Bapco "submits to the tax as imposed by that decree and made applicable to the Company" in a letter to the Ruler of Bahrain, CON-935, 29 November 1955.
44. C. R. Barkhurst, *letter* of agreement to the Ruler of Bahrain, CON-937, also dated 29 November 1955.
45. M. Lipp, letter to Gault, C/FA-326, 7 December 1958.
46. As from the date of the *Supplemental Agreement*, 8 December 1952, all monetary obligations designated by the *Lease*, as amended, in terms of rupees, including the present rate or royalty payable by the company, would "hereafter be due and payable in pounds sterling""at the rate of one shilling and sixpence (1/6) sterling for each rupee (Clause 2).
47. The "majors", a term which today often applies to the eight major oil companies: British Petroleum (BP), Chevron (in the 1950s, still Socal), Compagnie Française des Pétroles (CFP), Exxon (in the 1950s, still Standard Oil Company (New Jersey)), Gulf Oil, Mobil (in the late 1950s, Socony-Mobil), Shell and Texaco.
48. Clause 1 in Sections A, B and C of the *Amendatory Agreement*, 9 August 1965.
49. Clauses 2, 7 and 10 (Section A) and Clause 2 (Sections B and C) of the *Amendatory*

Agreement, 9 August 1965.

50. Clause 12, Section A, *Amendatory Agreement,* 9th August 1965.

51. W. A. Schmidt, *letter* to the Ruler of Bahrain, MISC. CON-25, 8 February 1961.

52. W. A. Schmidt, *letter* to the Ruler of Bahrain, MISC. CON-601, 27 October 1962.

53. W. A. Schmidt, *letter* to the Ruler of Bahrain, MISC. CON-792, 10 February 1963.

54. L. D. Josephson, *letter* to the Ruler of Bahrain, M-CON 188, 11 August 1964.

55. The term "affiliate" with respect to Bapco Limited meant any other company more than 50-percent of whose share capital was owned directly by Bapco or indirectly by those who owned Bapco directly (Caltex) or indirectly (Socal and Texaco, owners of Caltex). This redefinition, Clause 14 (g) of the letter to from Bapco to the Bahrain Government, M-CON 243, 8 March 1966, reflected a solution to the Head of Finance's concerns as expressed in his letter to Bapco's Vice-President, 2.40/37, dated 5 August 1964, pp. 2-3.

56. The term "non-affiliate" applied to any other company which was not an affiliate of Bapco Limited.

57. Clause 2 of the letter from L. D. Josephson to the Ruler of Bahrain, M-CON-188, 11 August 1964.

58. On 17 April 1966, L. D. Josephson, Vice-President and Director of Bapco [*letter* to the Head of Finance, Government of Bahrain, M-CON 246], submitted a proposed letter agreement to "accomplish" the accelerated royalty and tax schedule of payments. On 11 June 1966, W. O. Stolz, Director of Bapco [letter to the Ruler of Bahrain, M-CON 247], confirmed the accelerated payment of royalties and tax.

59. *Letter* from L. D. Josephson confirming the agreement between Bapco and the Ruler of Bahrain, M-CON 255, 24 April 1967.

60. This value had been determined in accordance with paragraph 9 of the letter agreement M-CON 243 between the Ruler of Bahrain and the Company, 8 March 1966.

61. *Letter* from L. D. Josephson confirming the agreement between Bapco Limited and the Ruler of Bahrain, MISC. CON 268, 3 August 1968, paragraphs 1 and 2.

62. *Letter* agreement as in footnote 61, paragraph 3.

63. On 16 October 1965, the Bahrain Dinar was created as Bahrain's unit of currency, equivalent to 10 rupees, and a series of bank notes inscribed "Bahrain Currency Board" were placed in circulation.

64. In October 1967, the UK devalued the pound sterling from US$2.8 to US$2.4.

65. *Letter* from L. D. Josephson confirming the agreement entered into between Bapco Limited and the Ruler of Bahrain, MISC. CON 265, 3 August 1968, paragraph A.

66. *Letter* from L. D. Josephson confirming the agreement entered into between Bapco Limited and the Ruler of Bahrain, MISC. CON 267, 3 August 1968, paragraphs 1 and 2.

67. Gulf Co-operation Council member countries: State of *Bahrain,* State of *Kuwait,* Sultanate of *Oman,* State of *Qatar,* Kingdom of *Saudi Arabia* and the Federation of the *United Arab Emirates.*

Chapter Ten

1. A "most favoured nation" clause in an international trade agreement is one which states that contracting parties to the agreement are bound to grant to each other treatment as favourable as they extend to any other country regarding the application of import and export duties and other trade regulations.

2. American Overseas Petroleum Limited (AMOSEAS), a non-profit service company owned equally by Socal (Chevron after 1984) and Texaco, was incorporated on 9 January 1952. Until 1 February 1957, and again after 1 January 1969, upon Mr L. D. Josephson's promotion, the Chairman of Caltex was usually the President and Chief Executive officer of Bapco Limited. During that twelve-year interim period, the Chairman of Amoseas was President of Bapco. Throughout this time, many of the same persons were members of the boards of directors of all three companies (Amoseas, Bapco and Caltex). On the occasion of board meetings, essentially the same group of persons would meet during one day for a continuous session, convened as the

Caltex board (or the Amoseas board) and then, after formally adjourning that meeting, declare open a meeting of the Bapco board. The decisions and other actions, however, were taken by the *separate* boards, even though the same persons might be directors of both boards.

3. The original members of OAPEC were: Algeria, Bahrain*, Egypt, Kuwait, Libya, Oman*, Qatar, Saudi Arabia, Syria* and the United Arab Emirates. (Non-OPEC members are indicated by a *.) Subsequently, Egypt's membership was suspended temporarily.

4. Article One, paragraph (i), *Agreement* between the Government of the State of Bahrain and the Bahrain Petroleum Company Limited, signed on 23 November 1974.

5. Paragraph (ii) of the above article.

6. The Bahrain Petroleum Company Limited was originally formed in January 1929 as a wholly-owned subsidiary of the Standard Oil Company of California (Socal). By the 1970s, Bapco was directly owned by Caltex. The company's place of registration remained Ottawa, Canada, whilst its head office was located in New York, USA. This situation continued until Bapco BSC (closed) was incorporated on 21 May 1981 and Bapco Limited was subsequently dissolved.

7. Article II, "Articles of Association", the Bahrain National Oil Company, translated from the *Official Gazette*, No. 1165, 4 March 1976.

8. Mr L. D. Josephson, Vice President and General Manager of Bapco, succeeded Mr Roy L. Lay. Mr. Josephson assumed his new position as President and Chief Executive Officer of Bapco on 1 November 1968, shortly before Mr Lay's retirement.

9. The supply contract was signed on 3 December 1968.

10. Source: The Bahrain Petroleum Company Limited 1968 *Annual Report*.

11. *Tonne* = metric ton, 1000 kilograms.

12. ASRY's shareholders in 1989: Bahrain, Kuwait, Qatar, Saudi Arabia and the United Arab Emirates, each holding 18.8358-per-cent of the equity (totalling 94.179-per-cent); Iraq holding 4.7180-per-cent and Libya 1.030-per-cent. (Algeria and Syria, the other OAPEC members, are not equity holders of ASRY.)

13. *Dead-weight tonnage* = the weight or displacement of cargo, fuel, crew/passengers carried by a ship. A ton, in this instance, is a unit of volume in shipping, equivalent to 2240 pounds (long ton) of total load, 35 cubic feet of water displaced, with the load-line just immersed.

14. *Ahead of Bahrain*: Hong Kong and Singapore (5 hours) and Tokyo (6 hours).
Behind Bahrain: Frankfurt (2 hours), London (2 hours in the summer, 3 hours in the winter) and New York (7 hours in the summer, 8 hours in the winter).

15. Clause 2.1, *Agreement* between the Government of the State of Bahrain and the Bahrain Petroleum Company Limited, 15 December 1979.

16. Clause 1 of the above *Agreement*.

17. Clause 6.1 of the above *Agreement*.

18. Caltex Bahrain Limited succeeded the Bahrain Petroleum Company Limited as a party to the participation agreements.

19. Article 3, "Legislative Decree" No. 25/1980, published in the *Official Gazette*, Issue 1408, 6 November 1980.

20. Membership of the Supreme Oil Council: His Highness the Prime Minister (Chairman), Their Excellencies the Ministers of Foreign Affairs, Development and Industry, Finance and National Economy, Works, Power and Water, *and* Labour and Social Affairs.

BIBLIOGRAPHY

Abbreviations

Many of the original documents quoted in this book are now stored in various archives around the world. These are identified in the bibliography in the following abbreviated form:

ARPGPR	*Administration Report on the Persian Gulf Political Residency and Maskat Political Agency*
BAPCO	The Bahrain Petroleum Company BSC (closed)
BSA	British State Archive
CCL	Chevron Corporate Library
CO	Colonial Office records
FO	(British) Foreign Office records
HMSO	His/Her Majesty's Stationery Office (London)
IOR	India Office Library and Records
RGIFD	Records of the Government of India Foreign Department

Listing Procedure

To avoid unnecessary complexity and confusion, the following convention has been adopted. No attempt has been made to categorize the bibliography. Latin expressions have been omitted. Authors and sources without credited originators are listed alphabetically in capitals.

Authors are listed according to family name. The works of authors who are listed more than once appear in chronological order. This practice accommodates the identification of correspondence and cables with otherwise identical information.

For the purposes of this bibliography, authors are defined as the originators of both published and unpublished books, cables, letters, memoranda, minutes, papers and theses.

Sources, in most entries, are highlighted in *italics* according to usual bibliographical practice.

Works with no credited author – such as annual reports, legal agreements, manuals, media articles and year books – are listed alphabetically according to document-title. To facilitate searches in these instances, both titles and dates of issue and/or signature have been set in *italics*.

Primary Source Notes

For readers wishing to pursue further study, the following information may be helpful.

1. Whilst direct quotations from administrative and/or annual reports are identified with individual source notations in the bibliography, collections of annual reports are in addition listed according to the corporate or government body to which they refer, e.g. *BAHRAIN GOVERNMENT ANNUAL REPORTS, BAHRAIN PETROLEUM COMPANY ANNUAL REPORTS* and so on.

2. Classifications of India Office records and Government of India Foreign Department records archived in the British Library and researched for the purposes of this book, are listed in the bibliography under *INDIA OFFICE LIBRARY AND RECORDS.* Specific documents from this source which are quoted in the narrative are listed in the bibliography in their own right.

3. Classifications of original documents now in the *BRITISH STATE ARCHIVE* maintained at the Public Record Office in London, have been treated similarly. These include the Admiralty, Colonial Office, Foreign and Commonwealth Office and War Office records.

4. With the approval of the Ministry of Information, State of Bahrain, *The Bahrain Government Annual Reports 1924-70* have been published in 8 volumes by Archive Editions, Gerrards Cross, Buckinghamshire, England. Reports for the years 1924-46 are reprinted by agreement with the British Library from originals in the India Office Library and Records, London. Reports for the years 1947-55 (except 1951) are reprinted from originals in the possession of Mr Robert Belgrave. Reports for 1951 and 1956 are reprinted by courtesy of the Foreign and Commonwealth Office Library. Reports for the years 1957-70 are reprinted courtesy of the Centre for Arab Gulf Studies, University of Exeter. (As the 1962 report was withdrawn by the Bahrain Government it is not included in the Archive Editions.)

5. By agreement with the British Library, Archive Editions have published the following: the original *Annual Administration Reports of the Persian Gulf Political Residency 1873-1947*, reprinted and bound in 10 volumes entitled *The Persian Gulf Administration Reports*; Lorimer's *Gazetteer of the Persian Gulf, Oman and Central Arabia*, first printed at Calcutta in 1908 and 1915, reprinted and bound in 9 volumes; *The Persian Gulf Précis*, first printed in 18 volumes at Calcutta and Simla (India) between 1903 and 1908, reprinted and bound in 8 volumes. The original works of all these volumes comprise part of the India Office Library and Records, now stored in the British Library.

6. Non-UK residents may wish to know also that certain international research centres maintain Archive Editions' publications in their reading rooms.

7. Residents of the Arabian Gulf region may find it helpful to know that copies of many of the original documents now housed in the British Library and the British State Archive (Public Record Office) are available for reference on microfiche and/or hard copy at the Centre of

Documentation and Research, Ministry of Foreign Affairs, PO Box 2380, Abu Dhabi, United Arab Emirates.

8. Bapco records are maintained in the Finance and Legal Division, The Bahrain Petroleum Company BSC (closed), Bahrain Refinery, State of Bahrain.

9. The Chevron Corporate Library is maintained at 225 Bush Street, San Francisco, California 94104, USA.

10. T. E. Ward's original documents and correspondence are maintained at the Petroleum History and Research Center of the University of Wyoming, Laramie, Wyoming 82070, USA.

It may be assumed that when a library or a private collection is not attributed, the related document is maintained in the Bapco archive.

A

ACCOUNTING GUIDE, signed on *15 December 1979* by Hassan A. Fakhro for the Bahrain National Oil Company (Banoco) and William E. Tucker for the Bahrain Petroleum Company Limited (Bapco), agreed for the Government of the State of Bahrain by Y. A. Shirawi and attached to the *Interim Operating Agreement* between Banoco and Bapco. (See also *AGREEMENT*, same date, and *SIDE LETTER No. 3 – INTERIM OPERATING AGREEMENT*)

ADAMS, H. T. (Eastern and General Syndicate Secretary), letter to T. E. Ward, 20 April 1927, T. E. Ward Jr's personal papers

ADAMS, H. T., letter to T. E. Ward, 9 June 1927, T. E. Ward Jr's personal papers

ADAMS, H. T., letter to T. E. Ward, 16 June 1927, T. E. Ward Jr's personal papers

ADAMS, H. T., letter to T. E. Ward, 14 September 1927, T. E. Ward Jr's personal papers

ADAMS, H. T., letter to T. E. Ward, 4 October 1927, T. E. Ward Jr's personal papers

ADAMS, H. T., letter to T. E. Ward, 7 October 1927, T. E. Ward Jr's personal papers

ADAMS, H. T., letter to T. E. Ward, 5 November 1927, T. E. Ward Jr's personal papers

ADAMS, H. T., letter to T. E. Ward, 15 November 1927, T. E. Ward Jr's personal papers

ADAMS, H. T., letter to T. E. Ward, 1 December 1927, T. E. Ward Jr's personal papers

ADAMS, H. T., letter to T. E. Ward, 23 December 1927, T. E. Ward Jr's personal papers

ADAMS, H. T., letter to T. E. Ward, 2 January 1928, T. E. Ward Jr's personal papers

ADAMS, H. T., letter to T. E. Ward, 18 April 1928, T. E. Ward Jr's personal papers

ADAMS, H. T., letter to T. E. Ward, 4 June 1928, T. E. Ward Jr's personal papers

ADAMS, H. T., letter to T. E. Ward, 26 July 1928, T. E. Ward Jr's personal papers

ADAMS, H. T., letter to T. E. Ward, 22 August 1928, T. E. Ward Jr's personal papers

ADAMS, H. T., letter to the Under Secretary of State, Colonial Office, London, 22 October 1928, CCL; classification CO C.59115/28 [No. 11], BSA and Ward, 1965, p. 153

ADAMS, H. T., cable to T. E. Ward in Chicago, 3 December 1928, Ward, 1965, pp. 119-20

ADAMS, H. T., letter to the Under Secretary of State, Colonial Office, London, 19 December 1928, F. B. Loomis' private file: *Bahrein 1928-1929*, CCL; classification CO C.59115/28 [No. 25], BSA and Ward, 1965, pp. 141-5

ADAMS, H. T., memorandum to T. E. Ward, 10 January 1929, Ward, 1965, pp. 151-2

ADAMS, H. T., letter from the Eastern and General Syndicate to the Colonial Office, 8 April 1929, classification CO C.69035/29 [No. 23], BSA

ADAMS, H. T., letter from the Eastern and General Syndicate to the Colonial Office, 24 September 1929, classification CO C.69035/29 [No. 140], BSA

ADAMS, H. T., letter to William T. Wallace, 5 November 1929, CCL

ADAMS, H. T., letter to William T. Wallace, 15 November 1929, CCL

ADAMS, H. T., letter from the Eastern and General Syndicate to the Colonial Office, 30 January 1930, classification CO C.79035/30 [No. 9], BSA, enclosing Harry G. Davis' letter of 29 January 1930

ADAMS, H. T., letter to Standard Oil Company of California, 21 February 1930, CCL

ADAMS, H. T., letter to F. A. Leovy, Eastern Gulf Oil Company, 21 February 1930, CCL

ADAMS, H. T., letter from the Eastern and General Syndicate Limited to the Colonial Office, 16 September 1930, confirming the formal assignation of the Bahrein Oil Concession to the Bahrein Petroleum Company Limited on 1 August 1930, classification CO C.79035/30, BSA

ADDITIONAL BAHREIN OPTION, Letter from Eastern Gulf Oil Company to the Eastern and General Syndicate Limited, 28 May 1928 (accepted by the Eastern and General Syndicate Limited on 28 November 1928)

AGREED PRINCIPLES governing the basic terms of the acquisition by the Government of the State of Bahrain and the future operations of the Bahrain Petroleum Company Limited's refinery, signed on the 26 June 1980 by Y. A. Shirawi (Minister of Development and Industry) and William E. Tucker

AGREEMENT between the Eastern and General Syndicate Limited (119 Finsbury Pavement, City of London, UK) and Eastern Gulf Oil Company (Pittsburgh, Pennsylvania, USA) re Kuwait Concession, *6 November 1933*, CCL

AGREEMENT between the Sheikh of Bahrein and the Bahrein Petroleum Company Limited, Modifying Mining Lease of 29 December 1934, as to Tax Exemption and increasing Guaranteed Minimum Royalty, *3 June 1936* [otherwise known as the first Supplemental Agreement/ Deed]. (See also *MINING LEASE* 29 December 1934, *DEED OF FURTHER MODIFICATION* 19 June 1940 and *SUPPLEMENTAL AGREEMENT* 8 December 1952)

AGREEMENT between Eastern and General Syndicate (Bahamas) Limited and the Bahrain Petroleum Company Limited [relating to overriding royalties], *29 March 1961*

AGREEMENT between Eastern and General Syndicate (Bahamas) Limited and the Bahrain Petroleum Company Limited [relating to EGS' voluntary liquidation], *11 March 1965*

AGREEMENT between Eastern and General Investment Company Limited (formerly Eastern and General Syndicate Limited), Berry Wiggins and Company Limited and the Bahrain Petroleum Company Limited regarding overriding royalty payments, *8 March 1973*

AGREEMENT between the Government of the State of Bahrain and the Bahrain Petroleum Company Limited, signed on *23 November 1974*, whereby the Government acquired 60% of Bapco's Exploration and Producing Rights, Operations and Facilities and Bapco's related production in Bahrain, effective from 1 January 1974

AGREEMENT between the Government of the State of Bahrain and the Bahrain Petroleum Company Limited whereby the Government acquired Bapco's remaining 40% interest in the Exploration and Producing Rights, Operations and Facilities and Bapco's related production in Bahrain, signed on the 25th day of Muharram 1400, corresponding to *15 December 1979*, by Y. A. Shirawi for the Government of Bahrain and William E. Tucker for the Bahrain Petroleum Company Limited. (*See also ACCOUNTING GUIDE*, same date, and *SIDE LETTER NO. 3 – INTERIM OPERATING AGREEMENT*)

AGREEMENT FOR SERVICES between the Bahrain National Oil Company and the Bahrain Petroleum Company BSC (closed), made on *1 March 1982*, signed by Hassan A. Fakhro (Chairman and Managing Director, Banoco) and Donald Frank Hepburn (Chief Executive, Bapco)

AITCHISON, C. U. (Under Secretary to the Government of India in the Foreign Department), compiler of *A Collection of Treaties, Engagements and Sanads relating to India and Neighbouring Countries, Volume XI* containing The Treaties etc. relating to Aden and the South Western coast of Arabia, the Arab Principalities in the Persian Gulf, Muscat (Oman), Baluchistan and the North-West Frontier Province, revised and continued up to the end of 1930 under the authority of the Government of India. [Part II (2) Bahrain section.] Published in Delhi, 1933

AL-AHRAM (Cairo newspaper), 24 January 1939, also in CCL

AL-KHALIFAH, Isa bin Ali, Sheikh of Bahrain, letter to Major A. P. Trevor, (British Political Agent in Bahrain), 18 Jamadi II 1332, corresponding to 14 May 1914. For English translation *see* AITCHISON, Vol. XI, p. 239

AL-KHALIFAH, His Excellency Sheikh Hamad bin Isa, letter to the Bahrein Petroleum Company Limited (at 225 Bush Street, San Francisco, California), sealed by the Ruler of Bahrain and British Political Agent, 27 January 1935 [re amendment to the Mining Lease]

AL-KHALIFAH, Sulman bin Hamad, Ruler of Bahrain, letter to the Chief Local Representative, Bahrain Petroleum Company, 18 April 1950 [re revision of oil royalty, acknowledging Bapco's letter CON-456]

AL-KHALIFAH, Sulman bin Hamad, Ruler of Bahrain, letter to the Chief Local Representative, Bahrain Petroleum Company, 22 December 1959 [acknowledging Bapco's letter CON-643]. (*See also* LIPP, CON-643, 1 December 1959)

AL-KHALIFAH, Sulman bin Hamad, Ruler of Bahrain, letter to Chief Local Representative, Bapco, 8 February 1961. (*See also* SCHMIDT, MISC. CON-25)

AL-KHALIFAH, Isa bin Sulman, Ruler of Bahrain, letter to Bapco's Chief Local Representative, referring to MISC. CON-601, 27 October 1962 and acknowledged on 10 November 1962, 2 December 1962. (*See also* SCHMIDT, MISC. CON-601)

AL-KHALIFAH, Isa bin Sulman, Ruler of Bahrain, letter to Bapco's Chief Local Representative, re financial issues, 25 November 1963

AL-KHALIFAH, Khalifa bin Sulman, Head of Finance, Bahrain Government Finance Department (Oil Affairs Bureau), letter to the Vice President in charge of Bapco, re channel for all routine communications between the Government of Bahrain and Bapco and a newly proposed pricing letter, 5 August 1964

AL-OTAIBA, HE Mana Saeed, *OPEC and the Petroleum Industry*, Croom Helm, London, 1985

AMENDATORY AGREEMENT between His Highness Shaikh Isa bin Sulman Al-Khalifa, Ruler of Bahrain and Its Dependencies, and the Bahrain Petroleum Company, Sections A to C, signed on 9 August 1965

AMERY, L. S., letter from the Secretary of State to the PRPG, 19 June 1928, classification CO C.59115/28 [No. 2], BSA

AMIRI DECREE (English translation) enacting Law No. 9/1976 in respect of the incorporation of the Bahrain National Oil Company in accordance with the attached Articles of Association, published in the Official Gazette, No. 1165, 4 March 1976

ANDERSON, Irvine H., *ARAMCO, The United States and Saudi Arabia: A Study of the Dynamics of Foreign Oil Policy 1933-1950*, Princeton University Press, Princeton, New Jersey, USA/Guildford, Surrey, UK, 1981

ANDERSON, William D., "Oil Policies of the Gulf Countries", Ch. 5, pp. 60-78, in *Conflict and Cooperation in the Persian Gulf*, ed. Mohammed Mughisuddin, Praeger, 1977

ANDERSON, W. P. (Chief Local Representative), letter CON-153 to E. B. Wakefield, (British Political Agent), re aviation gasoline, 14 July 1942, R/15/2/428, IOR

ANDERSON, W. P. confidential letter to Bird [no initials given], 28 August 1945, classification FO 371/45189 XL141812, BSA

ARAB ECONOMIST, "Bahrain: All Set for Another Boom", (no credited correspondent), No. 137, Vol. 13, February 1981, pp. 23-8

ARIKAT, Harby Moh'd Mousa, *The Economic Environment of the Arab World*, Economic Research

Paper No. 12, Centre for Middle Eastern and Islamic Studies, University of Durham, England, 1985

ARIKAT, Dr Harby Mohammad Mousa, *The Arab Gulf Economy: A Demographic and Economic Profile*, Economic Research Paper No. 17, Centre for Middle Eastern and Islamic Studies, University of Durham, UK, (undated)

ARPGPR Ch. VIII – "Administration Report of the Political Agency, Bahrein, for the year 1927", submitted by the PRPG on 19 June 1928, RGIFD, 1928

ARPGPR for the year 1878-79, RGIFD, Foreign Dept. Press, Calcutta 1879, Table 15, p. 88

ARPGPR for the year 1879-80, RGIFD, Foreign Dept. Press, Calcutta 1880, Table 20, p. 118

ARTICLES OF ASSOCIATION, the Bahrain Aviation Fuelling Company BSC (Closed)

ARTICLES OF ASSOCIATION, the Bahrain National Oil Company BSC

ASHE, Mary, letter from the Treasury (ref. FO 243/208/01) to T. E. Rogers at the Foreign Office, 25 November 1949, classification FO 371/75021, BSA

ASHE, Mary, letter from the Treasury (ref. FO 243/208/01) to A. Leavett at the Foreign Office, 16 January 1950, classification FO 371/75021, BSA

ASSIGNMENT OF BAHREIN ISLAND CONCESSION from the Eastern and General Syndicate Limited to the Bahrein Petroleum Company Limited, 1 August 1930

ATIYAH BIN ALI, *A Dialogue between Pearl Diving and Oil Wells*, a narrative poem in Arabic, written by a learned Bahraini in 1353 AH (corresponding to 1934), unpublished

AYOOB, Mohammed, "Oil, Arabism and Islam: The Persian Gulf in World Politics", Ch. 5, pp. 118-35, in *The Middle East in World Politics*, ed. Mohammed Ayoob, Croom Helm, London, 1981

AZZAM, Henry T., *The Gulf Economies in Transition*, Macmillan Press, UK, 1988

AZZAM, Henry T., editor, *Gulf Financial Markets*, Gulf International Bank BSC, Bahrain, 1988

B

BAHRAIN: ECONOMIC DEVELOPMENT STUDY 1967 REPORT, the work of the Bapco-Caltex Bahrain Study Group, submitted in May 1967

BAHRAIN EMPLOYED PERSONS COMPENSATION ORDINANCE, 1957, Government of Bahrain, November 1957

BAHRAIN GOVERNMENT ANNUAL REPORTS 1924-70, Archive Editions, Gerrards Cross, Buckinghamshire, England, 1986 (Vols. I-V) and 1987 (Vols. VI-VIII)

BAHRAIN INCOME TAX REGULATION 1955, Queen's Regulation made under Article 82 of the Bahrain Orders, 1952 and 1953, No. 8 of 1955, made by B. A. B. Burrows, Her Majesty's Political Resident in the Persian Gulf, published in Bahrain on 28 November 1955

BAHRAIN INCOME TAX REGULATION 1955, DRAFT AMENDMENT, contained in letter M-CON 243 (16 pages, plus Attachments A and B), from the Bahrain Petroleum Company Limited to His Highness Shaikh Isa bin Sulman Al-Khalifah, dated *8 March 1966*, "which agreement terminated the agreements recorded in the Company's letter concerning prices dated 8 December 1952, and in the Company's letter M-CON 188 dated 11 August 1964". (*See also* JOSEPHSON, M-CON 241, also dated 8 March 1966)

BAHRAIN INCOME TAX (AMENDMENT) DECREE 1966 (Decree No. 11 (Finance) of 1966), 15 June 1966

BAHREIN ISLAND CONCESSION from Shaikh Hamad bin Isa Al-Khalifa to the Eastern and General Syndicate Limited, *2 December 1925* (Articles I to X, followed by the *First Schedule* (privileges to be enjoyed by the Company under the exploration licence), *Second Schedule* (privileges to be enjoyed by the company under the prospecting licence) and the *Third Schedule* (Mining Lease)

BAHRAIN ISLANDER, THE, bi-monthly publication of the Bahrain Petroleum Company Limited, Vol. 27, No. 49, 12 July 1967

BAHRAIN LABOUR ORDINANCE, 1957, THE, issued by the Government of Bahrain, November 1957

BAHRAIN NEWSPAPER, THE, "Education Department Report", 13 February 1940

BAHRAIN OIL FIELD – PHASE-IN OF BANOCO PERSONNEL INTO OPERATIONS, see FILE MEMORANDUM and MINUTES OF MEETINGS)

BAHRAIN PETROLEUM COMPANY ANNUAL REPORTS, 1937-89

BAHRAIN REFINERY PARTICIPATION AGREEMENT, between the Government of the State of Bahrain and the Bahrain Petroleum Company Limited, signed by Y. A. Shirawi (Minister of Development and Industry) and William E. Tucker (a Director of the Bahrain Petroleum Company Limited) on *19 July 1980*. (*See also* AGREED PRINCIPLES)

BAHRAIN REFINERY PARTICIPATION AGREEMENTS, between the Government of the State of Bahrain and the Bahrain Petroleum Company Limited, signed on the 30th day of Jumada Al Akhira 1401, corresponding to *4 May 1981*:

1. *Participants' and Operating Agreement* signed for the Government of the State of Bahrain by Y. A. Shirawi and for the Bahrain Petroleum Company Limited (to be succeeded as a Party by Caltex Bahrain Limited) by R. N. Trackwell. Additionally, the Bahrain Petroleum Company BSC (closed) and Caltex Bahrain Limited were parties to and bound by the Agreement

2. *Accounting Guide – Supplemental Agreement to the Participants' and Operating Agreement* signed for the Government of the State of Bahrain by Y. A. Shirawi and the Bahrain Petroleum Company Limited by R. N. Trackwell. The Bahrain Petroleum Company BSC (closed) was also party to and bound by the Agreement

3. Annex A, *The Bahrain Petroleum Company BSC (closed), A Bahrain Joint Stock Company, Memorandum of Association*, signed for the Government of the State of Bahrain by Y. A. Shirawi (Minister of Development and Industry) and for Caltex Bahrain Limited by R. N. Trackwell (President)

4. Annex B, *The Bahrain Petroleum Company BSC (closed), A Bahrain Joint Stock Company, Articles of Association*, with the same signatories as for Annex A. Section I, Incorporation of Company [Articles 1-4]; Section II, Capital of the Company [Articles 5-13]; Section III, Alteration of Capital [Articles 14-15]; Section IV, Administration of the Company [Articles 16-28]; Section V, The General Assembly [Articles 29-40]; Section VI, Accounts of the Company [Articles 41-48]; Section VII, Dissolution and Liquidation of the Company [Articles 49-53]

5. *Processing Agreement*, between the Government of the State of Bahrain and the Bahrain Petroleum Company Limited (to be succeeded as a Party by Caltex Bahrain Limited), signed by Y. A. Shirawi and R. N. Trackwell. The Bahrain Petroleum Company BSC (closed) was a party to the Agreement. This document comprised 12 paragraphs accompanied by five additional documents: First Schedule, *Definitions* (refers to paragraph 1.2); Second Schedule, *Loading and Discharge Conditions* (refers to paragraph 8); Third Schedule, *Allocation of Differences between actual Productions and those of Retrospective Operating Programmes* (refers to paragraph 6.1 [c]); Fourth Schedule, *Allocation of Refinery Operating Costs Bases on the TSRV Method* (refers to paragraph 11); Fifth Schedule, *Allocation of Refinery Operating Costs using Systems Costing Method* (refers to paragraph 11).

6. *Technical and Manpower Services Agreement* between the Government of the State of Bahrain and the Bahrain Petroleum Company Limited (to be succeeded as a Party hereto by Caltex Bahrain Limited), signed by Y. A. Shirawi and R. N. Trackwell. The Bahrain Petroleum Company BSC (closed) was party to and bound by the Agreement

7. *Side Letter between Participants – Buyback Procedure* signed by Y. A. Shirawi and for the Bahrain Petroleum Company Limited by R. N. Trackwell. [This document refers to Annex B of the 19 July 1980 Agreement between the Government of the State of Bahrain and the Bahrain Petroleum Company Limited]

8. *Fuel and Gas Supply Agreement* between the Government of the State of Bahrain and the Bahrain Petroleum Company Limited (to be succeeded as a party by Caltex Bahrain Limited), signed by Y. A. Shirawi and R. N. Trackwell. The Bahrain Petroleum Company BSC (closed) was a party to the Agreement

BAHREIN: STANDARD OIL COMPANY OF CALIFORNIA AND THE TEXAS CORPORATION, a brief description of recent consolidations of production and marketing facilities of the Standard Oil Company of California and the Texas Corporation in the Far East, prepared and distributed by Dean Witter & Co. (Municipal and Corporation Bonds), members of the New York Stock Exchange and San Francisco Stock Exchange, 1937, document also in CCL

BALLANTYNE, H. R. (solicitor in charge of Bapco's UK office and the recognized channel of communication between the company and the British Government), letter to Colonial Office, 27 February 1931, classification CO C.89035/31 [No. 1], BSA

BALLANTYNE, H. R., letter to F. B. Loomis, 22 June 1932, CCL

BALLANTYNE, H. R., letter to F. B. Loomis, 4 October 1932, CCL

BALLANTYNE, H. R., letter to M. E. Lombardi, 4 November 1932, CCL

BALLANTYNE, H. R., telegram to F. B. Loomis, 20 March 1933, CCL

BALLANTYNE, H. R., letter to F. B. Loomis, 21 March 1933, CCL

BALLANTYNE, H. R., letter to W. F. Vane, 21 March 1933, CCL

BALLANTYNE, H. R., telegrams to W. F. Vane, 22 and 23 March 1933, CCL

BALLANTYNE, H. R., letter to F. B. Loomis, 31 March 1933, CCL

BALLANTYNE, H. R., letter to F. B. Loomis, 9 May 1933, CCL

BALLANTYNE, H. R., cable to E. A. Skinner in Bahrain, 9 May 1933, CCL

BALLANTYNE, H. R., letter to W. F. Vane, 12 May 1933, CCL

BALLANTYNE, H. R., cable to Eastern and General Syndicate representative in Bahrain, 12 May 1933, CCL

BALLANTYNE, H. R., cable to E. A. Skinner in Bahrain, 12 May 1933, CCL

BALLANTYNE, H. R., letter to Frank Holmes (Eastern and General Syndicate representative in Bahrain), 12 May 1933, CCL

BALLANTYNE, H. R., letter to M. E. Lombardi, 12 May 1933, CCL

BALLANTYNE, H. R., letter to E. A. Skinner, Bapco, Bahrain, 12 May 1933, CCL

BALLANTYNE, H. R., cable to W. H. Berg, 16 May 1933, CCL

BALLANTYNE, H. R., telegrams to W. F. Vane, 19, 24 and 27 May 1933, CCL

BALLANTYNE, H. R., telegram to W. F. Vane, 2 June 1933, CCL

BALLANTYNE, H. R., cables to M. E. Lombardi, 6, 7, 8, 9 and 16 June 1933, CCL

BALLANTYNE, H. R., two letters to W. F. Vane, 31 July 1933: [1] refers to Holmes, [2] refers to Laithwaite's activities on behalf of the India Office, CCL

BALLANTYNE, H. R., letter to W. F. Vane, 1 August 1933, CCL

BALLANTYNE, H. R., letter to J. G. Laithwaite, 1 August 1933, CCL

BALLANTYNE, H. R., letter to F. B. Loomis, 22 August 1933, CCL

BALLANTYNE, H. R., letter to the Under Secretary of State for India, India Office, London, 22 August 1933, CCL

BALLANTYNE, H. R., telegram to W. F. Vane, 24 August 1933, CCL

BALLANTYNE, H. R., cable to W. F. Vane, 8 September 1933, CCL

BALLANTYNE, H. R., cable to M. E. Lombardi, 9 September 1933, CCL

BALLANTYNE, H. R., cable to M. E. Lombardi, 12 September 1933, CCL

BALLANTYNE, H. R., cable to M. E. Lombardi, 15 September 1933, CCL

BALLANTYNE, H. R., letter to F. B. Loomis, 26 September 1933, CCL

BALLANTYNE, H. R., letter to F. B. Loomis, 17 October 1933, CCL

BALLANTYNE, H. R., letter to L. N. Hamilton, 5 February 1934, CCL

BALLANTYNE, H. R., letter to L. N. Hamilton, 20 February 1934, CCL

BALLANTYNE, H. R., letter to E. W. Janson, 28 February 1934, CCL

BALLANTYNE, H. R., letter to the Secretary, Eastern and General Syndicate, 9 November 1934, T. E. Ward Jr's personal papers

BALLANTYNE, H. R., letter to the Bahrein Petroleum Company Limited, 10 April 1935, T. E.

Ward Jr's personal papers

BALLANTYNE, H. R., letter to Mr Henry J. Kiernan (Attorney to Bapco Limited, 130 East 43rd Street, New York City, USA) from No. 135 Cliffords Inn, London EC4), 21 January 1944, B. M. van Benschoten private collection. (*See also* DIXON, 18 September 1943)

BALLANTYNE, H. R., and HAMILTON, L. N., cable to M. E. Lombardi, 15 October 1932, CCL

BALLANTYNE, H. R., and HAMILTON, L. N., cable to M. E. Lombardi, 17 November 1933, CCL

BARKHURST, C. R. (Bapco's Chief Local Representative), letter CON-926 to the Ruler of Bahrain, 31 December 1953, classification FO 371/109899, BSA

BARKHURST, C. R., letter CON-935 to the Ruler of Bahrain, acknowledging Decree No. 80 issued on 23 November 1955, superseding Decree No. 8 of 6 December 1952

BARKHURST, C. R., letter CON-936 to the Ruler of Bahrain, re Clause 1 of the Supplemental Agreement 8 December 1952 having no effect after 1 January 1955, 29 November 1955

BARKHURST, C. R., letter CON-937 to the Ruler of Bahrain, re aggregate oil prices and computation of corporate taxes, 29 November 1955

BARRETT, C. C. J., "Administration Report of the Political Agency, Bahrein, for the year 1928", *ARPGPR*, RGIFD, Simla, 1928

BARRETT, C. C. J. (Political Agent, Bahrein), letter to Eastern and General Syndicate Representative, Bahrain [Holmes], 3 May 1928, Ward, 1965, p. 74

BARRETT, Lt-Col. C. C. J., (PRPG), letter from the British Residency and Consulate-General (Bushire) to the Secretary of State (office not mentioned), 19 July 1929, CO, C.69035/29 [No. 111], enclosing a copy of letter No. 90 dated 15 July 1929 from the Political Agent (Bahrein) to the PRPG, re the Landing Ground and Seaplane Station at Bahrein and Eastern and General Syndicate's Exploration Licence

BEATTY, Jerome Jr, "Is John Bull's Face Red", *American Magazine,* January 1939

BELGRAVE, C. D., letter to the British Political Agent (Bahrein), ref. 691-210, 17 January 1949, classification FO 371/75021 and 371/82047, BSA

BELGRAVE, C. D., *Personal Column*, Librairie du Liban, Beirut, 1972

BELGRAVE, J. H. D., "Oil and Bahrain", *The World Today*, Vol. 7, No. 2, February 1951, pp. 76-83

BELING, Willard A., "Recent Developments in Labor Relations in Bahrayn", *The Middle East Journal,* Spring 1959, pp. 156-69

BERG, W. H. (Vice President, Standard Oil Company of California), letter to Thomas E. Ward, 27 December 1928, Ward, 1965, p. 131

BERG, W. H., telegram to Judge Frank Feuille, 27 December 1928, Ward, 1965, p.132

BERG, W. H., cable to F. B. Loomis, 26 October 1929, CCL

BERG, W. H. (Vice-President, Bapco Limited), letter from 225 Bush Street, San Francisco to the Under-Secretary of State, Colonial Office, London, 17 July 1930, enclosure 67 of Duncan Smith's letter to the Colonial Office, 2 August 1930, classification CO C.79035/30 [No.100], BSA

BERG, W. H. (Vice-President, Bapco Limited), letter from 225 Bush Street, San Francisco to the Under-Secretary of State, Colonial Office, London, 10 February 1931, enclosure in H. R. Ballantyne's letter to the Colonial Office, 27 February 1931, classification CO C.89035/31 [No.1], BSA

BERG, W. H., telegram to M. E. Lombardi, 20 March 1933, CCL

BERG, W. H., cable to M. E. Lombardi, 12 May 1933, CCL

BERG, W. H., telegram to L. N. Hamilton, 9 February 1937, CCL

BIBBY, G., *Looking for Dilmun: The Search for a Lost Civilization,* first published by Collins, 1970; reprinted by Penguin Books, UK, 1984

BIDWELL, Dr Robin, "Introduction – Middle East Politics before 1916: the background to the Arab Bulletin", *The Arab Bulletin*, Vol. I, 1916

BIRKS, J. S., and RIMMER, J. A., *Developing Education Systems in the Oil States of Arabia: Conflicts of Purpose and Focus,* Occasional Paper No. 21, Manpower and Migration Series No. 3, Centre for Middle Eastern and Islamic Studies, University of Durham, UK, 1984

BISCOE, Lt-Col. H. V. (Political Resident in the Persian Gulf), letter to the Secretary of State, Colonial Office, 30 June 1930, enclosing copy of a letter from the local representative of the Eastern and General Syndicate (Bahrein) to the Political Agent (Bahrein), dated 17 June 1930, classification CO C.79035/30 [No. 96], BSA

BISCOE, Lt-Col. H. V., letter to the Secretary of State, 9 July 1930, enclosing a copy of Major Frank Holmes' letter to the Political Agent (Bahrein), dated 5 July 1930, classification CO C.79035/30 [No. 97], BSA

BISCOE, Lt-Col. H. V., letter to the Secretary of State, re Eastern and General Syndicate's Application for a further Concession in Bahrein, 15 February 1932, File C.98035/32 [No. 1], classification CO 935/7, BSA

BREWER, William, "Yesterday and Tomorrow in the Persian Gulf", *Middle East Journal*, Vol. 23, No. 2, Spring 1969, pp. 149-58

BRITISH STATE ARCHIVE at the PUBLIC RECORD OFFICE, LONDON:

The following records, listed in numerical sequence, comprise document classifications relevant to this book. Note that the *numerical chronology* of references does *not correspond* to a *chronology of dates.*

CO 732/52/5 – Telegram and letter, 11 October and 27 December 1932

CO 732/59/9 – Cabinet conclusions, memoranda and letters, 9 June to 29 July 1933

CO 732/62/12 – Despatch and letters, 31 January to 8 June 1933 re customs and duties

CO 935/3 – CORRESPONDENCE PRINTED FOR THE USE OF THE COLONIAL OFFICE, 1926 TO 1931, *Persian Gulf: Concessions in Bahrein, Kuwait etc.*, Middle East No. 32. References:

C.7171/26

C.59115/28 – Nos. 1-2, 11, 14, 19, 25, 27

C.69035/29 – Nos. 7, 15, 17, 21-4, 37, 43, 45, 48-9, 53, 64-8, 75, 81, 83, 86, 86A, 91A, 92-3, 101-2, 104, 110-1, 114, 121-2, 124-5, 137-8, 140, 149, 181

C.79035/30 – Nos. 1, 5, 9, 12-9, 30, 41, 61, 72, 78, 80, 82, 84, 86-7, 89, 90-1, 96-100, 115, 117, 122, 128, 140-2, 155

C.89035/31 – Nos. 1-2, 9-10

CO 935/7 – CORRESPONDENCE PRINTED FOR THE USE OF THE COLONIAL OFFICE, 1932 TO July, 1933, *Persian Gulf: Concessions in Bahrein, Kuwait etc.*, Middle East No. 49 (in continuation to Middle East No. 32). References:

18135/33 – Nos. 4, 7, 18, 22, 50, 55-56, 76, 84-6, 95, 97, 99, 100-5, 117, 122-3, 126-9, 138-141, 178, 185, 190-7, 202, 207, 213, 222-6, 236-7, 241, 260-2, 271-2, 277, 290, 295

18301/33 – Nos. 1-4, 33-43, 51A, 52

C.98035/32 – Nos. 1, 13, 15, 16, 27, 31, 33, 46-8, 51-4, 56, 62, 66-9, 70, 79-87, 92, 96, 106, 109-110, 116, 130-1, 138, 140, 144-7, 155, 161-2, 167-74, 180-5, 190

FO 371/1242 – Translation of an agreement signed by the Ruler of Bahrain, 22 December 1880

FO 371/18911 – Secret Intelligence Report, May 1935

FO 371/19965 – Letters re nationality of Bapco employees, December 1935-August 1936; first Supplemental Deed, 3 June 1936; memos re Bapco employee disturbances, April 1946

FO 371/19977 – Correspondence re forms of address, 1936

FO 371/20771 – Letters, telegrams, minutes, 16 August to 27 October 1937 re Petroleum Concessions Ltd and the Additional Area

FO 371/20777 – Despatch re Bapco's oil production, 22 January 1937; letters re Caltex and marketing, 16 April to 10 July 1937; news clip, 11 August 1937

FO 371/21822 – Telegram re Holmes' movements, March 1938; letter re sale of fuel oil to Japanese navy, March 1938

FO 371/24542 – Memorandum and letters re concession, June-August 1940

FO 371/31317 – Confidential letters, New Delhi to India Office re Bahrein refinery output, November to December 1941; Secret Cipher Telegrams, re "scorched earth policy" for Bahrain, between Bahrain, Baghdad and War Office, London, during 1942

FO 371/31318 – Letters re Bapco's Chief Local Representative, February and March 1942

FO 371/34899 – Letters re war damage in Bahrain, April 1943

FO 371/34900 – Letters and telegrams re US State Department request to establish an American consulate in Bahrain, 8 February to 20 July 1943; cipher telegrams re US Air Corps in Bahrain, November to December 1943

FO 371/39906 – Secret memorandum re Americans and British in Arabia, 18 March 1944

FO 371/45189 – Letters re Bahrain refinery, August to September 1945

FO 371/52266 – Memorandum on Bahrein (42 pages and appendices) re history of Bahrain from the late 18th to early 20th centuries, 13 January 1947

FO 371/61422 – Letter, telegrams, despatch, re agricultural problems in Bahrain, 22 July to 25 August 1947

FO 371/61426 – Cabinet conclusions, memoranda and letters, 9 June to 29 July 1933; note of a meeting, 14 May 1947

FO 371/61428 – Letters and Memoranda, 13 September to 15 October 1947

FO 371/61441 – Telegrams and reports re drilling and surveys, February to December 1947

FO 371/68333 – Minute, letters, re Ward's inspection of Bapco's royalty oil, 25 November to 6 December 1948

FO 371/69330 – Despatch, 14 January 1948; letter, 30 January 1948

FO 371/75009 – Letter to the Foreign Office, 17 May 1949

FO 371/75021 – Letters, 17 January, 3 June, 4 July, 25 November 1949; 6 January 1950; three undated memoranda, all re oil imports, US$ sales of oil to US oil companies, oil royalties

FO 371/82012 – Confidential despatch on the Social and Political Effects of the Development of the Oil Industry in the Persian Gulf, 24 April 1950

FO 371/82027 – Despatch, 4 January 1950; Persian Gulf Economic Report, 7 August 1950

FO 371/82155 – Letter, Hay to FO, re 1941 Bahrain census and conscription, 1 May 1950

FO 371/91281 – Top secret correspondence re assurances to the Ruler of Bahrain of protection against aggression, June 1951

FO 371/91323 – Despatch re Bapco's deepwater

pier, March 1951; letters on same subject, May to October 1951

FO 371/91264 – Letters re employee discontent, June 1951; report from *The Bahrein Islander*, "Twenty-Five Years in Bahrein", 12 April 1951

FO 371/98398 – Letter re cost of living in Bahrein; despatch, 10 January 1952; letter, 23 May 1952

FO 371/98407 – Income Tax Regulation, 1952; departmental ciphers between Sir R. Hay (Bahrain) and Foreign Office (London), January 1952; Legal Adviser's communications to Minister of State and Secretary of State, 1952

FO 371/98428 – Memoranda re Bapco's oil revenue, January 1952

FO 371/98429 – Memo re guaranteed minimum royalties, July 1952; letters re 50-50 tax, June to December 1952

FO 371/98466 – Correspondence re Bapco's labour relations, 1952; memoranda, including one dated 4 February 1952

FO 371/104336 – Despatch, 22 December 1953; reports, 14 December 1953 re supplies of water and natural gas to Bapco

FO 371/104346 – Despatch and letters, 8 July to 26 August 1953

FO 371/104357 – Memo re cost of living in Bahrain, 1951

FO 371/104400 – Despatch re offshore prospecting beyond 100,000 acres, 7 December 1953

FO 371/104449 – Correspondence re Bapco's training and wages, 1952-3; report, 23 May 1953; despatch, letters, minutes, Translations, 6 May to 21 July 1953

FO 371/109866 – Minutes, 31 March 1954 to 9 January 1956

FO 371/109867 – Despatch and letters, 14 July to 26 August 1954

FO 371/109899 – Correspondence, December 1953; February 1954

FO 371/114600 – Study by Research Department re Britain's Treaty Relations with Bahrain, 4 May 1955

FO 371/114714 – PRPG despatch to Foreign Office on 50/50 profit, February 1955; minutes, March 1955; paper for Middle East Oil

Committee, March 1955; PRPG despatch to Foreign Office on oil prices, April 1955

FO 371/126957 – Press Release, 4 March 1957

FO 371/126997 – Bahrain letter re training and education, 26 August 1957

FO 371/127006 – Despatches, letters, minutes, 6 March to 27 June 1957

FO 371/127014 – Despatch, report, minutes, April 1957 to 3 June 1957; reports re census

FO 406/60 – References E4499/8184/91

FO 406/65 – Further correspondence respecting Eastern Affairs, Part XXVI, January to June 1930

FO 406/73 – Further correspondence respecting Eastern Affairs, Part XXXVII, July to December 1935. References: E 7262/452/91; E 6564/452/91

FO 406/77 – Further correspondence respecting Eastern Affairs, Part XLIV, January to June 1939. Reference: E 3750/2670/91

FO 406/78 – Further correspondence respecting Eastern Affairs, Part XLVI, January to December 1940

POWE 33/194 – Paper by G. C. Gester on oil development in Bahrein, 10 March 1934; memorandum re Bapco finding a market for its crude oil, October 1934

POWE 33/195 – Letter, 4 February 1938; *World Petroleum* extract, July 1938; letter, 16 February 1947; news clip, 2 May 1947

POWE 33/199 – Note on the Bahrein drilling moratorium, July 1942

POWE 33/1699 – Despatch by Hay, 19 March 1951; notes on Lease and Deed of Further Modification; report on Bahrain refinery, Saudi Arabian oil and the AB pipeline, July 1948

POWE 33/1919 – Letters re granting leases and licences to companies registered in the UK, October 1952

BROWN, Neville, "Britain and the Gulf – don't go just yet please!: the wisdom of withdrawal reconsidered", *New Middle East*, No. 24, September 1970, pp. 43-6

BROWN, R. M. (Bapco's Chief Local Representative), letter C/PA-271 to the British Political Agent, Manama, 4 January 1947, B. M. Van Benschoten's personal papers. (*See also* GAL-LOWAY, 10 December 1946)

BROWN, R. M., letter No. CON-456 to His Highness Shaikh Sulman bin Hamad Al-Khalifah, Ruler of Bahrain, 12 April 1950

BROWN, R. M., letter No. C/PA-472 to the British Political Agent (Pelly), 28 June 1951, B. M. Van Benschoten's personal papers

BROWN, R. M., letter CON-722 to the Ruler of Bahrain, re oil revenue payments, 28 June 1951, B. M. Van Benschoten's personal papers. (*See also* PELLY, 4 July 1951, and BROWN, 21 July 1951)

BROWN, R. M., letter CON-768 to the Ruler of Bahrain, re voluntary payment of Rps. 500,000 per month, 21 July 1951, B. M. Van Benschoten's personal papers

BROWN, R. M., letter (unnumbered) to the Ruler of Bahrain, re weighted average of crude oil prices, 8 December 1952

BROWN, R. M., letter CON-118 to the Ruler of Bahrain, confirming the Supplemental Agreement of 8 December 1952 and referring to $2\,^1/_4$ pence import fee, signed 8 December 1952. (*See also SUPPLEMENTAL AGREEMENT*)

BROWN, R. M., letter CON-119 to the Ruler of Bahrain, re provision to review the situation should States bordering Bahrain in which oil is being produced receive "substantially better terms than does Your Highness", 8 December 1952

BROWN, R. M., letter CON-120 to the Ruler of Bahrain (responding to Decree No. 8 issued by His Highness on 6 December 1952 "imposing a 50-50 tax upon income from the sales of crude petroleum etc"), 16 December 1952

BROWN, R. M., letter CON-121 to the Ruler of Bahrain (re Article 12 of Decree No. 8 dated 6 December 1952), 16 December 1952

BROWN, R. M., letter CON-306 to the Ruler of Bahrain re computation of taxation, 16 December 1952

BROWN, R. M., letter C/FA-90 to the Adviser to the Bahrain Government, re Bapco's issue of the 1952 Annual Report in accordance with the Ruler's "expressed wishes", 1 July 1953

BROWN, R. M., letter C/PA-59 to the British Political Agent (Bahrain), re Bapco's 1952 Annual

Report, 1 July 1953

BROWN, R. M., letter to the British Political Agent (re Bapco's 1952 Annual Report in accordance with Article 18 of the Lease dated December 29 1934), 1 July 1953

BURROWS, B. A. B., despatch no. 107 (1087/10/53), from the British Residency, Bahrain, to the Rt Hon. Anthony Eden, MC, MP, Principal Secretary of State for Foreign Affairs, Foreign Office, London, 7 December 1953, classification FO 371/104400, BSA

BURROWS, B. A. B., letter 1531/2/2/55 from the British Residency, Bahrain, to L. A. C. Fry, Eastern Department, Foreign Office, London, 22 February 1955, classification FO 371/114714, BSA

BURROWS, B. A. B., confidential despatch no. 41 (1531/2/8/55) from the British Residency, Bahrain, to Sir Anthony Eden, KG, MC, MP, Foreign Office London, re increase of oil revenues for the Bahrain Government, 7 April 1955, classification FO 371/114714, BSA

BUSCH, Briton Cooper, *Britain and the Persian Gulf 1894-1914*, University of California Press, Berkeley and Los Angeles, 1967

C

CALCOTT GASKIN, J., "Report on the Trade and Commerce of the Bahrein Islands for 1900", *ARPGPR for 1900-1901*, RGIFD, Calcutta, 1901

CALCOTT GASKIN, J., "Report on the Trade and Commerce of the Bahrein Islands for 1901", *ARPGPR for 1901-1902*, RGIFD, Calcutta, 1902, Part VII

CALCOTT GASKIN, J., "Report on the Trade of the Bahrein Islands for the year 1902", *ARPGPR for 1902-1903*, RGIFD, Calcutta, 1903

CALCOTT GASKIN, J., "Report on the Trade of the Bahrein Islands for the Year 1903", *ARPGPR for 1903-1904*, RGIFD, Calcutta, 1904, Part VII

CALCOTT GASKIN, J., holograph, 23 April 1904, classification R/15/1/317, IOR

CALIFORNIA STANDARD OIL COMPANY LIMITED (London office), telegram to W. H. Berg, (California Standard Oil Company Limited), 9 February 1937, CCL

CALIFORNIA STANDARD OIL COMPANY LIMITED (London office), telegram to W. H. Berg, 22 and 31 March 1937, CCL

CALIFORNIA STANDARD OIL COMPANY LIMITED (London office), telegram to W. H. Berg, 1, 5 and 14 April 1937, CCL

CALIFORNIA STANDARD OIL COMPANY LIMITED (London office), telegram to W. H. Berg, 5 May 1937, CCL

CALIFORNIA STANDARD OIL COMPANY LIMITED (London office), letter to Eastern and General Syndicate, 27 September 1938, T. E. Ward Jr's personal papers

CALIFORNIA STANDARD OIL COMPANY LIMITED (London office), coded cable to M. E. Lombardi re cable sent by R. G. Wedemeyer to CASOC and to W. J. Lenahan, 22 March 1939, CCL

CALTEX STORY, THE, published by the Caltex Petroleum Corporation, December 1981

CAROE, Olaf, *Wells of Power: The Oilfields of South-Western Asia*, Macmillan, London, 1951. [Also available for reference in CCL]

CARRIERA, A. M. Caetano, "The significance of ASRY to the future industrial development of Bahrain", pp. 71-84 in *Engineering and Development in the Gulf*, published for the Bahrain Society of Engineers by Graham and Trotman, London, 1977

CATTAN, Henry, *The Evolution of Oil Concessions in the Middle East and North Africa*, published for the Parker School of Foreign and Comparative Law by Oceana Publications Inc., Dobbs Ferry, New York, 1967

CHADWICK, J. E., minute, 25 November 1948, classification FO 371/68333, BSA

CHADWICK, J. E., letter to Sir Rupert Hay, 31 December 1948, classification FO 371/68333, reference E 15009/327/91, BSA

CHARTER OF THE BAHREIN PETROLEUM COMPANY LIMITED, 11 January 1929

CHEVRON, *Oil Concessions in the Middle East*, Vols. I and II, Chevron Corporation, San Francisco, California, USA (undated)

CHEVRON WORLD, "Centennial 1879-1979", Standard Oil Company of California, 1979/Winter

CHEVRON MANAGEMENT NEWSLETTER, July 1984, No. 453

CHEVRON WORLD, "Caltex Celebrates a Golden Anniversary", Spring 1986

CHEVRON WORLD, "The Great Arabian Discovery: Chevron's landmark oil find in Saudi Arabia reverberates 50 years later", *Fall 1988*

CHIEF REPRESENTATIVE, the Bahrein Petroleum Company Limited, letters to His Britannic Majesty's Political Agent (C. G. Prior), 22 and 24 April 1932, CCL

CHISHOLM, Archibald H. T., CBE, *The First Kuwait Oil Concession Agreement: A Record of the Negotiations 1911-1934*, Frank Cass, London (UK) and Portland, Oregon, USA, 1975

CHRISTIAN SCIENCE MONITOR, "Sheba and El Segundo" (no credited author), 22 January 1934, CCL

CHRONICLE COMMUNICATIONS LONDON, *CHRONICLE OF THE 20TH CENTURY*, Longman, 1988

COLE, Capt. G. A., "Administration Report of the Bahrain Agency for the Year 1934", *ARPGPR for the Year 1934*, Government of India Press, Simla, 1935, Ch. VII

COLE, H. W., letter from the Director of the Mines Department to Colonial Office, 2 March 1929, CO, File C.69035/29 [No. 15]

COLE, H. W., letter from the Director of the Mines Department to Colonial Office, 9 April 1929, CO, File C.69035/29 [No. 24]

COLE, H. W., letter from the Director of the Mines Department to Colonial Office, 21 June 1929, CO, File C.69035/29 [No. 75]

COMMANDER-IN-CHIEF MIDDLE EAST, secret telegram to the War Office, London, 16 April 1942, classification FO 371/31317 XL 141812, BSA

COMMANDER-IN-CHIEF MIDDLE EAST, secret telegram to the War Office, London, 24 April 1942, classification FO 371/31317 XL 141812, BSA

COMMANDER-IN-CHIEF MIDDLE EAST, secret telegram to the War Office, London, 6 May 1942, classification FO 371/31317 XL 141812, BSA

COMMANDER-IN-CHIEF MIDDLE EAST, secret telegram to the War Office, London, 11 May 1942, classification FO 371/31317 XL 141812, BSA

COMMANDER-IN-CHIEF MIDDLE EAST, secret telegram to the War Office, London, 2 June 1942, classification FO 371/31317 XL 141812, BSA

COMMANDER-IN-CHIEF MIDDLE EAST, secret telegram to the War Office, London, 7 July 1942, classification FO 371/31317 XL 141812, BSA

CRANFIELD, John, "Problems of Downstream Operation", *Petroleum Economist*, March 1984, pp. 101-02

CROMBIE, J., unpublished personal notes, March 1990, Bapco archive

CUNLIFFE-LISTER, P., letter from the Secretary of State to the Political Resident, re employment of British subjects by Bapco, 31 January 1933, classification CO 935/7, file 18135/33 [No. 22], BSA

D

DAILY TELEGRAPH, THE, "Three-way deal gives Lex 20pc of Berry Wiggins", (no credited correspondent), 26 September 1972. (*See also FINANCIAL TIMES* entry on the same subject)

DAVIES, F. A., letter to Clark Gester, 24 May 1930, CCL

DAVIES, F. A., letter to Clark Gester, 31 May 1930, CCL

DAVIES, F. A., letter to Clark Gester, 14 June 1930, CCL

DAVIES, F. A., letter to Clark Gester (from Baghdad), 12 August 1930, CCL

DAVIES, F. A., *Supplementary Report on the Bahrein Islands*, 26 November 1930, CCL

DAVIES, F. A., and TAYLOR, W. F., letter to G. C. Gester and W. I. McLaughlin, Socal, 17 May 1930, CCL

DAVIS, Harry G., letter to William Wallace, 24 May 1929, F. B. Loomis *Bahrein 1928-1929* file, CCL

DAVIS, Harry G., letter to William Wallace, 27 May 1929, F. B. Loomis *Bahrein 1928-1929* file, CCL

DAVIS, Harry G., letter to William Wallace, 3 June

1929, CCL

DAVIS, Harry G., letter to E. W. Janson, Eastern and General Syndicate, 3 June 1929, CCL

DAVIS, Harry G., telegram to South American Gulf Oil Company, 19 July 1929, F. B. Loomis' private file: *Bahrein 1928-29* file, CCL

DAVIS, Harry G., telegram to South American Gulf Oil Company, 22 July 1929, F. B. Loomis' private file: *Bahrein 1928-29* file, CCL

DAVIS, Harry G., telegram to South American Gulf Oil Company, 1 August 1929, F. B. Loomis *Bahrein 1928-1929* file, CCL

DAVIS, Harry G., telegram to South American Gulf Oil Company, 2 August 1929, F. B. Loomis *Bahrein 1928-1929* file, CCL

DAVIS, Harry G., telegram to South American Gulf Oil Company, 9 August 1929, F. B. Loomis *Bahrein 1928-1929* file, CCL

DAVIS, Harry G., cable to William T. Wallace, Gulf Oil, 11 December 1929, CCL

DAVIS, Harry G., cable to William T. Wallace, Gulf Oil, 18 December 1929, CCL

DAVIS, Harry G., cable to Judge Frank Feuille, 25 January 1930, CCL

DAVIS, Harry G., letter to the Eastern and General Syndicate from the Hyde Park Hotel, Knightsbridge, London, dated 29 January 1930, enclosed in H. T. Adams' letter to the Colonial Office, 30 January 1930, classification CO C.79035/30 [No. 9], BSA

DEACON, C. W. (Bapco General Manager), circular to employees, 23 April 1938

DECREE No. 8 (English translation), issued by His Highness Sheikh Sir Sulman bin Hamad Al-Khalifah, Ruler of Bahrain, re Bapco's corporate income tax liability, *6 December 1952*

DECREE No. 80, (English translation), issued by His Highness Sheikh Sir Sulman bin Hamad Al-Khalifa, Ruler of Bahrain, superseding Decree No. 8 (above), *28 November 1955*

DEED OF FURTHER MODIFICATION of Lease dated 29 December 1934, made between His Highness Shaikh Hamad bin Isa Al-Khalifa and the Bahrein Petroleum Company Limited, and being Supplemental to the Agreement dated the 3rd day of June 1936 (later known as the first Supplemental Agreement/Deed), signed on 19 June 1940. (*See also MINING LEASE* 29 December 1934, *AGREEMENT* 3 June 1936 and *SUPPLEMENTAL AGREEMENT* 8 December 1952)

DEEGAN, Charles J., and BURNS, Warren W., "World Oil", *The Oil and Gas Journal*, 28 December, 1946, pp. 154-9

DIXON, Capt. M. G. (British Political Agent), letter C/1244, to Bapco Chief Local Representative, re moratorium in respect of the company's obligations under Article 5 of the *Deed of Further Modification* [19 June 1940], dated 18 September 1943, B. M. van Benschoten private collection. (*See also* BALLANTYNE, 21 January 1944)

DRAKE, Waldo, "Bahrein Island Voyage Starts" (re *El Segundo* tanker), *Los Angeles Times*, 31 December 1933

DRAKE WELL MUSEUM pamphlet, Drake Well Museum, Venango Co., Pennsylvania, USA (undated)

DUFF, Dahl M., "Refining Expansion outside U.S. to reach 4,000,000 barrels by 1951", *The Oil and Gas Journal*, 27 December 1947, pp. 214, 216, 219 and 242-6

DUNCAN SMITH, D., letter to the Colonial Office, 2 August 1930, classification CO, C.79035,30 [No. 100], BSA

DURAND, Capt. E.L., "Notes on the Pearl Fisheries of the Persian Gulf", *ARPGPR for the year 1877-78*, RGIFD, Calcutta 1878, Appendix A to Part II

DURAND, Capt. E. L., "Description of the Bahrain Islands", *ARPGPR for the Year 1878-79*, RGIFD, Calcutta, 1880

DURAND, Capt. E. L., "The Islands and Antiquities of Bahrain", *Journal of the Royal Arabic Society* (New Series), XII, Part II, 1880, pp. 189-227

E

EASTERN AND GENERAL SYNDICATE, radiogram to Ward, Oilfield, New York, 20 April 1927, T. E. Ward Jr's personal papers

EASTERN AND GENERAL SYNDICATE, radiogram to T. E. Ward, 28 October 1927, T. E. Ward Jr's personal papers

EASTERN AND GENERAL SYNDICATE, radiogram to T. E. Ward, 9 November 1927, T. E. Ward Jr's personal papers

EASTERN AND GENERAL SYNDICATE, radiogram to Ward, Oilfield, New York, 22 November 1927, T. E. Ward Jr's personal papers

EASTERN AND GENERAL SYNDICATE, radiogram to T. E. Ward, 23 November 1927, T. E. Ward Jr's personal papers

EASTERN AND GENERAL SYNDICATE, radiogram to T. E. Ward, 25 November 1927, T. E. Ward Jr's personal papers

EASTERN AND GENERAL SYNDICATE, radiogram to T. E. Ward, 27 November 1927, T. E. Ward Jr's personal papers

EASTERN AND GENERAL SYNDICATE, radiogram to T. E. Ward, 28 November 1927, T. E. Ward Jr's personal papers

EASTERN AND GENERAL SYNDICATE, letter to Messrs Thomas Cook and Sons, Marseilles, France, 23 December 1927, T. E. Ward Jr's personal papers

EASTERN AND GENERAL SYNDICATE, letter to the British Political Agent (Bahrein), 3 May 1928, T. E. Ward Jr's personal papers

EASTERN AND GENERAL SYNDICATE, letter to the British Political Agent (Bahrein), 28 May 1928, T. E. Ward Jr's personal papers

EASTERN AND GENERAL SYNDICATE, letter to the Political Agent (Bahrein), 5 June 1928, T. E. Ward Jr's personal papers

EASTERN AND GENERAL SYNDICATE, radiogram to Ward, Oilfield, New York, 12 August 1928, T. E. Ward Jr's personal papers

EASTERN AND GENERAL SYNDICATE, cable to Thomas E. Ward, 16 November 1928, Ward, 1965, p. 109

EASTERN AND GENERAL SYNDICATE, radiogram to T. E. Ward, New York, 26 February 1929, T. E. Ward Jr's personal papers

EASTERN AND GENERAL SYNDICATE, radiogram to F. B. Loomis, 15 November 1929, CCL

EASTERN AND GENERAL SYNDICATE, letter to Judge Frank Feuille, 19 November 1929, CCL

EASTERN AND GENERAL SYNDICATE, letter to H. R. Ballantyne, 2 October 1934, T. E. Ward Jr's personal papers

EASTERN AND GENERAL SYNDICATE, letter to T. E. Ward, 12 November 1936, T. E. Ward Jr's personal papers

EASTERN AND GENERAL SYNDICATE, letter to T. E. Ward, 18 May 1937, T. E. Ward Jr's personal papers

EASTERN GULF OIL COMPANY, letter to the Eastern and General Syndicate Ltd (London), 3 October 1928, T. E. Ward Jr's personal papers

ELLIS JONES, Peter, *Oil: A Practical Guide to the Economics of World Petroleum*, Woodhead-Faulkner Ltd, Cambridge, UK, and Nichols Publishing Company, New York, USA, 1988

ELWELL-SUTTON, L. P., *Persian Oil: A Study in Power Politics*, Lawrence and Wishart Ltd, London, 1955

EVANS, John, *OPEC, Its Member States and the World Energy Market*, Longman, UK, 1986; distributed in the USA and Canada by Gale Research Company, Book Tower, Detroit, Michigan, USA

EWING, J. S., interview with H. D. Collier on 13 March 1958, CCL

F

FAKHRO, Hassan A., General Manager, the Bahrain National Oil Company, *Financial Communication (in letter form) to the President, the Bahrain Petroleum Company*, 13 May 1976

FARID, Abdul Majid, editor, *Oil and Security in the Arabian Gulf*, Croom Helm, London, 1981

FAROUGHY, Dr Abbas, *The Bahrein Islands (750-1951): A Contribution to the Study of Power Politics in the Persian Gulf*, Verry, Fisher and Co., New York, 1951, available in CCL

FEUILLE, Judge Frank, letter to Thomas E. Ward, 28 December 1928, Ward, 1965, pp. 132-3

FEUILLE, Judge Frank, letter to W. F. Vane, 31 January 1929, CCL

FEUILLE, Judge Frank, letter to F. B. Loomis, 5 June 1929, CCL

FEUILLE, Judge Frank, letter to F. B. Loomis, 14 June 1929, F. B. Loomis *Bahrein 1928-1929* file, CCL

FEUILLE, Judge Frank, telegram to F. B. Loomis, 28 September 1929, CCL

G

GALLOWAY, A. C. (Political Agent, Bahrain), letter C/1370 to Bapco's Chief Local Representative, 10 December 1946, B. M. Van Benschoten's personal papers. (*See also* BROWN, 4 January 1947)

GAULT, C. A. (Political Residency, Bahrain), letter to D. M. H. Riches, Foreign Office, London, 28 August 1958, classification CO 935/3, BSA

GAULT, C. A., letter to A. R. Walmsley, Foreign Office, London, 11 October 1958, classification CO 935/3, BSA

GESTER, G. C., "Petroleum Developments in Bahrein Island", *The Petroleum Times*, 10 March 1934. Also filed under classification POWE 33/194, BSA

GIBSON, J. P., letter to H. R. Ballantyne, re Bapco Chief Local Representative's role under wartime conditions, 17 May 1941

GIDDENS, Paul H., "The Significance of the Drake Well", *Oil's First Century*, Papers Given at the Centennial Seminar on the History of the Petroleum Industry, Harvard Graduate School of Business Administration, 1960

GIDDENS, Paul H., "Edwin L. Drake and the birth of the Petroleum Industry", *Historic Pennsylvania Leaflet*, No. 21, Pennsylvania Historical and Museum Commission, Harrisburg, 1975

GILMOUR, John R. (Petroleum Dept), secret letter to E. W. Lumby (India Office), 8 May 1942, classification FO FO371/31318 XL 141812, BSA

GORNALL, John, *Some Memories of Bapco*, Awali, Bahrain, May 1965

GOVERNMENT OF BAHRAIN, letter No. 691-210 from the Office of the Adviser, 17 January 1949 (signed by C. Dalrymple Belgrave), to the British Political Agent, classification FO 371/75021, BSA

GOVERNMENT OF BAHRAIN, letter No. 626-20 to Bapco's Chief Local Representative, 26 December 1949

GOVERNMENT OF BAHRAIN, letter No. 3170-20B to Bapco's Chief Local Representative, acknowledging Barkhurst's letter CON-935, 29 November 1955

GOVERNMENT OF INDIA, External Affairs Dept, secret telegram to Secretary of State for India, 23 April 1942, classification FO 371/31317 XL 141812, BSA

GOVERNMENT OF INDIA, External Affairs Dept, telegram to Secretary of State for India, 20 April 1943, classification FO 371/34900 139378, BSA

GREER, James M., letter to T. E. Ward, 6 October 1928, T. E. Ward Jr's personal papers

GREER, James M. (Law Department, Gulf Company New York), letter to Thomas E. Ward, 17 December 1928, Ward, 1965, pp. 122-3

GREER, James M., letter to Judge Frank Feuille, 1 March 1929, CCL

GREER, James M., memorandum re contracts between Eastern and General Syndicate and Eastern Gulf Oil Company dealing with the Bahrein islands and Kuwait territory (14 pages + 13 enclosures), dated March 1929, CCL

GREER, James M., Western Union cable to Francis B. Loomis, Standard Oil Company of California, 27 May 1929, CCL

GREER, James M., letter to Judge Frank Feuille (typed on the letterhead of the law department, South American Gulf Oil Company), 13 June 1929, CCL

GREER, James M., letter to Judge Frank Feuille, 17 July 1929, F. B. Loomis *Bahrein 1928-29* file, CCL

GREER, James M. telegram to Judge Frank Feuille, 20 July 1929, F. B. Loomis *Bahrein 1928-29* file, CCL

GREER, James M., letter to Judge Frank Feuille, 24 September 1929, CCL

GREER, James M., letter to Judge Frank Feuille, 4 November 1929, CCL

GREER, James M., letter to Judge Frank Feuille, 14 January 1930, CCL

GREER, James M., letter to Judge Frank Feuille, 15 January 1930, CCL

GREER, James M., letter to William T. Wallace, 29 January 1930

GREER, James M., letter to Judge Frank Feuille, 25 February 1930, CCL

GREER, James M., cable to F. B. Loomis, 27 February 1930, CCL

GREER, James M., letter to Judge Frank Feuille, 7 March 1930, CCL

GREER, James M., letter to Judge Frank Feuille, 9 May 1930, CCL

GREER, James M., letter to F. B. Loomis, 2 January 1931, CCL

GREER, James M., letter to F. B. Loomis, 20 February 1934, CCL

GRILL, N.C., *Urbanisation in the Arabian Peninsula*, Occasional Papers Series No. 25 (1984), Centre for Middle Eastern and Islamic Studies, University of Durham, UK, 1984

H

HALL, J. H., letter from the Colonial Office to the Foreign Office, India Office and the Petroleum Department, 17 June 1929, classification CO C.69035555/29 [Nos. 64-66], BSA

HALL, J. H., letter from the Colonial Office to the Foreign Office, 3 August 1929, classification CO C.69035/29 [No. 92], BSA

HALL, J. H., letter from the Colonial Office to the India Office, 3 August 1929, classification CO C.69035/29 [No. 93], BSA

HAMILTON, L. N., cables to M. E. Lombardi, 26 and 28 October 1932, CCL

HAMILTON, L. N., cable to M. E. Lombardi, 11 November 1932, CCL

HAMILTON, L. N., cable to M. E. Lombardi, 15 December 1932, CCL

HAMILTON, L. N., telegrams to M. E. Lombardi, 6, 11 and 17 July 1933, CCL

HAMILTON, L. N., letter to F. B. Loomis, 12 August 1933, CCL

HAMILTON, L. N., cable to W. F. Vane, 21 August 1933, CCL

HAMILTON, L. N., letter to W. F. Vane, 8 February 1934, CCL

HAMILTON, L. N., letter to W. F. Vane, 23 February 1934, re minimum annual return to the Ruler under the Bahrein Concession, CCL

HAMILTON, L. N., telegram to W. F. Vane from London, 25 June 1934, CCL

HAMILTON, L. N., telegram to W. F. Vane from London, 1 February 1935, CCL

HAMILTON, L. N., telegram to W. F. Vane from London, 2 April 1935, CCL

HAMILTON, L. N., cables to M. E. Lombardi, 13 and 17 September 1935, CCL

HAMILTON, L. N., telegram to H. D. Collier, California Standard Oil Co. Ltd, 8 April 1936, CCL

HAMILTON, L. N., telegram to W. H. Berg, 20 June 1936, CCL

HAMILTON, L. N. (for California Standard Oil Company Limited), letter to William Pocock, Eastern and General Syndicate, 17 November 1938, T. E. Ward Jr's personal papers

HAMILTON, L. N., confidential memorandum F-302 to M. E. Lombardi (from 6 Lothbury, London EC2), copied to F. W. Ohliger, Casoc, Khobar (No. 268), W.J. Lenahan, Casoc, Djedda [Jeddah] (No. 194) and H. R. Ballantyne, 9 March 1939, CCL

HAMILTON, L. N., confidential cable F-408 to M. E. Lombardi, copied to F. W. Ohliger, W. J. Lenahan and H. R. Ballantyne, 12 April 1939, CCL

HAMILTON, L. N., confidential Cable F-416 to M. E. Lombardi, copied to Ohliger, Lenahan and Ballantyne, 13 April 1939 (from London), CCL

HARTSHORN, J. E., *Politics and World Oil Economics: An Account of the International Oil Industry in its Political Environment*, Frederick A. Praeger, New York, 1962

HARTSHORN, J. E., *Oil Companies and Governments: An Account of the International Oil Industry in its Political Environment*, Faber and Faber, London, 1967

HAY, Sir Rupert, "The Impact of the Oil Industry on the Persian Gulf Shaykhdoms", *Middle East Journal*, Vol. 9, No. 4, Autumn 1955, pp. 361-72

HEARD-BEY, Dr Frauke, *From Trucial States to United Arab Emirates*, Longman, London and New York, 1982

HEARINGS before a Special Committee Investigating Petroleum Resources, United States Senate, Seventy-Ninth Congress, First Session, *27-8 June 1945*, "American Petroleum Interests in Foreign Countries", Library of Congress, Washington DC.

HEARINGS before a Special Committee Investigating Petroleum Resources, United States Senate, Seventy-Ninth Congress, First Session, *28-30 November 1945*, "Wartime Petroleum Policy Under the Petroleum Administration for War", Library of Congress, Washington DC.

HERRON, H. M., coded cable to California Standard Oil Company, London, UK, from California Texas Oil Company Limited, New York, 24 March 1939, CCL

HERRON, H. M., coded cable from New York to Anderson, Bapco, Bahrain, 22 August 1945, classification FO FO371/45189 XL141812, BSA

HEWINS, Ralph, *Mr Five Per Cent – The Story of Calouste Gulbenkian*, New York, 1958

HOLMES, Major Frank, cable to EASGENSYND London, 28 September 1926, T. E. Ward Jr's personal papers

HOLMES, Major Frank, cable to EASGENSYND London re Holmes' power of attorney, 28 September 1926, T. E. Ward Jr's personal papers

HOLMES, Major Frank, cable to EASGENSYND London, 29 September 1926, T. E. Ward Jr's personal papers

HOLMES, Major Frank, cable to Wallace, Gulf Refining Company, Pittsburgh, re Farisan concession, 19 October 1926, T. E. Ward Jr's personal papers

HOLMES, Major Frank, cable to William T. Wallace, Gulf Refining Company, Pittsburgh (second cable) re Farisan and Bahrein concessions, 19 October 1926, T. E. Ward Jr's personal papers

HOLMES, Major Frank, letter to T. E. Ward, 5 November 1927, T. E. Ward Jr's personal papers

HOLMES, Major Frank, letter to His Majesty, Abdul Aziz bin Abdul Rehman bin Faisal bin Saud, King of Hedjas and Sultan of Nejd and Dependencies, 23 December 1927, T. E. Ward Jr's personal papers

HOLMES, Major Frank, c/o A. M. Yateem Bros, Manama, letter to William T. Wallace, 9 February 1928, T. E. Ward Jr's personal papers and Ward, 1965, pp. 45-6

HOLMES, Major Frank, c/o A. M. Yateem Bros, Manama, letter to W. T. Wallace, 12 March 1928, T. E. Ward Jr's personal papers

HOLMES, Major Frank, letter to William T. Wallace, 25 March 1928, CCL

HOLMES, Major Frank, letter to William T. Wallace, 16 April 1928, Ward, 1965, pp. 59-60

HOLMES, Major Frank, letter to R. O. Rhoades, 19 April 1928, Ward, 1965, p. 67

HOLMES, Major Frank, letter to R. O. Rhoades, 20 April 1928, Ward, 1965, p.68

HOLMES, Major Frank, letter to William T. Wallace, 20 April 1928, Ward, 1965, pp. 60-6

HOLMES, Major Frank, letter to William T. Wallace, 6 May 1928, Ward, 1965, pp. 69-74

HOLMES, Major Frank, letter to T. E. Ward, 13 May 1928, T. E. Ward Jr's personal papers

HOLMES, Major Frank, letter to William T. Wallace, 9 June 1928, T. E. Ward Jr's personal papers

HOLMES, Major Frank, letter to the Secretary, Eastern and General Syndicate (London) from Cairo, 12 August 1928, T. E. Ward Jr's personal papers

HOLMES, Major Frank, letter to William T. Wallace, 20 September 1929, CCL and Ward, 1965, pp. 167-8

HOLMES, Major Frank, telegram to M. E. Lombardi, 5 January 1930, CCL

HOLMES, Major Frank, letter to the Political Agent, Bahrein, 5 July 1930, (enclosed in Biscoe's letter to the Secretary of State, 9 July 1930), classification CO C.79035/30 [No. 97], BSA

HOLMES, Major Frank, letter to F. B. Loomis, 9 September 1930, CCL

HOLMES, Major Frank, letter to British Political Agent, Bahrain, 11 September 1930, classification CO C.79035/30 [No. 117], (also in CCL)

HOLMES, Major Frank, telegram to M. E. Lombardi, 12 March 1931, CCL

HOLMES, Major Frank, letter to Thomas E. Ward, 18 March 1931, T. E. Ward Jr's personal papers

HOLMES, Major Frank, telegram to M. E. Lombardi, 19 March 1931, CCL

HOLMES, Major Frank, telegram to F. B. Loomis, 21 April 1931, CCL

HOLMES, Major Frank, letter to Ward, 2 November 1931, T. E. Ward Jr's personal papers

HOLMES, Major Frank, letter to Ward, 3 April 1932, T. E. Ward Jr's personal papers

HOLMES, Major Frank, cable to Taylor, Socal, 11 April 1932, CCL

I

R/15/2/419 – Communication between Russell Brown, Bapco and British Political Agent re South African employees, August 1945

R/15/2/422 – PRPG Report re possible pipeline to be laid by Casoc from Saudi Arabia to Bahrain, June 1935

R/15/2/423 – Communications re code and ciphers during war, 1943

R/15/2/425 – Bapco Royalties 1950

R/15/2/428 – Letter, July 1942; telegram August 1944 and 1945 on aviation plant construction

R/15/2/435 – Letters re Bapco, particularly Sitra wharf, March to May 1945

R/15/2/440 – Correspondence between the Bahrain Government Adviser and the British PA, July-August 1945

R/15/2/443 – Ministry of Fuel and Power/India Office communications re Texas Oil Co. and Bapco, April to June 1945

R/15/2/444 – Letters, July 1945

R/15/2/450 – Anderson endeavours to keep Bapco's competitors in the dark, 1944-6

R/15/2/453 – Letters re Bapco and water supply, November 1948 to January 1949; April-May 1950

R/15/2/460 – Documents relating to Bapco and "dollar oil", December 1948 to June 1949

R/15/2/499 – Documents relating to possibility of selling Bapco's petroleum coke to Iraq, 1940

R/15/2/808 – Report on active co-operation between Government of Bahrain and Bapco, 1940

R/15/2/838 – PRPG and PA correspond re Bapco and technical training, March 1945; report on Bahrain's School, January 1948

R/15/2/1340 – Brown seeks permission to arm guards transferring Bapco's cash, 4 June 1949

INFORMATION FOR EMPLOYEES ABOUT BAHREIN, first edition, 1 September 1938, pamphlet produced by the Bahrein Petroleum Company Limited, CCL

INITIAL AGREEMENTS, see AGREED PRINCIPLES and *BAHRAIN REFINERY PARTICIPATION AGREEMENT*, 1980

INSTITUTE OF PETROLEUM REVIEW (editor George Sell), "Twenty Five years of Middle East Oil – Bapco Operations in Bahrain", Volume 12, No. 142, October 1958, pp. 333-6

INTERIM OPERATING AGREEMENT, see SIDE LETTER NO. 3

INTERNATIONAL PETROLEUM CARTEL, THE, Staff Report to the Federal Trade Commission submitted to the Subcommittee on Monopoly of the Select Committee on Small Business [Part II Development of Joint Control Over the International Petroleum Industry – deals with IPC, Gulbenkian, Red Line Agreement; Part III Production and Marketing Agreements – deals with Achnacarry Agreement, "As Is" etc.], United States Senate, 22 August 1952

ISSAWI, Charles (Bayard Dodge Professor of Near Eastern Studies, Princeton University), *The 1973 Oil Crisis and After*, Princeton Near East Paper No.27, published by the Program in Near Eastern Studies, Princeton University, 1979.

J

JAMES, Marquis, *The Texaco Story: The First Fifty Years 1902-1952*, written for and published by the Texas Company, CCL

JANSON, E. W., letter to T. E. Ward, 19 August 1926, T. E. Ward Jr's personal papers

JANSON, E. W., letter to T. E. Ward, 23 February 1927, T. E. Ward Jr's personal papers

JANSON, E. W., letter to T. E. Ward, 28 November 1927, T. E. Ward Jr's personal papers

JANSON, E. W., letter to T. E. Ward, 15 December 1927, Ward 1965, p. 43

JANSON, E. W., letter to William T. Wallace, Gulf Exploration Company, 31 July 1930, CCL

JANSON, E. W., letter to T. E. Ward, 19 August 1932, T. E. Ward Jr's personal papers

JANSON, E. W., letter to T. E. Ward, 31 March 1933, T. E. Ward Jr's personal papers

JANSON, E. W., letter to T. E. Ward, 10 August 1933, T. E. Ward Jr's personal papers

JANSON, E. W., letter to T. E. Ward, 27 March 1935, T. E. Ward Jr's personal papers

JANSON, Jonathan (Chairman of Eastern and General Investment Company Limited), letter to J.M. Voss (at Bapco's Head Office in Madison Avenue, New York), 29 September 1972, T.

E. Ward Jr's personal papers

JANSON, Jonathan, letter to T. E. Ward Jr, 29 September 1972, T. E. Ward Jr's personal papers

JOSEPHSON, L. D. (Bapco's Chief Local Representative), letter C/PA-563 to the British Political Agent in Bahrain, re communication with the Ruler of Bahrain, 10 August 1964

JOSEPHSON, L. D., letter M-CON188 to the Ruler of Bahrain, confirming an agreement between His Highness and Bapco re realization prices, 11 August 1964

JOSEPHSON, L. D., letter C/PA-572 to the British Political Agent in Bahrain, re communication with the Ruler of Bahrain, 25 August 1964

JOSEPHSON, L. D., letter to the Secretary of the Bahrain Government, re communication with the Bahrain Government, 2 September 1964

JOSEPHSON, L. D., personal letter to Roy L. Lay, re Realization Principles, 14 September 1964

JOSEPHSON, L. D., letter C/PA-582 to the British Political Agent in Bahrain re amendments to the 29 June 1940 *Political Agreement* between the British Government and Bapco, 18 November 1964. (S*ee also* OLDFIELD, 19 August 1964)

JOSEPHSON, L. D., letter M-CON 203 to His Highness Shaikh Isa bin Sulman Al-Khalifah, Ruler of Bahrain, re deletions and alterations to avoid terminology and provisions in agreements which His Highness considers inconsistent with current circumstances in Bahrain, 29 March 1965

JOSEPHSON, L. D., letter M-CON 228 to His Highness Shaikh Isa bin Sulman Al-Khalifah, Ruler of Bahrain, re definition of exploratory wells, intangible costs and computation of corporate taxation, dated 27 September 1965

JOSEPHSON, L. D., letter M-CON 239 to His Excellency Shaikh Khalifah bin Sulman Al-Khalifah, Head of Finance, Government of Bahrain, re draft documents, dated 8 March 1966

JOSEPHSON, L. D., letter M-CON 240 to His Highness Shaikh Isa bin Sulman Al-Khalifah, Ruler of Bahrain, confirming the agreement entered into re export of Bahrain crude oil and referring to letter M-CON 243, dated *8 March*

1966. (*See also BAHRAIN INCOME TAX REGULATION 1955, DRAFT AMENDMENT*, M-CON 243, 8 March 1966)

JOSEPHSON, L. D., letter M-CON 241 to His Highness Shaikh Isa bin Sulman Al-Khalifa, Ruler of Bahrain, confirming that, for the implementation of Bapco's letter M-CON 243 of 8 March 1966, His Highness and Bapco entered into a further agreement for the purposes of reporting Bapco's income tax, dated 8 March 1966

JOSEPHSON, L. D., letter M-CON 242 to His Excellency Shaikh Khalifah bin Sulman Al-Khalifah, Head of Finance, Government of Bahrain (confirming the company's understanding to paragraph 12 of the agreement set forth in the company's letter M-CON 243 of 8 March 1966), dated 8 March 1966. (*See also BAHRAIN INCOME TAX REGULATION 1955, DRAFT AMENDMENT*, 8 March 1966)

JOSEPHSON, L. D., letter M-CON 246 to His Excellency Shaikh Khalifah bin Sulman Al-Khalifah, Head of Finance, Government of Bahrain, re proposal to accelerate payment of corporate taxation, 17 April 1966

JOSEPHSON, L. D., letter M-CON 247 to His Highness Shaikh Isa bin Sulman Al-Khalifah, Ruler of Bahrain, confirming agreement concerning changes to corporate taxation, 11 June 1966. (*See also BAHRAIN INCOME TAX (AMENDMENT) DECREE 1966*)

JOSEPHSON, L. D., letter M-CON 255 to His Highness Shaikh Isa bin Sulman Al-Khalifah, Ruler of Bahrain, confirming the agreement to amend royalty rates, 27 April 1967

JOSEPHSON, L. D., letter MISC. CON 265 to His Highness Shaikh Isa bin Sulman Al-Khalifah, Ruler of Bahrain, confirming an amendment to the company's letter M-CON 243 dated 8 March 1966, whereby one half of a US cent substitutes the words three sevenths pence sterling, dated 3 August 1968

JOSEPHSON, L. D., letter MISC. CON 267 to the Ruler of Bahrain, confirming an agreement between His Highness Shaikh Sulman bin Hamad Al-Khalifah and the Bahrain Petroleum Company Limited re payments related to finished and semi-finished products processed by Bapco utilizing imported crude oil, dated 3 August 1968

JOSEPHSON, L. D., letter MISC. CON 268 (to the Ruler of Bahrain), confirming method of computation and payment of Bapco's income tax and royalty obligations to His Highness, dated 3 August 1968

JOSEPHSON, L. D., letter MISC. CON 276 (to the Ruler of Bahrain) superseding the agreement recorded in Bapco's letter MISC. CON 267 of 3 August 1968, dated 1 April 1969

JOSEPHSON, L. D., letter MISC. CON 278 to the Ruler of Bahrain, confirming agreements into which His Highness and Bapco entered re marketing allowances, 1 April 1969

K

KELSEY, E. O., "Standard Oil to enter Asiatic Markets with Bahrein Crude Products", *San Francisco Chronicle*, 15 November 1935

KELLOGG, The Hon. Frank B. (US Secretary of State), telegram to the US Chargé d'Affaires, Atherton, at the US Embassy (London), 28 March 1929, Ward, 1965, pp. 159-60

KRUG, A. E., *Memorandum* [re dissolution of Caltex Bahrain Limited], 3 June 1983, Caltex Petroleum Corporation

KUBURSI, Atif A., *Oil, Industrialization and Development in the Arab Gulf States*, Croom Helm, London, UK; Sydney, Australia and Dover, New Hampshire, USA, 1984

KUBURSI, Atif A., and NAYLOR, Thomas (editors), *Co-operation and Development in the Energy Sector: The Arab Gulf States and Canada*, Proceedings of a Symposium on the Energy Sector co-sponsored by the Petroleum Information Committee of the Arab Gulf States and McMaster University, Canada, and held at McMaster University 16-17 May 1984, published by Croom Helm, London, Sydney and Dover, New Hampshire, 1985

L

LANGENKAMP, Robert D., editor, *The Illustrated Petroleum Reference Dictionary*, Third Edition, PennWell Books, Tulsa, Oklahoma

LANGER, William L., *An Encyclopedia of World History*, Houghton Mifflin Company, Boston, Massachusetts, USA, 5th edition, 1980

LAWLESS, R. I., editor, *The Gulf in the Early 20th Century: Foreign Institutions and Local Responses*, Occasional Papers Series No. 31 (1986), Centre for Middle Eastern and Islamic Studies, University of Durham, UK, 1986

LEASE between His Excellency Sheikh Hamad bin Shiekh Issa Al Khalifah Sheikh of Bahrein and the Bahrein Petroleum Company Limited, sealed by the Ruler and signed by J. M. Russell and F. A. Davies (Attorneys-in-Fact) on 29 December 1934, before G. Loch, British Political Agent in Bahrein. (Copies are maintained by Bapco and in CCL)

LEGISLATIVE DECREE No. 25/1980 with respect to establishing the Supreme Oil Council [State of Bahrain], Official Gazette, Issue 1408, 6 November 1980

LENAHAN, W. J., (of California Arabian Standard Oil Company), confidential memo, No. 364 [re German Legation in Saudi Arabia], to California Standard Oil Company's representative in London, 22 February 1939, Jeddah, Saudi Arabia, CCL

LENAHAN, W. J., confidential telegram to Ohliger in Dhahran, sent from Jeddah on 5 April 1939, CCL

LEOVY, F. A. (Vice-President, Eastern Gulf Oil Company), letter to T. E. Ward, 4 December 1927, Ward, 1965, p. 188

LEOVY, F. A., letter to the Eastern and General Syndicate, 28 May 1928, T. E. Ward Jr's personal papers

LEOVY, Frank A., letter to Thomas E. Ward, 28 May 1928, Ward, 1965, pp. 82-4

LEOVY, F. A. , letter to T. E. Ward, 26 December 1928, Ward, 1965, p. 127

LEOVY, F. A., letter to the Eastern and General Syndicate, 24 October 1929, Enclosure No. 45 in H. T. Adams' letter to the Colonial Office, 6 November 1929, classification CO C.69035/29 [No. 149], BSA and CCL

LEY, R. F., letter to William T. Wallace, 5 March 1929, T. E. Ward Jr's personal papers

LIPP, M. H. (Bapco's Chief Local Representative), letter CON-810 to C. Dalrymple Belgrave, re safety of Bahrein geological plans and drilling logs, should the field "fall into the hands of the enemy", 17 May 1941

LIPP, M. H., letter C/PA-317 to C. A. Gault, British Political Agent (Bahrain), re Bapco's letter C/PA-326 of 7 December 1958 to the Secretary of the Bahrain Government, 7 December 1958

LIPP, M. H., letter C/FA-326 to G.W.R. Smith, Secretary to the Bahrain Government, re increase of petrol supplied to the Government to 200,000 imperial gallons annually, 7 December 1958

LIPP, M. H., letter C/PA-359 to the British Political Agent (Bahrain), re extension of applicable oil royalty for 10 more years, 1 December 1959

LIPP, M. H., letter CON-643 to the Ruler of Bahrain, re Letters: No. 626-20 of 26 December 1949 (*see* Government of Bahrain entry for that date), No. CON-456 to the Ruler of 12 April 1950, the Ruler's letter to Bapco of 18 April 1950, by which the royalty rate was established for the 10-year period ending 1 January 1960, and now proposed to continue this for an additional 10-year period, 1 December 1959. (*See also* AL-KHALIFAH, 22 December 1959)

LIPP, M. H., letter C/PA-360 to the British Political Agent (Bahrain), re CON-645 of 1 December 1959, 1 December 1959

LIPP, M. H., letter CON-645 to the Ruler of Bahrain, re possible removal of a portion of the offshore area under the Ruler's dominion from the scope of the *Mining Lease* (29 December 1934) and the *Deed of Further Modification* (19 June 1940), 1 December 1959.

LIPP, M. H., letter C/PA-384 to the British Political Agent (Bahrain), re Lipp's retirement and appointment of W. A. Schmidt as Resident Vice President of Bapco effective 1 August 1960, 15 June 1960. (*See also* WILTSHIRE, 1533/13/60)

LITTLE, J. E. R. (Acting Political Agent), letter 153010/6/53/G to Bapco's Chief Local Representative, re Bapco's 1952 Annual Report, 9 July 1953

LOCH, Lt-Col. G., (British Political Agent in Bahrain) "Administration Report of the Bahrain Agency for the Year 1932", *ARPGPR for the Year 1932*, Government of India Press, Simla, 1933, Ch. VIII

LOCH, Lt-Col. G., "Administration Report of the Bahrain Agency for the Year 1933" *ARPGPR for the Year 1933*, Government of India Press, Simla, 1934, Ch. VIII

LOCH, Lt-Col. G., "Administration Report of the Bahrain Agency for the Year 1935", *ARPGPR for the Year 1935*, Government of India Press, New Delhi, 1936, Ch. VII

LOCH, Lt-Col. G., "Administration Report of the Bahrain Agency for the Year 1936", *ARPGPR for the Year 1936*, Government of India Press, New Delhi, 1937, Ch. VI

LOMBARDI, M. E., letter to F. B. Loomis, 20 May 1929, CCL

LOMBARDI, M. E., night letter to F. B. Loomis at the Chicago Club, Chicago, Illinois, 4 December 1929, CCL

LOMBARDI, M. E., telegram to F. B. Loomis, 13 December 1929, CCL

LOMBARDI, M. E., letter to F. B. Loomis, 6 June 1930, F. B. Loomis *Bahrein 1930* file, CCL

LOMBARDI, M. E., letter to Francis B. Loomis, 11 June 1930, F. B. Loomis *Bahrein 1930* file, CCL

LOMBARDI, M. E., cable to Judge Frank Feuille, 16 July 1930, CCL

LOMBARDI, M. E., letter to William T. Wallace, 12 August 1930, CCL

LOMBARDI, M. E., letter to Frank Holmes, 12 January 1931, CCL

LOMBARDI, M. E., letter to Frank Holmes, 12 March 1932, CCL

LOMBARDI, M. E., cables to Frank Holmes, c/o EASGENSYND, London, 19 and 21 April 1932, CCL

LOMBARDI, M. E., letter to F. B. Loomis, 2 June 1932, CCL

LOMBARDI, M. E., letter to Major Frank Holmes, 21 June 1932, CCL

LOMBARDI, M. E., letter to F. B. Loomis, 21 June 1932, CCL

LOMBARDI, M. E., letter to F. B. Loomis, 3 August 1932, CCL

LOMBARDI, M. E., cable to E. A. Skinner, Bahrein, 7 October 1932, CCL

LOMBARDI, M. E., cable to Major Frank Holmes, Kuwait, 7 October 1932, CCL

LOMBARDI, M. E., cable to L. N. Hamilton, London, 1 November 1932, CCL

LOMBARDI, M. E., letter to L. N. Hamilton, London, 21 November 1932, CCL

LOMBARDI, M. E., letter to Major Frank Holmes in Bahrein, 22 November 1932, CCL

LOMBARDI, M. E., letter to H. R. Ballantyne, 23 November 1932, CCL

LOMBARDI, M. E., letter to W. H. Berg (from Bahrain), 12 March 1933, CCL

LOMBARDI, M. E., cable to W. H. Berg (from Kuwait), 17 March 1933, CCL

LOMBARDI, M. E., cable to W. H. Berg (from Cairo), 28 March 1933, CCL

LOMBARDI, M. E., letter to W. F. Taylor (from Cairo), 28 March 1933, CCL

LOMBARDI, M. E., telegram to W. H. Berg, 3 April 1933, CCL

LOMBARDI, M. E., telegram to W. H. Berg, 12 May 1933, CCL

LOMBARDI, M. E., telegrams to H. R. Ballantyne, 5-7, 9 and 15 June 1933, CCL

LOMBARDI, M. E., cable to H. R. Ballantyne, 8 September 1933, CCL

LOMBARDI, M. E., letter to F. B. Loomis, 12 February 1934, enclosing memorandum re Bahrein Title Documents of the same date, CCL

LOMBARDI, M. E., letter to F. B. Loomis, 15 February 1934, enclosing list of Iraq Petroleum Company Documents, CCL

LOMBARDI, M. E., letter to F. B. Loomis, 15 February 1934, re location of Bahrein title documents, CCL

LOMBARDI, M. E., cable to F. B. Loomis at the Metropolitan Club, Washington DC, 28 May 1934, CCL

LOMBARDI, M. E., cable to J. A. Moffett, Standard Oil Co. of California, 30 Rockefeller Center, New York City, 28 May 1934, CCL

LOMBARDI, M. E., letter to the Secretary of State for India, India Office, Whitehall, London, 15 April 1935, CCL

LOMBARDI, M. E., telegram to L. N. Hamilton, 1 November 1935, CCL

LOMBARDI, M. E., memorandum to K. R. Kingsbury, 2 August 1937, CCL

LOMBARDI, M. E., coded cable to California Standard Oil Company, London, UK, 22 March 1939

LONGRIGG, Stephen Hemsley, *Oil in the Middle East: Its Discovery and Development*, issued under the auspices of the Royal Institute of International Affairs by Oxford University Press, London/New York/Toronto, Canada, 3rd edition, 1968

LOOMIS, Francis B., letter to Paul T. Culbertson, Department of State, Washington DC, 6 February 1929, CCL

LOOMIS, Francis B., letter to Judge Edward C. Finney, First Assistant Secretary, Dept. of the Interior, Washington DC, 13 February 1929, F. B. Loomis *Bahrein 1928-1929* file, CCL

LOOMIS, Francis B., letter from the Carlton Hotel, London, to M. E. Lombardi, 15 April 1929, F. B. Loomis *Bahrein 1928-1929* file, CCL

LOOMIS, Francis B., letter to Judge Frank Feuille, 21 June 1929, F. B. Loomis *Bahrein 1928-1929* file, CCL

LOOMIS, Francis B., letter to W. H. Berg (Vice President of Socal), 24 October 1929, F. B. Loomis B*ahrein 1928-1929* file, CCL

LOOMIS, Francis B., letter to Judge Frank Feuille, 14 November 1929, F. B. Loomis *Bahrein 1928-1929* file, CCL

LOOMIS, Francis B., telegram to Judge Frank Feuille, 26 December 1929, F. B. Loomis *Bahrein 1928-1929* file, CCL

LOOMIS, Francis B., telegram to Judge Frank Feuille, 27 December 1929, F. B. Loomis *Bahrein 1928-1929* file, CCL

LOOMIS, Francis B., letter to Judge Frank Feuille, 8 January 1930, F. B. Loomis *Bahrein 1930* file, CCL

LOOMIS, Francis B., telegram to William T. Wallace, 23 January 1930, F. B. Loomis *Bahrein 1930* file, CCL

LOOMIS, Francis B., telegram to William T. Wallace, 24 January 1930, F. B. Loomis *Bahrein 1930* file, CCL

LOOMIS, Francis B., telegram to William T. Wallace, 25 January 1930, F. B. Loomis *Bahrein 1930* file, CCL

LOOMIS, Francis B., letter to Felix T. Smith, 1 February 1930, CCL

LOOMIS, Francis B., letter to Major Frank Holmes, 6 April 1931, CCL

LOOMIS, Francis B., letter to Lombardi, 10 April 1933, CCL

LOOMIS, Francis B., letter to H. R. Ballantyne, 5 October 1933, CCL

LOOMIS, Francis B., cable to M. E. Lombardi, 28 May 1934, CCL

LOOMIS, Francis B., letter to M. E. Lombardi sent from The Hague, 19 June 1934, CCL

LOOMIS, Francis B., air mail letter to Hon. Cordell Hull, Secretary of State, Washington DC, 25 April 1939, CCL

LOOMIS, Francis B., air mail letter to Hon. George S. Messersmith, Assistant Secretary of State, Washington DC, 25 April 1939, CCL

LOOMIS, Francis B., letter to Hon. George S. Messersmith, Assistant Secretary of State, Washington DC, 16 May 1939, CCL

LORIMER, J. G., "German interests in Bahrain, 1895-1904", *Gazetteer of the Persian Gulf, Oman and Central Arabia*, Vol. I (Historical), Part I, Ch. V, 1908 and 1915, Calcutta, and reprinted by Archive Editions, 1986

LOS ANGELES HERALD, "Bahrein Island Oil Cleared by Standard for Japan" (no credited author), 18 July 1934, also in CCL

LOS ANGELES HERALD, "Standard Oil Bahrein Is. Right Clear" (no credited author), 2 August 1934, CCL

LOS ANGELES TIMES, 31 December 1933, p. 6, re *El Segundo*

LOS ANGELES TIMES, "Standard Active in Persian Gulf" (no credited author), 4 February 1934, also in CCL

LUCIANI, Giacomo, *The Oil Companies and the Arab World*, Croom Helm, London and Canberra, Australia/St Martin's Press, New York, 1984

M

MABRO, Robert, "The Present Oil Crisis: The Causes and Implications", *Arab Affairs*, London, Vol. 1, No. 1, Summer 1986, pp. 23-9

MADGWICK, T. G., letter to E. W. Janson, 10 August 1926, T. E. Ward Jr's personal papers

MADGWICK, Professor T. George, *Report*, 23 September 1926, Ward, 1965, pp. 29-32

MADGWICK, Professor T. George, letter to T. E. Ward, 4 December 1928, Ward, 1965, p. 139

MARTIN, Esmond Bradley, and MARTIN, Chryssee Perry, *Cargoes of the East: The Ports, Trade and Culture of the Arabian Seas and Western Indian Ocean*, Elm Tree Books, London, 1978

MATTHEWS, H. Freeman (Chargé d'Affaires, US Embassy), letter to C. W. Baxter (Foreign Office), 8 February 1943, classification FO 371/34900 139378, BSA

McLAUGHLIN, W. I., letter to W. F. Taylor, c/o Ottoman Bank, Baghdad, Iraq, 27 May 1930, CCL

MEMORANDUM of a Conference held at the Colonial Office, London, on 10 May 1929, F. B. Loomis *Bahrein 1928-1929* file, CCL

MEMORANDUM re L. N. Hamilton's cable to E. A. Skinner of 28 October 1932, repeated to M. E. Lombardi, written in San Francisco on 29 October 1932 ,CCL

MEMORANDUM of a discussion at the India Office, (Confidential P.Z. 6687/34) on 26 October 1934, between Mr Fraser of the Anglo-Persian Oil Company, and Messrs Walton and Laithwaite, POWE 33/194, BSA

MEMORANDUM re *Political Agreement* between His Majesty's Government, UK, and the Bahrein Petroleum Company Limited, dated the second day of August one thousand nine hundred and forty [2 August 1940], signed sealed and delivered by Hamilton Richard Ballantyne on behalf of Bapco Limited

MEMORANDUM OF ASSOCIATION, the Bahrain Petroleum Company BSC (closed), A Bahrain Joint Stock Company, made in accordance with the approval of The Registry of Commerce Control as per letter issued on 28 March 1981 under No. SH/18/81/1. (*See also* BAHRAIN REFINERY PARTICIPATION AGREEMENTS, 4 May 1981)

MERCER, Derrik, editor, *Chronicle of the 20th Century*, published for Chronicle Communications London by Longman, UK, 1988

MIDDLE EAST ECONOMIC SURVEY, Vol. XXIII, No. 6, 26 November 1979 and Vol. XXX, No. 32, 18 May 1987, Middle East Economic Survey, Cyprus

MIKDASHI, Zuhayr M., CLELAND, Sherrill, SEYMOUR, Ian, editors, *Continuity and Change in the World Oil Industry*, The Middle East Research and Publishing Center, Beirut, Lebanon, 1970

MILLER, R. P., letter to G. C. Gester, weekly report No. 2, summarizing drilling operations at Jebel Ad-Dukhan No. 1 well for the week ending 2 June 1932

MINING LEASE between Sheikh Hamad bin Sheikh Issa Al Khalifah, Sheikh of Bahrein and the Bahrein Petroleum Company Ltd, (covering 100,000 acres of Bahrein Island), Articles I to XXIII, 29 December 1934. (*See also AGREEMENT* 3 June 1936, *DEED OF FURTHER MODIFICATION* 19 June 1940 and *SUPPLEMENTAL AGREEMENT* 8 December 1952)

MINUTES OF MEETINGS held on Monday, 2 November 1981 at Ministry of Development and Industry Conference Room, Government House, Manama; Wednesday, 11 November 1981 in the Awali Management Conference Room of Bapco BSC (closed); Monday, 30 November 1981, at Awali, re Bahrain Oil Field – Phase-In of Banoco Personnel Into Operations. (*See also* FILE MEMORANDUM, 13 December 1981)

MOBIL ANNUAL REPORTS for 1975, 1979, Mobil Corporation, New York

MOFFETT, J. A. (Vice President, Standard Oil Company of California), cable to M. E. Lombardi, 23 May 1934, CCL

MONTEAGLE, Lord (for the Secretary of State, Foreign Office, London), letter to Ray Atherton, US Chargé d'Affaires in London, 29 May 1929, CCL

MONTEAGLE, Lord, letter from the Foreign Office to the Colonial Office, 2 April 1929, classification CO C. 69035/29 [No. 21], BSA

MONTEAGLE, Lord, letter from the Foreign Office to the Colonial Office, 8 August 1929, classification CO C.69035/29 [No. 101], BSA

MOODY'S INDUSTRIAL MANUAL, 1987/Vol. 2, 1988/ Vol. 1 A-I/Vol. 2 J-Z, published by Moody's Investors Service Inc. (a company of Dun and Bradstreet Corporation), New York, USA

MOORE, Frederick Lee, *Origin of American Oil Concessions in Bahrein, Kuwait and Saudi Arabia*, an unpublished thesis presented to the School of Politics and International Affairs, Princeton University, 1948. (Also available for reference in CCL)

MOORE, W. F., "Yesterday – A Sun-Baked Desert; Today – A $140,000,000 Industry; Tomorrow – A 600,000-barrel daily production", *The Oil and Gas Journal*, 27 December 1947, pp. 162-5, 233-4 and 236

MORRIS, Richard B., editor, *Encyclopedia of American History*, Harper & Row, New York, London and Sydney, 6th edition, 1982

MOSLEY, Leonard, *Power Play: The Tumultuous World of Middle East Oil 1890-1973*, Weidenfeld and Nicolson, London, 1973

MURRAY, Wallace (Chief, Division of Near Eastern Affairs, Department of State, Washington DC), letter to F. B. Loomis, 4 January 1935, CCL

MURRAY, Wallace, letter to F. B. Loomis, 30 April 1935, enclosing abstracts from the British Parliamentary Debates, House of Commons, 8 April 1935, London, columns 775-6, CCL

N

NAWWAB, Ismail I., SPEARS, Peter C., and HOYE, Paul F., editors, *Aramco and its World*, Arabian American Oil Company, Washington DC, 1980

NEW YORK HERALD TRIBUNE, "Bahrain: The Enchanting Isles of the Arabian Gulf", 16 May 1965 (16-page feature)

NEW YORK HERALD TRIBUNE, "Far East Trade in Oil Sought by Texas Corp.", 12 June 1936, also in CCL

NEW YORK TIMES, "Will Tap New Oil Field" (no credited author), 1 January 1934, also in CCL

NEW YORK TIMES, "Tanker El Segundo Sails for

Bahrein" (no credited author), 14 January 1934, also in CCL

NEW YORK TIMES, "Persian Gulf Oil Finds New Outlet" (no credited author), 8 September 1935, also in CCL

NEW YORK TIMES, "Edward Skinner, Oil Official Dies", 25 November 1956

NIBLOCK, Tim, and LAWLESS, Richard, editors, *Prospects for the World Oil Industry*, Proceedings of a Symposium on the Energy Economy co-sponsored by the Petroleum Information Committee of the Arab Gulf States and the University of Durham, UK, and held in Durham, 9-10 May 1984, published by Croom Helm, London, Sydney and Dover, New Hampshire, 1985

NOTE OF A CONFERENCE between the Colonial Office and the Eastern and General Syndicate Limited, 8 August 1929, classification CO C.69035/29 [No. 102], BSA

NOTE OF AN INTERVIEW at the Colonial Office with the Representatives of the Eastern and General Syndicate, 6 February 1930, classification CO C.79035/30 [Nos. 13-19], BSA

NOTE OF AN INTERVIEW with the Representatives of Messrs Holmes, Son and Pott and Messrs. Freshfields, 17 February 1930, classification CO C.79035/30 [Nos. 13-19], No. 52, BSA

NUGENT, Jeffrey B., and THOMAS, Theodore H., editors, *Bahrain and the Gulf: Past Perspectives and Alternative Futures*, Croom Helm, London/Sydney, 1985

NUTTALL, W. L. F. (Ministry of Fuel and Power), letter to Chadwick, 6 December 1948, classification FO 371/68333, reference E 15009/327/91, BSA

O

OAKLAND TRIBUNE, "Japan Gets More Oil" (no credited correspondent), 20 April 1939

OAPEC NEWS BULLETIN, Vol. 3, No. 12, December 1977, special supplement (unpaginated): "The Arab Shipbuilding and Repair Yard Co. on the occasion of the inauguration of the A.S.R.Y. drydock December 15, 1977"

ODELL, Peter R., *Oil and World Power*, Penguin Group, London/New York/Ontario/Sydney/

Auckland, eighth edition, 1986

OIL AND GAS JOURNAL, THE, "O'Mahoney Selected to Head Senate Oil Group", (correspondent not credited), 17 February 1945

OIL AND GAS JOURNAL, THE, "O'Mahoney Committee to Hear Impressive List of Witnesses" (correspondent not credited), 16 June 1945, p. 91

OIL AND GAS JOURNAL, THE, 30 June 1945. (*See also* RALPH, Henry D., report published in the same issue)

OIL AND GAS JOURNAL, THE, "O'Mahoney May Resume Hearings on Pipe Lines", (correspondent not credited), 14 July 1945

OIL AND GAS JOURNAL, THE, "Diplomatic Support Sought for Oil Operators Abroad" (correspondent not credited), 27 October 1945, p. 76

OIL AND GAS JOURNAL, THE, "PAW, Its Task Finished, Gives Account of Its Stewardship" (correspondent not credited), 1 December 1945, p. 48

OIL AND GAS JOURNAL, THE, 8 December 1945. (*See also* RALPH, Henry D., report published in the same issue)

OIL AND GAS JOURNAL, THE, "International Section", 29 December 1945

OIL AND GAS JOURNAL, THE, "Saudi Arabia and Kuwait – Expansion" (correspondent not credited), 28 December 1946, pp. 171, 175-6 and 213

OIL AND GAS JOURNAL, THE, "Survey of World Oil Refineries", 28 December 1946, pp. 197, 199, 201, 203, 207 and 209. (*See also* DEEGAN and BURNS, report published in the same issue)

OIL AND GAS JOURNAL, THE, "Survey of World Oil Refineries", 27 December 1947, pp. 200, 203, 206, 207, 209, 211 and 213 (*See also* DUFF, Dahl M. and MOORE, W. F., reports published in the same issue)

OIL AND GAS JOURNAL, THE, "Bahrain Halts Long Production Decline", (no credited correspondent), Middle East Report, 26 June 1978, pp. 132-42

OIL AND GAS JOURNAL, THE, "Arab Shipyard Marks 3 Years of Service", 10 November 1980, p.164

OIL AND PETROLEUM YEAR BOOKS

OIL CONCESSIONS IN THE MIDDLE EAST, Volumes I and II, CCL

OILFIELD EQUIPMENT CO. INC. (President [T. E. Ward]), letter to Major Frank Holmes, 14 October 1927, T. E. Ward Jr's personal papers

OLDFIELD, K. (British Political Agent in Bahrain), letter 1535/64 to L. D. Josephson, re amendments to the *Political Agreement* between the British Government and Bapco dated 29 June 1940, 19 August 1964.

O'MAHONEY COMMITTEE, see specified entries under *OIL AND GAS JOURNAL, THE,* and RALPH, Henry D.

ONE HUNDRED YEARS HELPING TO CREATE THE FUTURE [1879-1979], (no credited author), published by the Standard Oil Company of California, CCL

OPTION COVERING BAHREIN CONCESSION from Eastern and General Syndicate Limited to Eastern Gulf Oil Company, 30 November 1927

OWEN, Edgar Wesley, *Trek of the Oil Finders: A History of Exploration for Petroleum*, The American Association of Petroleum Geologists, Semicentennial Commemorative Volume, Tulsa, Oklahoma, USA, March 1975

P

PACIFIC COAST WALL STREET JOURNAL, "S. O. Developing Far East Field" (no credited author), 5 February 1934, also in CCL

PACIFIC COAST WALL STREET JOURNAL, "California Standard, After Four Years of Work, to Ship First Bahrein Crude Shortly" (no credited author), 18 April 1934, also in CCL

PACIFIC COAST WALL STREET JOURNAL, "Bahrein Oil Lease Attacked" (no credited author), 27 June 1934, also in CCL

PACIFIC COAST WALL STREET JOURNAL, "Standard Ships Bahrein Oil" (no credited author), 10 September 1934, also in CCL

PACIFIC COAST WALL STREET JOURNAL, "First of S.O.'s Bahrein [sic] Oil Coming Next Month" (no credited author), 25 May 1934, also in CCL

PACIFIC COAST WALL STREET JOURNAL, "Bahrein Oil Parley Seen Inconclusive" (no credited author), 1 November 1935, also in CCL

PACIFIC COAST WALL STREET JOURNAL, "Bahrein Oil Outlet Studied" (no credited author), 21 September 1936, also in CCL

PACIFIC COAST WALL STREET JOURNAL, "Italian Flyers Claim Record Raid" (Rome correspondent), 21 October 1940

PARSONS, Anthony, *They Say the Lion*, Jonathan Cape, London, 1986

PARTICIPANTS' AND OPERATING AGREEMENT, see BAHRAIN REFINERY PARTICIPATION AGREEMENTS, May 1981

PASSFIELD, Lord, letter from the Secretary of State to the PRPG, 15 January 1920, classification CO C.79035/30 [No. 5], BSA

PASSFIELD, Lord, letter from the Secretary of State to the PRPG, 13 March 1930, classification CO C.79035/30 [No. 41], enclosing a draft of a proposed Indenture between the Ruler of Bahrain and the Eastern and General Syndicate Limited

PASSFIELD, Lord, paraphrased telegram from the Secretary of State to the PRPG, 1 April 1930, classification CO C.79035/30 [No. 61], BSA

PASSFIELD, Lord, confidential telegram from the Secretary of State to the PRPG, 1 May 1930, classification CO C.79035/30 [No. 72], BSA

PATRICK, P. J. (India Office), letter to the Colonial Office, 23 August 1929, classification CO C.69035/29 [No. 122], BSA

PAYTON-SMITH, D. J., *Oil – A Study of Wartime Policy and Administration* (History of the Second World War), HMSO, London, 1971

PELLY, C. J. (British Political Agent), letter C10/1/11/51 to E. A. Skinner, 4 July 1951, B. M. Van Benschoten's personal papers

PENROSE, Edith T., *The Large International Firm in Developing Countries: The International Petroleum Industry*, Greenwood Press, Westport, Connecticut, USA, 1968

PENROSE, Edith, "Oil and State in Arabia", Ch. 13, pp. 271-85 of *The Arabian Peninsula: Society and Politics*, ed. D. Hopwood, Allen and Unwin, London, 1972

PETERSON, Maurice (Foreign Office), letter to

H. Freeman Matthews, Chargé d'Affaires, US Embassy, London, 18 March 1943, classification FO 371/34900 139378, BSA

PETROLEUM AND ENERGY INTELLIGENCE WEEKLY [PIW], "PIW Ranks World's Top 50 Oil Companies", PIW Special Supplement Issue, 12 December, 1988

PETROLEUM ECONOMIST, "Bahrain Dockyard Scheme Inaugurated" (no credited correspondent), January 1975, p. 30

PETROLEUM ECONOMIST, "Plans for Future Without Oil", (no credited correspondent), November 1976, pp. 427-9

PETROLEUM ECONOMIST, "Bapco Reports Progress" (no credited correspondent), August 1977, pp. 322-3

PETROLEUM ECONOMIST, "Looking to Gas for the Future" (no credited correspondent), January 1981, pp. 13-14

PETROLEUM ECONOMIST, news clip on Bahrain refinery shutdown during February, reported in the April 1983 edition (unpaginated)

PETROLEUM ECONOMIST, "Problems for Bapco Refinery" (no credited correspondent), November 1985, p. 412

PETROLEUM PRESS SERVICE, "The Search for Oil in the near and Middle East" (no credited correspondent), 30 June 1939, pp. 313-16

PETROLEUM PRESS SERVICE, "Bahrein as a Refining Centre" (no credited correspondent), November 1947, pp. 101-2

PETROLEUM PRESS SERVICE, "Middle East: Costs and Supplies" (no credited correspondent), December 1951, pp. 401-2

PETROLEUM PRESS SERVICE, "The Persian Gulf" (no credited correspondent), No. 5, May 1953, pp. 161-7

PETROLEUM PRESS SERVICE, "Prosperous Bahrain" (no credited correspondent), August 1954, pp. 298-9

PETROLEUM PRESS SERVICE, "The Story of Oil in the Middle East" (no credited correspondent), September 1954, pp. 331-4

PETROLEUM PRESS SERVICE, "New Plants at Bahrain Refinery" (no credited correspondent), August 1955, p. 302

PETROLEUM PRESS SERVICE, "Bahrain Refinery Expansion Modified" (no credited correspondent), September 1956, p. 347

PETROLEUM PRESS SERVICE, "Busy Bahrain" (no credited correspondent), August 1959, pp. 314-15

PETROLEUM PRESS SERVICE, "Bahrain: Trading Centre" (no credited correspondent), July 1958, pp. 268-9

PETROLEUM PRESS SERVICE, news clips on Bahrain: August 1960, p. 310; December 1960, p. 470; July 1961, p. 274; May 1962, p. 194; June 1963, p. 234; May 1964, p. 194

PETROLEUM PRESS SERVICE, "Small But Prosperous" (no credited correspondent), June 1965, p. 230

PETROLEUM PRESS SERVICE, "Shared Offshore Field in Production" (no credited correspondent), February 1966, pp. 66-7. News clips on Bahrain: July 1966, p. 274; June 1967, p. 234

PETROLEUM PRESS SERVICE, "Bahrain and Qatar – Increased Production" (no credited correspondent), June 1968, p. 230

PETROLEUM PRESS SERVICE, "Aluminium and Refining" (no credited correspondent), December 1969, pp. 470

PETROLEUM PRESS SERVICE, "Concession for Superior" (no credited correspondent), February 1971, pp. 67 and 69

PETROLEUM PRESS SERVICE, "Bapco in 1970" (no credited correspondent), June 1971, p. 231. News clip, June 1972, p. 214

PETROLEUM PRESS SERVICE, "Dry-Dock Project Approved" (no credited correspondent), January 1973, p. 31

PETROLEUM RELATIONS IN BAHRAIN, Vol. I, 1925-1971 (Arabic text only), Ministry of Development and Industry, State of Bahrain

PETROLEUM TIMES, THE, "Bahrein Now an Isle of Opportunity for Britishers", 9 July 1938, pp. 37-9

PETROLEUM TIMES, THE, "World Refineries Survey", 1985

PILGRIM, Guy E., report, 9 June 1905, classification R/15/1/317, IOR

PILGRIM, Guy E., "The geology of the Persian

Gulf and the adjoining portions of Persia and Arabia", *Geological Survey of India Memoir*, 1908

PILLSBURY, MADISON & SUTRO (Socal lawyers), letter to F. B. Loomis, 13 March 1930, CCL

POCOCK, William (Secretary of the Eastern and General Syndicate), letter to T. E. Ward, 31 January 1934, T. E. Ward Jr's personal papers

POCOCK, William, letter to the Bahrein Petroleum Company Limited, 2 August 1934, T. E. Ward Jr's personal papers

POCOCK, William, letter to T. E. Ward, 15 November 1934, T. E. Ward Jr's personal papers

POCOCK, William, letter to H. R. Ballantyne (lawyer acting for Bapco Limited in London), 15 November 1934, T. E. Ward Jr's personal papers

POLITICAL AGENT (Bahrain), letter No. 90 to PRPG, 15 July 1929, enclosed in PRPG's letter to the Secretary of State, classification CO C.69035/29 [No.111], BSA

POLITICAL AGENT (Bahrain), secret telegram to Secretary of State for India, 11 May 1942, classification FO 371/31318 XL 141812, BSA

POLITICAL AGREEMENT between His Majesty's Government in the United Kingdom and the Bahrein Petroleum Company Limited, 29 June 1940

POLITICAL RESIDENT IN THE PERSIAN GULF, telegram to the Secretary of State for India, received in the India Office on 27 April 1933, copy received by the Colonial Office, 1 May 1933, classification CO 18135/33 [No. 84], BSA

POLITICAL RESIDENT IN THE PERSIAN GULF, secret telegram to the Secretary of State for India, 2 May 1942, classification FO 371/31318 XL 141812, BSA

POLITICAL RESIDENT IN THE PERSIAN GULF, telegram to Secretary of State for India, 27 April 1943, classification FO 371/34900 139378, BSA

PRIOR, C. G. (Political Agent, Bahrain), "Administration Report of the Bahrain Agency for the Year 1931", *ARPGPR for the Year 1931*, Government of India Press, Simla, 1932, Ch. VIII

PRIOR, C. G., letter to Bapco's Chief Local Representative, 23 April 1932, CCL

PRIOR, C. G., letter No. C/5 of 1932 to Political Resident Persian Gulf, re Eastern and General's application for an additional concession, classification CO 935/7, BSA

R

RALPH, Henry D., "Witnesses at O'Mahoney Hearing Plead for Removal of Controls and Federal Competition", *The Oil and Gas Journal*, 30 June 1945, pp. 76-8

RALPH, Henry D., "Industry Can Thrive Only Under Stimulus of Free Enterprise, Committee Is Told", *The Oil and Gas Journal*, 8 December 1945, pp. 56-8 and 75

RALPH, Henry D., "US Foreign Oil Policy", *The Oil and Gas Journal*, 27 December 1947, pp. 220-7, 275-8 and p. 281

RAYNER, Charles (Petroleum Adviser), *State Department Memorandum on US Petroleum Policy*, 10 February 1944, prepared for the Special Senate Committee (Truman Committee) investigating the National Defense Program Report No. 10, Part 15, 78th Congress, 2nd session, Government Printing Office, pp. 70-6

RECORD OF A DISCUSSION between the Colonial Office and the Representatives of the Eastern and General Syndicate held on 19 July 1929 and dated 20 July 1929, classification CO C.69035/29 [No. 86A], BSA

RECORD OF A DISCUSSION between the Colonial Office and Representatives of the Eastern and General Syndicate, 31 July 1929, classification CO C.69035/29 [No. 91A], BSA

REFINERY PARTICIPATION AGREEMENTS, (*see BAHRAIN REFINERY PARTICIPATION AGREEMENTS*)

RENDEL, G. W., letter from the Foreign Office to the Colonial Office, re the special treaty relations existing between His Majesty's Government and the Ruler of Bahrain, 4 September 1929, classification CO C.69035/29 [No. 124], BSA

REPORT OF AN INTERDEPARTMENTAL CONFERENCE, 7 June 1929, classification CO C.69035/29 [No. 68], BSA

REVUE PETROLIFÈRE, LA, English translation, "Oil Prospecting in the Iranian (Persian) Gulf Region", 9 November 1935, p. 1455, also in CCL

REVUE PETROLIFÈRE, LA, English translation, "Wells of the Bahrein Petroleum Company", 20 June 1936, p. 923, also in CCL

REVUE PETROLIFÈRE, LA, English translation, "Increase in Production", 20 March 1937, p. 390, also in CCL

REVUE PETROLIFÈRE, LA, English translation, "Oil Shipments from Bahrein Island", 12 November 1937, p. 1589, also in CCL

REVUE PETROLIFÈRE, LA, English translation, "Trip of One of Ibn Saud's Advisers to Germany", 7 July 1939, p. 913, also in CCL

RHOADES, R. O., *Report on Bahrein Island, Persian Gulf*, 1928, CCL

RHOADES, R. O., letter to William T. Wallace in New York, 10 April 1928, CCL

RHOADES, R. O., letter to Major Frank Holmes, 19 April 1928, Ward, 1965, pp. 67-8

RIHANI, Ameen, *Maker of Modern Arabia*, Houghton Mifflin Company, Boston, Massachusetts, USA, 1928

ROENIGK, Allison F., "Good Neighbor Idea Works Well in Bahrain", *World Petroleum*, September 1955, pp. 102-5

ROUNDABOUT, the magazine for Caltex Staff and their families, "The Caltex Connection" (no credited author), September 1986

RUSTOW, Dankwart A., and MUGNO, John F., *OPEC Success and Prospects*, A Council on Foreign Relations Book, New York University Press, New York, 1976

S

SALDANHA, J. A. "Précis of Bahrein Affairs, 1854-1904", *The Persian Gulf Précis, Volume IV*, first printed at Calcutta and Simla, 1903-8, reprinted by agreement with the British Library from original works in the India Office Library and Records, London, by Archive Editions, Gerrards Cross, Buckinghamshire, UK, 1986

SAMPSON, Anthony, *The Seven Sisters: The Great Oil Companies and the World They Made*, Viking Press, New York, 1975

SAN FRANCISCO CHRONICLE, "S.F. Ship Set to Link Persia Isle to World" (no credited author), 13 December 1933, CCL

SAN FRANCISCO CHRONICLE, "Standard Oil to Enter Asiatic markets with Bahrein Crude Products" (no credited author), 15 November 1935, also in CCL

SAN FRANCISCO EXAMINER, "Standard Oil Venture Pays" (no credited author), 3 February 1934, also in CCL

SANDARS, N. K., translator, *The Epic of Gilgamesh* (English version), Penguin Books, London, 1972

SCHMIDT, W. A. (Bapco's Chief Local Representative), letter MISC. CON-25 to the Ruler of Bahrain, re amendment of Bapco's agreement letter CON-937 of 29 November 1955 to substitute 8 1/2 pence sterling for 7 pence sterling, 8 February 1961. (*See also* AL-KHALIFAH acknowledgement, 8 February 1961)

SCHMIDT, W. A., letter MISC. CON-601 to the Ruler of Bahrain, re Bapco's study of "the means by which the Ruler might derive some additional revenue from the sale of products refined by the company for consumption in Bahrain without imposing an additional burden upon the people of Bahrain who consume those products", 27 October 1962

SCHMIDT, W. A., letter MISC. CON-602 to the Ruler of Bahrain, re proposal to accommodate the requirement in CON-601 (above), 27 October 1962

SCHMIDT, W. A., letter MISC. CON-792 to the Ruler of Bahrain, re letter MISC. CON-717 of 10 December 1962 which acknowledged receipt of the Ruler's letter dated 2 December 1962 and acceptance of the Company's proposal in its letter MISC. CON-602 dated 27 October 1962, 10 February 1963. (*See also* SCHMIDT, MISC. CON-601 and CON-602)

SCHMIDT, W. A., letter MISC. CON-793 to the Ruler of Bahrain, re taxation of refined products consumed in Bahrain, 10 February 1963

SCHMIDT, W. A., letter MISC. CON-796 to the Ruler of Bahrain, re relinquishment of certain offshore areas under the terms of the *Mining Lease* and the *Deed of Further Modification*, 10 February 1963

SCHMIDT, W. A., letter M-CON-170 to the Ruler of Bahrain, confirming an agreement re the prices at which Bapco makes sales of refined products for export, 25 November 1963

SECCOMBE, Ian, and LAWLESS, Richard, *Work Camps and Company Towns: Settlement Patterns and the Gulf Oil Industry*, Occasional Paper No. 36, Centre for Middle Eastern and Islamic Studies, University of Durham, UK, 1987

SECRETARY OF STATE (Foreign Office, London), letter from Secretary Amery to the Embassy of the United States of America, London (attention of Chargé d'Affaires, Ray Atherton), 29 May 1929, classification CO C.69035/29 [No. 53], BSA

SECRETARY OF STATE, secret telegram to PRPG, 26 April 1942, classification FO 371/31318 XL 141812, BSA

SECRETARY OF STATE TO THE GOVERNMENT OF INDIA, External Affairs Department, secret telegram to India Office, 30 April 1943, classification FO 371/34900 139378, BSA

SECRETARY OF STATE TO THE GOVERNMENT OF INDIA, External Affairs Department, secret telegram to India Office, 14 July 1943, classification FO 371/34900 139378, BSA

SELL, G., "Twenty-Five years of Middle East Oil: Bapco Operations in Bahrain", *Institute of Petroleum Review*, No. 142, Vol. 12, 1958, pp. 333-6

SEYMOUR, H. J. (Foreign Office), letter to the Colonial Office, 6 April 1929, enclosing a memorandum to the US Embassy (London) dated 3 April 1929, classification CO C.699035/29 [No. 22], BSA

SEYMOUR, Ian, *OPEC Instrument of Change*, Macmillan Press, London, 1980

SHAW, G. Howland, (Chief, Division of Near Eastern Affairs, Department of State, Washington), letter to William Wallace, 3 June 1929, CCL

SHIRAWI, May Al-Arrayed, *Education in Bahrain: Problems and Progress*, Ithaca Press, Oxford, 1989

SHIRAWI, Y. A. (in his capacity as Chairman of the Bahrain National Oil Company), letter to the President, the Bahrain Petroleum Company, 24 June 1976, Bapco archive

SHIRAWI, Y. A. (signing for the Government of Bahrain), letter from the Ministry of Development and Industry to the President of Bapco, re the 13.444% "anchor", 15 December 1979, Bapco archive

SHIRAWI, Y. A. (signing for the Government of Bahrain), letter from the Ministry of Development and Industry to the President of Bapco, re future consultation with the Government concerning the Bahrain refinery, 15 December 1979, Bapco archive

SHIRAWI, Y. A. (signing for the Government of Bahrain), letter from the Ministry of Development and Industry to the President of Bapco, re the agreement entered into that day whereby the Government acquired the remaining 40% interest in the Bahrain oilfield, 15 December 1979, Bapco archive

SHIRAWI, Y. A. (signing as Minister of Development and Industry), letter to the President of Bapco, re three areas requiring review (not stated in the letter), 31 December 1979, Bapco archive

SHIRAWI, Y. A. (in his capacity as Chairman of the Bahrain Petroleum Company BSC (closed)), memorandum to all personnel of the Producing and Development Division, 10 December 1981, informing them that effective on 1 January 1982, Banoco would assume full responsibility for the management and conduct of the onshore oil and gas producing operations in Bahrain carried out by Caltex Bahrain Limited

SIDE LETTER NO. 3 – INTERIM OPERATING AGREEMENT, SIDE LETTER – PRICING PROCEDURES, referring to Articles 3 and 4 of the *Agreement* between the Government of the State of Bahrain and the Bahrain Petroleum Company Limited dated 23 November 1974 and to the *Annex to the Side Letter – Outstanding Issues*, signed on the 1st day of December 1979 by Y. A. Shirawi for the Government of the State of Bahrain and R. N. Trackwell for the Bahrain Petroleum Company Limited. (*See also* STOLZ, 23 November 1974)

SKINNER, E. A., cables to M. E. Lombardi, 2 and 14 November 1932, CCL

SKINNER, E. A., letter to M. E. Lombardi, 28 November 1932, CCL

SKINNER, E. A., cable to M. E. Lombardi, 13 December 1932, CCL

SKINNER, E. A., letter to L. N. Hamilton, 20 December 1932, CCL

SKINNER, E. A., letter to H. R. Ballantyne, 21 December 1932, CCL

SKINNER, E. A., telegram to Bapco, London, received on 9 May 1933 (no record of despatch date), CCL

SKINNER, E. A., telegram to W. F. Vane, 12 May 1933, CCL

SKINNER, E. A., cables to M. E. Lombardi, 17 and 20 November 1933, CCL

SKINNER, E. A., letter to H. M. Herron, Bapco, New York, 23 March 1939.

SKINNER, Walter E., *Oil and Gas International Year Book* for the years 1974, 1975/76, 1976/77, 1977/78, published by the Financial Times, London

SKINNER, Walter E., *The Oil and Petroleum International Year Book*, 1971-2, M & O Ltd, Cannon Street, London

SKINNER, Walter E., *The Oil and Petroleum Manual* for 1923, 1927, M & O Ltd, Cannon Street, London

SKINNER, Walter E., *The Oil and Petroleum Year Book* for 1940, 1966, 1968, M & O Ltd, Cannon Street, London

SMITH, Felix T., (lawyer at Pillsbury, Madison & Sutro), letter to M. E. Lombardi, 9 December 1930, CCL

STANDARD OIL BULLETIN (published quarterly by the Standard Oil Company of California), maintained in CCL: "Oilfield Operations on Bahrein Island" (quotation from Manager E. A. Skinner), Vol. XXI, No. 3, July 1933, pp. 8-9; "Petroleum Prospects in the Persian Gulf" (no credited author), Vol. XXI, July 1933, pp. 3, 5, 14-16

STANDARD OIL BULLETIN, "Bahrein Development Approaching Commercial Stage" (no credited author), Vol. XXI, January 1934, No. 9, p. 1

STANDARD OIL BULLETIN, "And Now – A Pipe-Line for Bahrein", (no credited author), Vol. XXI, April 1934, No. 12, pp. 10-13

STANDARD OIL BULLETIN, "Bahrein Oil Starts to Market" (no credited author), August 1934, Vol. XXII, No. 4, August 1934, pp. 2-10, 16

STANDARD OIL BULLETIN, "Another Step in the Bahrein Development" (no credited author), Vol. XXIII, No. 7, November 1935, p. 1

STANDARD OIL BULLETIN, "Company Officers Advanced" (no credited author), Winter 1948

STANDARD OIL BULLETIN, "Bahrein, Treasure Island of the Persian Gulf" (no credited author), Vol. XXXIV, No. 6, April 1950, pp. 3-7, CCL

STANDARD OIL COMPANY OF CALIFORNIA, unsigned letter to Robert Skinner, US Ambassador to Turkey, 12 February 1935, CCL

STANDARD OILER, THE, (corporate magazine of the Standard Oil Company of California), "In Distant Places", Vol. 1, No. 6, May 1939, pp. 8-9, CCL

STANDARD OILER, THE, Vol. 1, No. 11, October 1939, p. 14, CCL

STANDARD OILER, THE, Vol. 1, No. 12, November 1939, p. 15, CCL

STANDARD OILER, THE, Vol. 9, No. 5, May 1947, p. 20, CCL

STANDARD OILER, THE, Vol. 9, No. 7, July 1947, inside back cover, CCL

STANDARD OILER, THE, Vol. 9, No. 8, August 1947, pp. 14, 19, 21, CCL

STANDARD OILER, THE, Vol. 10, Nos. 1-2, February 1948, pp. 1-3, CCL

STANDARD OILER, THE, "The Eastern Half of the Globe" (no credited author) Vol. 10, No. 3, March 1948, pp. 1-4, 21-4, CCL

STANDARD OILER, THE, Vol. 12, No. 1, January 1950, pp. 4, 14, 22, 24, CCL

STANDARD OILER, THE, "The Man Behind Bahrain's Discovery" [re Edward Allen Skinner], July 1957, Vol. 19, No. 7, pp. 17-19 and inside back cover, CCL

STANDARD OILER, THE, "The Oil Discovery that made World History", July 1957, Vol. 19, No. 7, pp. 14-17

STANDARD OILER, THE, "Oil Under that Desert? Ridiculous!", Vol. 20, No. 3, March 1958, pp. 12-16

STARLING, F. C. (Petroleum Department), letter to the Colonial Office, 11 October 1930, classification CO C.79035/30 [No. 128], BSA

STATEMENT FOR THE PRESS issued by Standard Oil Company of California, San Francisco on behalf of K. R. Kingsbury (President of Socal) and Captain T. Rieber (Chairman of the Texas Corporation), dated 26 June 1936, announcing the formation of The California Texas Oil Company Limited, CCL

STEVENS, Paul, editor, *Oil and Gas Dictionary*, Macmillan Press, London, and Nichols Publishing, New York, 1988

STOLZ, W. O. (President, Bapco Limited), letter [MISC. CON-558] to the Minister of Finance and National Economy (His Excellency Sayed Mahmoud Ahmed Al-Alawi), 23 November 1974, referring to the *Agreement* between the Government of the State of Bahrain and the Bahrain Petroleum Company Limited (same date) and enclosing 4 Side Letters concerning: Outstanding Issues, Gas, Definition and Refinery

STOLZ, W. O., letter to J. M. Voss, 13 April 1975, re Berry Wiggins overriding royalty. (*See also* VOSS, J. M., 8 April 1975, *and WAIVER OF NOTICE*, 25 March 1975)

STRATEGIC PLAN – BAHRAIN REFINERY MODERNISATION AUGUST 1986, Bapco BSC "A Refining Company"

SUN AND FLARE, "Board Chairman Marks 35th Year" [re Fred Davies], 15 May, 1957, Vol. XIII, No. 20 (published by Aramco in Dhahran)

SUNDAY TIMES, THE, "Japan Making Treaty with Arabs – Oil Concession Hopes", Cairo correspondent, 9 April 1939

SUPPLEMENTAL AGREEMENT to the *Mining Lease* dated 29 December 1934, the *Deed* dated 3 June 1936 (subsequently referred to as the first *Supplemental Deed*) and the *Deed of Further Modification* dated 19 June 1940, made between His Highness Sheikh Sir Sulman bin Hamad Al Khalifah, Ruler of Bahrain and the Bahrein Petroleum Company Limited, signed and sealed on *8 December 1952*. (*See also MINING LEASE* dated 29 December 1934, *AGREEMENT* Modifying Mining Lease of 29 December 1934 (the *Supplemental Deed* dated 3 June 1936), and *DEED OF FURTHER MODIFICATION*, 19 June 1940)

T

TARBUTT, Percy, letter to T. E. Ward, 1 July 1936, T. E. Ward Jr's personal papers

TAYLOR, W. F. (Bill), letters to W. I. McLaughlin, 7, 28 and 30 April 1930, CCL

TAYLOR, W. F. (Bill), letters to W. I. McLaughlin, 7 and 30 May 1930, CCL

TAYLOR, W. F. (Bill), letter to W. I. McLaughlin, 7 June 1930, CCL

TAYLOR, W. F. (Bill), letter to R. C. Stoner, 11 June 1930, CCL

TAYLOR, W. F. (Bill), letter to W. I. McLaughlin, 14 June 1930, CCL

TAYLOR, W. F. (Bill), letter to R. C. Stoner, 2 July 1930, CCL

TAYLOR, W. F. (Bill), and DAVIES, F. A. (Fred), telegram to W. I. McLaughlin, 15 May 1930, CCL

TAYLOR, William F., and DAVIES, Fred A, "We Travel 'East of Suez'", *Among Ourselves*, December 1930, p. 3, CCL

TELEGRAMS from the Viceroy, Foreign and Political Department to the Secretary of State for India, No. 1656S and No. 1657S, 12 May 1929, classification CO C.69035/29 [No. 45], BSA

TEXACO INC. ANNUAL REPORTS

TEXACO STAR, THE, "East of Suez" (no credited author), No. 3, 1936, pp. 15-16

TIETZE, Emil, *Die Mineralreichthümer Persiens*, K.K. Geol. Reichsanstalt Jahrb., v. 29, no. 4, 1879, p. 592

TOMPKINS, J. D., "Standard Oil of California in Bahrein Deal", *New York Herald Tribune*, 10 May 1936, CCL

TRIPP, J. P., a lecture series presented in 1966, and papers lent from the author's private collection:

"The Arab States of the Gulf since 1947: Their Evolution and Development with Particular Reference to their Neighbours, Arab Nationalism and the British presence"

"Economic Realities in the Middle East and the Importance of Oil to the Arabs and the West"

"The Historical Evolution of British Interests in the Persian Gulf"

"Political Progress and Development in the Persian Gulf since 1947"

"Saudi Arabia's Place in the Arab World"

TUCHMAN, Barbara W., *Practicing History: Selected Essays*, Ballantine Books (a division of Random House Inc. New York), 1981

TUCKER, E. Stanley, "Demand for Oil: How Responsive to Price Change?", *Petroleum Economist*, London, November 1987, Vol. LIV, No. 11, pp. 402-10

TURNER, Louis, *Oil Companies in the International System*, published for the Royal Institute of International Affairs (London) by George Allen & Unwin, London and Sydney; Allen & Unwin Inc., Winchester, Massachusetts, USA, 3rd edition, 1983

TUSON, Penelope, *The Records of the British Residency and Agencies in the Persian Gulf*, India Office Library and Records/Foreign and Commonwealth Office, London, 1979

TUTTLE, J. H., memorandum to H. D. Collier, 30 July 1940 [re Captain Rieber visiting Tuttle, Lombardi and Davies in San Francisco], CCL

U

US SENATE, Ninety-Third Congress, Second Session: *Hearings before the Subcommittee on Multinational Corporations of the Committee on Foreign Relations*, 20-21 February, and 27-28 March 1974, Part 7

US STATE DEPARTMENT, secret letter to the Rt Hon. Anthony Eden, MC, MP, Secretary of State for Foreign Affairs, Foreign Office, London, 12 April 1943, classification FO 371/34900 139378, BSA

US STATE DEPARTMENT, secret letter to the Rt Hon. Anthony Eden, MC, MP, Secretary of State for Foreign Affairs, Foreign Office, London, 24 May 1943, classification FO 371/34900 139378, BSA

V

VANE, W. F., memorandum to F. B. Loomis, 26 December 1929, CCL

VANE, W. F., memorandum to M. E. Lombardi, 19 August 1930, CCL

VANE, W. F., memorandum to M. E. Lombardi, 14 November 1930, personal papers of the late Lt-Col. Arnold Galloway, UK, now maintained by Angela Clarke

VANE, W. F., letter to F. B. Loomis, 17 May 1932, CCL

VANE, W. F., memorandum to M. E. Lombardi dated 23 November 1932, re Bahrein and mainland matters, which includes a summary of L. N. Hamilton's 11 November letter to M. E. Lombardi, together with accompanying correspondence: E. A. Skinner's 2 November letter to M. E. Lombardi and E. A. Skinner's two letters of 2 November to L. N. Hamilton, CCL. Accompanying correspondence discusses:

Extension of Prospecting Licence to 2 December 1933

Second Year's Extension of Prospecting Licence

Additional Concession

Amendment of Clause IX of the Prospecting Licence

Compliance with Condition "C"

Information to the Colonial Office

Possible Alterations to the Mining Lease

VANE, W. F., memorandum to M. E. Lombardi dated 26 November 1932, re Bahrein and mainland matters, which includes a summary of E. A. Skinner's 7 and 8 November letters to L. N. Hamilton, E. A. Skinner's 4 November letter to M. E. Lombardi, L. N. Hamilton's 15 November letter to M. E. Lombardi, and F. B. Loomis' 23 November letter to M. E. Lombardi, CCL. Accompanying correspondence discusses:

Extension of Prospecting Licence

Additional Concession – Bahrein Island

Alteration of Clause IX

Compliance with Condition "C"

Possible Negotiations with Ibn Saud for Mainland Concession

VANE, W. F., memorandum to R. C. Stoner, re Supplemental Agreement with the Ruler of Bahrein, 15 February 1933, CCL

VANE, W. F., memorandum to W. H. Berg, re relations with the Eastern and General Syndicate, 20 March 1933, CCL

VANE, W. F., letter to E. A. Skinner, re the Additional Area in Bahrain, 12 May 1933, CCL

VANE, W. F., telegram to H. R. Ballantyne, 19 May 1933, CCL

VANE, W. F., telegrams to H. R. Ballantyne, 1 and 2 June 1933, CCL

VANE, W. F., cable to L. N. Hamilton, 21 November 1933, CCL

VANE, W. F., telegram to L. N. Hamilton, 8 March 1934, CCL

VANE, W. F., memorandum to M. E. Lombardi, 10 October 1934, CCL

VANE, W. F., memorandum to M. E. Lombardi, 28 December 1934, CCL

VENN, Fiona, *Oil Diplomacy in the Twentieth Century*, Macmillan Education Ltd, London, 1986

VOLNEY LEWIS, J., letter to T. E. Ward, 27 May 1927, T. E. Ward Jr's personal papers

VOSS, H. A., letter to T. E. Ward (typed on South American Gulf Oil Company letterhead), 24 May 1928, T. E. Ward Jr's personal papers

VOSS, J. M. (a Director of Bapco), letter to W. O. Stolz, President of Bapco in Bahrain, 8 April 1975, re settlement with Berry Wiggins with respect to overriding royalty. (*See also* STOLZ, 13 April 1975 *and WAIVER OF NOTICE*, 25 March 1975)

W

WAIVER OF NOTICE of a special meeting of the Board of Directors, The Bahrain Petroleum Company Limited, at 3.00 p.m. at 380 Madison Avenue, New York, on 25 March 1975, re settlement with Berry Wiggins with respect to overriding royalty claims. (*See also* STOLZ, 13 April 1975, *and* VOSS, 8 April 1975)

WALL, J. W. (British Political Agent in Bahrain), letter (153009/2/53G) to Bapco's Chief Local Representative, 24 February 1953

WALL, J. W., confidential minute (1532/3/54), 28 February 1954, classification FO 371/109899, BSA

WALLACE, William T., letter to T. E. Ward (typed on South American Gulf Oil Company letterhead but written on behalf of Gulf Oil Corporation), 2 November 1927, T. E. Ward Jr's personal papers

WALLACE, William T., letter to T. E. Ward, 12 November 1927, T. E. Ward Jr's personal papers

WALLACE, William T., letter to T. E. Ward (typed on South American Gulf Oil Company letterhead), 22 November 1927, T. E. Ward Jr's personal papers

WALLACE, William T., letter to T. E. Ward (typed on South American Gulf Oil Company letterhead), 1 December 1927, T. E. Ward Jr's personal papers

WALLACE, William T., letter to T. E. Ward (typed on South American Gulf Oil Company letterhead), 3 December 1927, T. E. Ward Jr's personal papers

WALLACE, William T., letter to Major Frank Holmes, 30 March 1928, Ward, 1965, pp. 56-9

WALLACE, William T., letter to Major Frank Holmes, 22 May 1928, Ward, 1965, pp. 76-9

WALLACE, William T., letter to T. E. Ward (typed on South American Gulf Oil Company letterhead), 24 May 1928, T. E. Ward Jr's personal papers

WALLACE, William T., letter to T. E. Ward (typed on South American Gulf Oil Company letterhead), 4 June 1928 (refers to additional Bahrein concession), T. E. Ward Jr's personal papers

WALLACE, William T., letter to T. E. Ward, 12 July 1928, T. E. Ward Jr's personal papers

WALLACE, William T., letter to T. E. Ward, 11 August 1928, T. E. Ward Jr's personal papers

WALLACE, William T., telegram to T. E. Ward, Stevens Hotel, Chicago, 3 December 1928, Ward, 1965, pp. 117-18

WALLACE, William T., letter to T. E. Ward, 4 January 1929, Ward, 1965, pp. 145-150

WALLACE, William T., letter to T. E. Ward (typed on Venezuela Gulf Oil Company letterhead), 15 February 1929, T. E. Ward Jr's personal papers

WALLACE, William T., letter to Francis B. Loomis, 19 August 1929, F. B. Loomis *Bahrein 1928-1929* file, CCL

WALLACE, William T., letter to Judge Frank Feuille, 13 November 1929, CCL

WALLACE, William T., two letters to Francis B. Loomis, 22 November 1929, [1] re form of transfer of Bahrein concession, [2] re future status of Major Frank Holmes, F. B. Loomis *Bahrein 1928-1929* file, CCL

WALLACE, William T., letter to Judge Frank Feuille, 23 November 1929, CCL

WALLACE, William T., letter to F. B. Loomis, 11 January 1930, CCL

WALLACE, William T., telegram to F. B. Loomis, 11 January 1930, CCL

WALLACE, William T., letter to Harry G. Davis (in London), 14 January 1930, CCL

WALLACE, William T., letter to Francis Loomis, 14 January 1930, CCL

WALLACE, William T., letter to Judge Frank Feuille, 17 January 1930, CCL

WALLACE, William T., cable to Harry G. Davis, 17 January 1930, CCL

WALLACE, William T., letter to M. E. Lombardi, 8 August 1930, CCL

WALLACE, William T., letter to M. E. Lombardi, 24 March 1931, CCL

WALTON, J. C., (Under Secretary of State, Foreign Office), letter No.18 from the India Office to the Colonial Office, 16 May 1929, classification COC.69035/29 [No. 45], enclosing:

1. Paraphrase telegram from Secretary of State for India to Viceroy, Foreign and Political Department, Nos. 1512 and 1513, 10 May 1929

2. Paraphrase telegram from Viceroy, Foreign and Political Department, to Secretary of State for India, No. 1656 S, 12 May 1929

3. Paraphrase telegram from Viceroy, Foreign and Political Department to Secretary of State for India, No. 1657 S, 12 May 1929

4. Letter from J. C. Walton, dated 16 May 1929

WALTON, J. C., letter to H. R. Ballantyne, 21 August 1933, CCL

WAR OFFICE, most secret cipher telegram to the Commander-in-Chief, Middle East, 17 June 1942, classification FO 371/31318 XL 141812, BSA

WARD, T. E., letter to Major Frank Holmes, 29 October 1926, T. E. Ward Jr's personal papers

WARD, T. E., letter to Dr J. Volney Lewis, 29 October 1926, T. E. Ward Jr's personal papers

WARD, T. E., letter to Major Frank Holmes, 5 November 1926, T. E. Ward Jr's personal papers

WARD, T. E., letter to H. T. Adams (Eastern and General Syndicate), 21 January 1927, T. E. Ward Jr's personal papers

WARD, T. E., letter to H. T. Adams, 23 May 1927, T. E. Ward Jr's personal papers

WARD, T. E., letter to Dr J. Volney Lewis, 24 May 1927, T. E. Ward Jr's personal papers

WARD, T. E., letter to Dr J. Volney Lewis, 28 May 1927, T. E. Ward Jr's personal papers

WARD, T. E., letter to H. T. Adams, 1 June 1927, T. E. Ward Jr's personal papers

WARD, T. E., letter to William T. Wallace, 16 June 1927, T. E. Ward Jr's personal papers

WARD, T. E., letter to William T. Wallace, 20 June 1927, T. E. Ward Jr's personal papers

WARD, T. E., letter to C. W. Hamilton (Gulf Oil Corporation), 23 September 1927, T. E. Ward Jr's personal papers

WARD, T. E., letter to William T. Wallace (Gulf Oil Corporation), 7 October 1927, T. E. Ward Jr's personal papers

WARD, T. E., letter to H. T. Adams, 21 October 1927, T. E. Ward Jr's personal papers

WARD, T. E., letter to H. T. Adams, 26 October 1927, T. E. Ward Jr's personal papers

WARD, T. E., letter to E. W. Janson, 26 October 1927, T. E. Ward Jr's personal papers

WARD, T. E., letter to H. T. Adams, 28 October 1927, T. E. Ward Jr's personal papers

WARD, T. E., letter to Major Frank Holmes, 28 October 1927, T. E. Ward Jr's personal papers

WARD, T. E., radiogram to EASGENSYND London, 29 October 1927, T. E. Ward Jr's personal papers

WARD, T. E., letter to William T. Wallace, 31 October 1927, T. E. Ward Jr's personal papers

WARD, T. E., radiogram to EASGENSYND London, 1 November 1927, T. E. Ward Jr's personal papers

WARD, T. E., letter to William T. Wallace, 2 November 1927, T. E. Ward Jr's personal papers

WARD, T. E., letter to H. T. Adams, 2 November 1927, T. E. Ward Jr's personal papers

WARD, T. E., letter to William T. Wallace (addressed to the Venezuelan Gulf Oil Company), 3 November 1927, T. E. Ward Jr's personal papers

WARD, T. E., letter to William T. Wallace, 12 November 1927, T. E. Ward Jr's personal papers

WARD, T. E., letter to William T. Wallace, 15 November 1927, T. E. Ward Jr's personal papers

WARD, T. E., letter to H. T. Adams, 22 November 1927, T. E. Ward Jr's personal papers

WARD, T. E., letter to William T. Wallace, 22 November 1927, T. E. Ward Jr's personal papers

WARD, T. E., letter to H. T. Adams, 23 November 1927, T. E. Ward Jr's personal papers

WARD, T. E., radiogram to EASGENSYND, 23 November 1927, T. E. Ward Jr's personal papers

WARD, T. E., radiogram to EASGENSYND, 25 November 1927, T. E. Ward Jr's personal papers

WARD, T. E., letter to C. W. Hamilton, 28 November 1927, T. E. Ward Jr's personal papers

WARD, T. E., letter to William T. Wallace, 30 November 1927, T. E. Ward Jr's personal papers

WARD, T. E., radiogram to EASGENSYND, 1 December 1927, T. E. Ward Jr's personal papers

WARD, T. E., letter to H. T. Adams, 3 December 1927, T. E. Ward Jr's personal papers

WARD, T. E., letter to William T. Wallace, 3 December 1927, T. E. Ward Jr's personal papers

WARD, T. E., letter to H. W. Adams, 14 December 1927, T. E. Ward Jr's personal papers

WARD, T. E., letter to C. W. Hamilton, 15 December 1927, T. E. Ward Jr's personal papers

WARD, T. E., letter to H. T. Adams, 23 December 1927, T. E. Ward Jr's personal papers

WARD, T. E., letter to C. W. Hamilton, 23 December 1927, T. E. Ward Jr's personal papers

WARD, T. E., letter to Major Frank Holmes, 23 December 1927, T. E. Ward Jr's personal papers

WARD, T. E., letter to Major Frank Holmes, 18 January 1928, T. E. Ward Jr's personal papers

WARD, T. E., letter to H. T. Adams, 29 March 1928, T. E. Ward Jr's personal papers

WARD, T. E., letter to William T. Wallace, 3 April 1928, T. E. Ward Jr's personal papers

WARD, T. E., letter to H. T. Adams, 4 April 1928, T. E. Ward Jr's personal papers

WARD, T. E., letter to H. T. Adams, 17 May 1928, T. E. Ward Jr's personal papers

WARD, T. E., letter to H. T. Adams, 31 May 1928, T. E. Ward Jr's personal papers

WARD, T. E., letter to H. T. Adams, 5 June 1928, T. E. Ward Jr's personal papers

WARD, T. E., letter to H. T. Adams, 8 June 1928, T. E. Ward Jr's personal papers

WARD, T. E., letter to William T. Wallace, 14 June 1928, Ward, 1965, p. 92

WARD, T. E., letter to Major Frank Holmes, 15 June 1928, T. E. Ward Jr's personal papers

WARD, T. E., letter to H. T. Adams, 29 June 1928, T. E. Ward Jr's personal papers

WARD, T. E., letter to Eastern and General Syndicate, 9 July 1928, Ward, 1965, pp. 90-91

WARD, T. E., letter to H. T. Adams, 8 August 1928, T. E. Ward Jr's personal papers

WARD, T. E., letter to William T. Wallace, 8 August 1928, T. E. Ward Jr's personal papers

WARD, T. E., letter to William T. Wallace, 13 August 1928, T. E. Ward Jr's personal papers

WARD, T. E., letter to H. T. Adams, 14 August 1928, T. E. Ward Jr's personal papers

WARD, T. E., letter to William T. Wallace, 14 August 1928, T. E. Ward Jr's personal papers

WARD, T. E., letter to William T. Wallace, 23 August 1928, Ward, 1965, pp. 104-5

WARD, T. E., letter to H. T. Adams, 4 October 1928, T. E. Ward Jr's personal papers

WARD, T. E., letter to Major Frank Holmes, c/o Yateem Bros, Bahrain, 2 November 1928, Ward, 1965, pp. 108-9

WARD, T. E., letter to William T. Wallace, 16 November 1928, T. E. Ward Jr's personal papers

WARD, T. E., cable to Eastern and General Syndicate, 23 November 1928, Ward, 1965, pp. 109-10

WARD, T. E., letter to Eastern and General Syndicate, 24 November 1928, Ward, 1965, pp. 110-12

WARD, T. E., cable to Eastern and General Syndicate, 27 November 1928, Ward, 1965, p. 112

WARD, T. E., cable to Eastern and General Syndicate, 30 November 1928, Ward, 1965, pp. 113-14

WARD, T. E., letter to H. T. Adams, 1 December 1928, Ward, 1965, pp. 114-17

WARD, T. E., telegram to William T. Wallace, New York, 4 December 1928, Ward, 1965, pp. 118-19

WARD, T. E., letter to H. T. Adams, 11 December 1928, Ward, 1965, pp. 119-121

WARD, T. E., letter to H. T. Adams, 13 December 1928, Ward, 1965, pp. 121-2

WARD, T. E., letter to H. T. Adams, 18 December 1928, T. E. Ward Jr's personal papers

WARD, T. E., letter to H. T. Adams, 21 December 1928, Ward, 1965, p. 125

WARD, T. E., letter to William T. Wallace, 21 December 1928, Ward, 1965, p. 126

WARD, T. E., letter to H. T. Adams, re Bahrein Option assignment, 28 December 1928, Ward, 1965, pp. 134-5

WARD, T. E., letter to Eastern Gulf Oil Company, Pittsburgh, PA, 28 December 1928, Ward, 1965, pp. 130-1

WARD, T. E., letter to Judge Frank Feuille, 28 December 1928, Ward, 1965, pp. 133-4

WARD, T. E., letter to William T. Wallace, 3 January 1929, CCL

WARD, T. E., letter to W. H. Berg, 5 January 1929, Ward, 1965, pp. 136-7

WARD, T. E., letter to William T. Wallace, 15 January 1929, T. E. Ward Jr's personal papers

WARD, T. E., letter to H. T. Adams, 23 January 1929, T. E. Ward Jr's personal papers

WARD, T. E., letter to Judge Frank Feuille, 30 January 1929, CCL

WARD, T. E., letter to William T. Wallace, 30 January 1929, CCL

WARD, T. E., letter to H. T. Adams, 31 January 1929, T. E. Ward Jr's personal papers

WARD, T. E., letter to H. T. Adams, 8 February 1929, T. E. Ward Jr's personal papers

WARD, T. E., letter to Judge Frank Feuille, 13 February 1929, T. E. Ward Jr's personal papers

WARD, T. E., letter to H. T. Adams, 15 February 1929, T. E. Ward Jr's personal papers

WARD, T. E., radiogram to EASGENSYND, 15 February 1929, T. E. Ward Jr's personal papers

WARD, T. E., letter to H. T. Adams, 22 February 1929, T. E. Ward Jr's personal papers

WARD, T. E., letter to William T. Wallace, 19 July 1929, T. E. Ward Jr's personal papers

WARD, T. E., letter to William T. Wallace, 30 September 1929, T. E. Ward Jr's personal papers

WARD, T. E., letter to H. T. Adams, 1 October 1929, Ward, 1965, p. 171

WARD, T. E., cable to the Eastern and General Syndicate, 1 October 1929, Ward, 1965, p. 172

WARD, T. E. (Attorney-in-Fact for Eastern and General Syndicate), letter to William T. Wallace, 21 November 1929, T. E. Ward Jr's personal papers

WARD, T. E., letter to E. W. Janson, 20 March 1931, T. E. Ward Jr's personal papers

WARD, T. E., letter to E. W. Janson, 28 April 1931, T. E. Ward Jr's personal papers

WARD, T. E., letter to Major Frank Holmes, 8 October 1931, T. E. Ward Jr's personal papers

WARD, T. E., letter to Major Frank Holmes, 9 May 1932, T. E. Ward Jr's personal papers

WARD, T. E., letter to E. W. Janson, 24 February 1933, T. E. Ward Jr's personal papers

WARD, T. E., letter to E. W. Janson, 19 May 1933, T. E. Ward Jr's personal papers

WARD, T. E., letter to E. W. Janson, 22 July 1933, T. E. Ward Jr's personal papers

WARD, T. E., letter to E. W. Janson, 5 December 1933, T. E. Ward Jr's personal papers

WARD, T. E., letter to E. W. Janson, 14 February 1934, T. E. Ward Jr's personal papers

WARD, T. E., letter to Eastern and General Syndicate (for the attention of the Secretary, William Pocock), 14 February 1934, T. E. Ward Jr's personal papers

WARD, T. E., letter to E. W. Janson, 22 March 1935, T. E. Ward Jr's personal papers

WARD, T. E., letter to E. W. Janson, 23 April 1935, T. E. Ward Jr's personal papers

WARD, T. E., letter to Percy Tarbutt, 14 December 1935, T. E. Ward Jr's personal papers

WARD, T. E., letter to Percy Tarbutt, 16 June 1936, T. E. Ward Jr's personal papers

WARD, T. E., letter to F. A. Leovy, 2 December 1937, Ward, 1965, pp. 185-8

WARD, Thomas E., *Negotiations for Oil Concessions in Bahrain, El Hasa (Saudi Arabia), the Neutral Zone, Qatar and Kuwait*, private publication, New York, 1965

WEBSTER, J. A., letter to the Air Ministry from the Colonial Office, 13 May 1929, CO, File C.69035/29 [No.43]

WEIGHTMAN, H., "Administration Report of the Bahrain Agency for the Year 1937", *ARPGPR for the Year 1937*, Government of India Press, New Delhi, 1938, Ch. VI

WEIGHTMAN, H., "Administration Report of the Bahrain Agency for the Year 1938", *ARPGPR for the Year 1938*, Government of India Press, New Delhi, 1939, Ch. VI

WEIGHTMAN, H., "Administration Report of the Bahrain Agency and the Trucial Coast for the Year 1939", *ARPGPR for the Year 1939*, Government of India Press, Simla, 1940, Ch. VI

WEIGHTMAN, H., "Administration Report of the Bahrain Agency and the Trucial Coast for the Year 1940", *ARPGPR for the Year 1940*, Government of India Press, Simla, 1941, Ch. VI

WHITE, Gerald T., *Formative Years in the Far West: A History of Standard Oil Company of California and Predecessors Through 1919*, Appleton-Century-Crofts, Division of Meredith Publishing Co., New York

WILLIAMS, Maynard Owen, "Bahrain: Port of Pearls and Petroleum", *National Geographic Magazine*, Vol. 89, No. 2, February 1946, pp. 195-210

WILLIAMS, O. G. R. (Under Secretary of State, Colonial Office, London), letter to the Board of Trade, 8 November 1928, classification CO C.59115/28 [No. 14], No. 5, BSA

WILLIAMS, O. G. R., memorandum to Eastern and General Syndicate, 23 November 1928, CCL,

CO, File C.59115/28 [No. 19] and Enclosure No. 6 (Draft Nationality Clause) and Ward, 1965, pp. 153-4, BSA

WILLIAMS, O. G. R., memorandum to the Board of Trade enclosed in letter of 12 January 1929, classification CO C.59115/28 [No. 27], BSA

WILLIAMS, O. G. R., letter to the Admiralty, 20 March 1929, classification CO C.69035/29 [No. 17], BSA

WILLIAMS, O. G. R., letter to Eastern and General Syndicate, 30 May 1929, classification CO C.69035/29 [No. 49], BSA; also in CCL

WILLIAMS, O. G. R., letter to the Foreign Office re special treaty relations existing between the British Government and the Ruler of Bahrein, 29 August 1929, classification CO C.69035/29 [No. 114], BSA

WILLIAMS, O. G. R., letter to the Eastern and General Syndicate, 16 September 1929, classification CO C.69035/29 [No. 125], BSA

WILLIAMS, O. G. R., letter to the Eastern and General Syndicate, 3 January 1930, classification CO C.69035/29 [No. 181], BSA; also in CCL

WILLIAMS, O. G. R., letter to the Colonial Office (copied to the Foreign Office, India Office, Petroleum Dept., Air Ministry, War Office, Board of Trade and Admiralty), 8 February 1930, classification CO C.79035/30 [Nos. 13-19], BSA

WILLIAMS, O. G. R., letter to the Secretary of the Eastern and General Syndicate, 8 February 1930, CCL

WILLIAMS, O. G. R., letter to H. R. Ballantyne, 28 July 1933, CCL

WILSON, Arnold, "A Periplus of the Persian Gulf", *Geographical Journal*, Vol. = 69, No. 3, March 1927, pp. 235-59

WILSON, C. O., "The Middle East – Its Present and Future", *The Oil and Gas Journal*, 29 December 1945, pp. 183-94

WILSON, D., "Memorandum Respecting the Pearl Fisheries in the Persian Gulf", *Geographical Journal*, III (1883)

WILTSHIRE, E. P. (British Political Agent, Bahrain), letter 1537/12/59/G to Lipp, 30 December 1959. (*See also* LIPP, C/PA 359 and C/PA 360)

WILTSHIRE, E. P., letter 1533/13/60 to Lipp, 19 June 1960. (*See also* LIPP, C/PA-384)

WORLD OIL, news clips on Bahrain, published on 15 August each year: 1973, p. 158; 1974, p. 160; 1975, p. 173; 1976, p. 175; 1977, p. 170; 1978, pp. 208-9; 1979, p. 218; 1980, p. 250; 1981, p. 277; 1982, pp. 280 and 282

WORLD PETROLEUM, Vol. VIII, October 1937, Number 10: "Bahrein's Place in World's Oil Picture" (editorial), p. 33; "Industrial City on a Tropical Island – The Story of Bahrein" (no credited correspondent), pp. 34-41

WORLD PETROLEUM, "Bahrain Petroleum Co. Ltd" (no credited correspondent), September 1954, p. 93

Z

ZAHLAN, A. B., editor, *The Arab Brain Drain: Proceedings of a Seminar organised by the Natural Resources, Science and Technology Division of the United Nations Economic Commission for Western Asia*, Beirut, 4-8 February 1980, published for the United Nations by Ithaca Press, London, 1981

ZIWAR-DAFTARI, May, editor, *Issues in Development: The Arab Gulf States*, MD Research and Services Ltd, London, 1980

ZWEMER, Samuel M., *Cradle of Islam*, New York, 1900

Postscript

It may be of further interest to readers that, during 1989, Archive Editions (an imprint from Archive International Group) published *Arabian Gulf Oil Concessions 1911-1953*. Two of the twelve volumes relate to Bahrain. The documents contained therein have been reproduced from originals in the India Office Library and Records, London, by agreement with the British Library.

INDEX